Applied and Numerical Harmonic Analysis

Ole Christensen

Frames and Bases

An Introductory Course

Birkhäuser
Boston • Basel • Berlin

Ole Christensen
Technical University of Denmark
Department of Mathematics
2800 Lyngby
Denmark

ISBN: 978-0-8176-4677-6   e-ISBN: 978-0-8176-4678-3

Library of Congress Control Number: 2007942994

Mathematics Subject Classification: 41-01, 42-01

Printed on acid-free paper.

9 8 7 6 5 4 3 2 1

www.birkhauser.com

To Karen

ANHA Series Preface

The *Applied and Numerical Harmonic Analysis (ANHA)* book series aims to provide the engineering, mathematical, and scientific communities with significant developments in harmonic analysis, ranging from abstract harmonic analysis to basic applications. The title of the series reflects the importance of applications and numerical implementation, but richness and relevance of applications and implementation depend fundamentally on the structure and depth of theoretical underpinnings. Thus, from our point of view, the interleaving of theory and applications and their creative symbiotic evolution is axiomatic.

Harmonic analysis is a wellspring of ideas and applicability that has flourished, developed, and deepened over time within many disciplines and by means of creative cross-fertilization with diverse areas. The intricate and fundamental relationship between harmonic analysis and fields such as signal processing, partial differential equations (PDEs), and image processing is reflected in our state-of-the-art *ANHA* series.

Our vision of modern harmonic analysis includes mathematical areas such as wavelet theory, Banach algebras, classical Fourier analysis, time-frequency analysis, and fractal geometry, as well as the diverse topics that impinge on them.

For example, wavelet theory can be considered an appropriate tool to deal with some basic problems in digital signal processing, speech and image processing, geophysics, pattern recognition, biomedical engineering, and turbulence. These areas implement the latest technology from sampling methods on surfaces to fast algorithms and computer vision methods. The underlying mathematics of wavelet theory depends not only on classical Fourier analysis, but also on ideas from abstract harmonic analysis, including von Neumann algebras and the affine group. This leads to a study of the Heisenberg group and its relationship to Gabor systems, and of the metaplectic group for a meaningful interaction of signal decomposition methods. The unifying influence of wavelet theory in the aforementioned topics illustrates the justification for providing a means for centralizing and disseminating information from the broader, but still focused, area of harmonic analysis. This will be a key role of *ANHA*. We intend to publish with the scope and interaction that such a host of issues demands.

Along with our commitment to publish mathematically significant works at the frontiers of harmonic analysis, we have a comparably strong commitment to publish major advances in the following applicable topics in which harmonic analysis plays a substantial role:

Antenna theory	*Prediction theory*
Biomedical signal processing	*Radar applications*
Digital signal processing	*Sampling theory*
Fast algorithms	*Spectral estimation*
Gabor theory and applications	*Speech processing*
Image processing	*Time-frequency and*
Numerical partial differential equations	*time-scale analysis*
	Wavelet theory

The above point of view for the *ANHA* book series is inspired by the history of Fourier analysis itself, whose tentacles reach into so many fields.

In the last two centuries Fourier analysis has had a major impact on the development of mathematics, on the understanding of many engineering and scientific phenomena, and on the solution of some of the most important problems in mathematics and the sciences. Historically, Fourier series were developed in the analysis of some of the classical PDEs of mathematical physics; these series were used to solve such equations. In order to understand Fourier series and the kinds of solutions they could represent, some of the most basic notions of analysis were defined, e.g., the concept of "function." Since the coefficients of Fourier series are integrals, it is no surprise that Riemann integrals were conceived to deal with uniqueness properties of trigonometric series. Cantor's set theory was also developed because of such uniqueness questions.

A basic problem in Fourier analysis is to show how complicated phenomena, such as sound waves, can be described in terms of elementary harmonics. There are two aspects of this problem: first, to find, or even define properly, the harmonics or spectrum of a given phenomenon, e.g., the spectroscopy problem in optics; second, to determine which phenomena can be constructed from given classes of harmonics, as done, for example, by the mechanical synthesizers in tidal analysis.

Fourier analysis is also the natural setting for many other problems in engineering, mathematics, and the sciences. For example, Wiener's Tauberian theorem in Fourier analysis not only characterizes the behavior of the prime numbers, but also provides the proper notion of spectrum for phenomena such as white light; this latter process leads to the Fourier analysis associated with correlation functions in filtering and prediction problems, and these problems, in turn, deal naturally with Hardy spaces in the theory of complex variables.

Nowadays, some of the theory of PDEs has given way to the study of Fourier integral operators. Problems in antenna theory are studied in terms of unimodular trigonometric polynomials. Applications of Fourier analysis abound in signal processing, whether with the fast Fourier transform (FFT), or filter design, or the adaptive modeling inherent in time-frequency-scale methods such as wavelet theory. The coherent states of mathematical physics are translated and modulated Fourier transforms, and these are used, in conjunction with the uncertainty principle, for dealing with signal reconstruction in communications theory. We are back to the raison d'être of the *ANHA* series!

John J. Benedetto
Series Editor
University of Maryland
College Park

Contents

Preface

The aim of this book is to present the central parts of the theory for bases and frames. The content can naturally be split into two parts: Chapters 1–5 describe the theory on an abstract level, and Chapters 7–11 deal with explicit constructions in L^2-spaces. The link between these two parts is formed by Chapter 6, which introduces B-splines and their main properties.

Some years ago, I published the book *An Introduction to Frames and Riesz Bases* [10], which also appeared in the ANHA series. So, what are the reasons for another book on the topic? I will give some answers to this question.

Books written by mathematicians are usually focused on characterizations of various properties and the search for sufficient conditions for a desired conclusion to hold. Concrete constructions often play a minor role. The book [10] is no exception. During the past few years, frames have become increasingly popular, and several explicit constructions of frames of various types have been presented. Most of these constructions were based on quite direct methods rather than the classical sufficient conditions for obtaining a frame. With this in mind, it seems that there is a need for an updated version of the book [10], which moves the focus from the classical approach to a more constructive one.

Frame theory is developed in constant dialogue between mathematicians and engineers. Again, compared with [10], this is reflected in the current book by several new sections on applications and connections to engineering. The hope is that these sections will help the mathematically oriented readers to see where frames are used in practice — and the engineers to find the chapter containing the mathematical background for applications in their field.

The third main change compared with [10] is that the current book is meant to be a textbook, which should be directly suitable for use in a graduate course. We focus on the basic topics, without too many side-remarks; in contrast, [10] tried to cover the entire area, including the research aspects. The chapters from [10] dealing with research topics have been removed (or reduced: for example, parts of Chapter 15 about perturbation results now appear in Section 5.6). We frequently mention the names of the people who first proved a given result, but for the parts of the theory that can be considered classical, we do not state a reference to the original source. A professional reader might miss all the hints to more advanced literature and open problems; however, the hope is that the more streamlined writing makes it easier for students to follow the presentation.

For use in a graduate course, a number of exercises is included; they appear at the end of each chapter. Some of the removed material from [10] now appears in the exercises.

Let us describe the chapters in more detail. Chapter 1 gives an introduction to frames in finite-dimensional vector spaces with an inner product. This enables a reader with a basic knowledge of linear algebra to understand the idea behind frames without the technical complications in infinite-dimensional spaces. Many of the topics from the rest of the book are presented here, so Chapter 1 can also serve as an introduction to the later chapters.

Chapter 2 collects some definitions and conventions concerning infinite-dimensional vector spaces. Some standard results needed later in the book are also stated here. Special attention is given to the Hilbert space $L^2(\mathbb{R})$ and operators hereon. We expect the reader to be familiar with this material, so most of the results appear without proof. The exceptions are the sections about pseudo-inverse operators and some special operators on $L^2(\mathbb{R})$, which play a key role in Gabor theory and wavelet analysis; these subjects are not treated in classical analysis courses and are therefore described in detail.

Chapter 3 deals with the theory for bases in Hilbert spaces and Banach spaces. The most important part of the chapter is formed by a detailed discussion of Bessel sequences and Riesz bases. The chapter also contains sections on Fourier analysis and wavelet theory, which motivate the constructions in Chapters 7–11.

Chapter 4 highlights some of the limitations on the properties one can obtain from bases. Hereby, the reader is provided with motivation for considering the generalizations of bases studied in the rest of the book.

Chapter 5 contains the core material about frames in general Hilbert spaces. It gives a detailed description of frames with full proofs, relates frames and Riesz bases, and provides various ways of constructing frames.

Chapter 6 introduces B-splines and their main properties. We do not aim at a complete description of splines but concentrate on the properties that play a role in the current context.

Chapters 7–11 deal with frames having a special structure. A central part concerns theoretical conditions for obtaining dual pairs of frames and explicit constructions hereof. The most fundamental frames, namely frames consisting of translates of a single function in $L^2(\mathbb{R})$, are discussed in Chapter 7. In Chapter 8, these considerations are extended to frames generated by translations of a collection of functions rather than a single function. These frames naturally lead to Gabor frames in $L^2(\mathbb{R})$, which is the subject of Chapter 9. We provide characterizations of such frames, as well as explicit constructions of frames and some of their dual frames. The discrete counterpart in $\ell^2(\mathbb{Z})$ is treated in Chapter 10; in particular, it is shown how one can obtain Gabor frames in $\ell^2(\mathbb{Z})$ by sampling of Gabor frames in $L^2(\mathbb{R})$. Wavelet frames are introduced in Chapter 11. The main part of the chapter is formed by explicit constructions via multiscale methods, but the chapter also contains a section about general wavelet frames.

Most readers of the second part of the book will mainly be interested in either Gabor systems or wavelet systems. For this reason, Chapters 7–11 are to a large extent independent of each other. The most notable exception from that rule is that some of the fundamental results in Gabor analysis are based on results derived in the chapter about shift-invariant systems. In general, careful cross-references (and, if necessary, repetitions) between Chapters 7–11 are provided.

Depending on the level and specific interests of the students, a graduate course based on the book can proceed in various ways:

- Readers with a limited background in functional analysis (and readers who just want to get an idea about the topic) are encouraged to read Chapter 1. It will provide the reader with a good understanding for the topic, without all the technical complications in infinite–dimensional vector spaces.

- A short course on frames and Riesz bases in Hilbert spaces can be based on Sections 3.1–3.3 and Sections 5.1–5.2; these sections will make the reader able to proceed with most of the other parts of the book and with a large part of the research literature concerning abstract frame theory.

- A theoretical graduate course on bases and frames could be based on Chapter 2, Chapter 3, and Chapter 5. It would be natural to continue with one or more chapters on concrete frame constructions in $L^2(\mathbb{R})$.

- For a course focusing on either Gabor analysis or wavelets, the detailed analysis of frames in Chapter 5 is not necessary. It is enough to read Chapter 2, Section 3.5 (or Section 3.6), Chapter 4, Section 5.1, and parts of Chapter 6 before continuing with the relevant specialized chapters.

I would like to acknowledge the various individuals and institutions who have helped me during the process of writing this book. First, I wish to thank the Department of Mathematics at the Technical University of Denmark for giving me enough freedom to realize the book project, e.g., via a semester without teaching obligations. Some weeks of that semester were used to visit other departments in order to get inspiration and concentrate on the work with the book for several weeks; I thank my colleagues Hans Feichtinger (NuHAG, University of Vienna) as well as Rae Young Kim (Yeungnam University, South Korea) and Jungho Yoon (EWHA Woman University, South Korea) for hosting me during these visits.

Thanks are also due to Martin McKinnon Edwards, Jakob Jørgensen, and Sumi Jang for help with the figures. Finally, I would like to thank Richard Laugesen and Azita Mayeli for correcting parts of the material, as well as Henrik Stetkær and Kil Kwon for several suggestions concerning the presentation of the material.

I also thank the staff at Birkhäuser, especially Tom Grasso, for assistance and support.

Ole Christensen
Kgs. Lyngby, Denmark
November 2007

Frames and Bases

1
Frames in Finite-dimensional Inner Product Spaces

In the study of vector spaces, one of the most important concepts is that of a basis. In fact, a basis provides us with an expansion of all vectors in terms of "elementary building blocks" and hereby helps us by reducing many questions concerning general vectors to similar questions concerning only the basis elements. However, the conditions to a basis are very restrictive: we require that the elements are linearly independent, and very often we even want them to be orthogonal with respect to an inner product. This makes it hard or even impossible to find bases satisfying extra conditions, and this is the reason that one might wish to look for a more flexible tool.

Frames are such tools. A frame for a vector space equipped with an inner product also allows each vector in the space to be written as a linear combination of the elements in the frame, but linear independence between the frame elements is not required. Intuitively, one can think about a frame as a basis to which one has added more elements. In this chapter, we present frame theory in finite-dimensional vector spaces. This restriction makes part of the theory much easier, and it also makes the basic idea more transparent. Our intention is to present the results in a way that gives the reader the right feeling about the infinite-dimensional setting as well. This also means that we sometimes use unusual words in the finite-dimensional setting. For example, we will frequently use the word "operator" for a linear map.

There are other reasons for starting with a chapter on finite-dimensional frames. Every "real-life" application of frames has to be performed in a finite-dimensional vector space, so even if we want to apply results from

O. Christensen, *Frames and Bases*. DOI: 10.1007/978-0-8176-4678-3_1,
© Springer Science+Business Media, LLC 2008

the infinite-dimensional setting, the frames will have to be confined to a finite-dimensional space at some point.

Most of the chapter can be fully understood with an elementary knowledge of linear algebra. In order not to make the proofs too cumbersome, we will at a few points use some results from analysis, mainly about norms in vector spaces.

This chapter is organized as follows. Section 1.1 contains the basic properties of frames. For example, it is proved that every set of vectors $\{f_k\}_{k=1}^m$ in a vector space with an inner product is a frame for $\mathrm{span}\{f_k\}_{k=1}^m$. We prove the existence of coefficients minimizing the ℓ^2-norm of the coefficients in a frame expansion and show how a frame for a subspace leads to a formula for the orthogonal projection onto the subspace. In Section 1.2 and Section 1.3, we consider frames in $\mathbb{C}^n$. In particular, we prove that the vectors $\{f_k\}_{k=1}^m$ in a frame for $\mathbb{C}^n$ can be considered as the first n coordinates of some vectors in $\mathbb{C}^m$ constituting a basis for $\mathbb{C}^m$, and that the frame property for $\{f_k\}_{k=1}^m$ is equivalent to certain properties for the $m \times n$ matrix having the vectors f_k as rows. In Section 1.4, we prove that the canonical coefficients from the frame expansion arise naturally by considering the pseudo-inverse of the pre-frame operator, and we show how to find the coefficients in terms of the singular value decomposition. Finally, in Section 1.5, we discuss applications of frames in the context of data transmission.

1.1 Basic frames theory

Let $V \neq \{0\}$ be a finite-dimensional vector space. As standing assumption we will assume that V is equipped with an inner product $\langle \cdot, \cdot \rangle$, which we choose to be linear in the first entry. Recall that a sequence $\{e_k\}_{k=1}^m$ in V is a *basis* for V if the following two conditions are satisfied:

(i) $V = \mathrm{span}\{e_k\}_{k=1}^m$;
(ii) $\{e_k\}_{k=1}^m$ is *linearly independent*, i.e., if $\sum_{k=1}^m c_k e_k = 0$ for some scalar coefficients $\{c_k\}_{k=1}^m$, then $c_k = 0$ for all $k = 1, \ldots, m$.

As a consequence of this definition, every $f \in V$ has a unique representation in terms of the elements in the basis, i.e., there exist unique scalar coefficients $\{c_k\}_{k=1}^m$ such that

$$f = \sum_{k=1}^m c_k e_k. \tag{1.1}$$

Sometimes, in particular in high-dimensional vector spaces, it is cumbersome to find the coefficients $\{c_k\}_{k=1}^m$. But if $\{e_k\}_{k=1}^m$ is an *orthonormal basis*, i.e., a basis for which

$$\langle e_k, e_j \rangle = \delta_{k,j} = \begin{cases} 1 & \text{if } k = j \\ 0 & \text{if } k \neq j, \end{cases}$$

then the coefficients $\{c_k\}_{k=1}^m$ are easy to find: taking the inner product of f in (1.1) with an arbitrary e_j gives

$$\langle f, e_j \rangle = \langle \sum_{k=1}^m c_k e_k, e_j \rangle = \sum_{k=1}^m c_k \langle e_k, e_j \rangle = c_j,$$

so

$$f = \sum_{k=1}^m \langle f, e_k \rangle e_k. \tag{1.2}$$

We now introduce frames. In Theorem 1.1.5 below, we prove that a frame $\{f_k\}_{k=1}^m$ also leads to a representation of the type (1.1).

Definition 1.1.1 *A countable family of elements $\{f_k\}_{k \in I}$ in V is a frame for V if there exist constants $A, B > 0$ such that*

$$A \|f\|^2 \leq \sum_{k \in I} |\langle f, f_k \rangle|^2 \leq B \|f\|^2, \quad \forall f \in V. \tag{1.3}$$

The numbers A, B are called *frame bounds*. They are not unique. The *optimal lower frame bound* is the supremum over all lower frame bounds, and the *optimal upper frame bound* is the infimum over all upper frame bounds. Note that the optimal frame bounds actually are frame bounds. The frame is *normalized* if $\|f_k\| = 1$, $\forall k \in I$.

In a finite-dimensional vector space, it is somehow artificial (though possible) to consider frames $\{f_k\}_{k \in I}$ consisting of infinitely many elements. Therefore, we will only consider finite families $\{f_k\}_{k=1}^m$, $m \in \mathbb{N}$. With this restriction, Cauchy–Schwarz' inequality shows that

$$\sum_{k=1}^m |\langle f, f_k \rangle|^2 \leq \sum_{k=1}^m \|f_k\|^2 \|f\|^2, \quad \forall f \in V,$$

i.e., the upper frame condition is automatically satisfied. However, one can often find a smaller upper frame bound than $\sum_{k=1}^m \|f_k\|^2$. Corollary 1.1.13 will show that it is important to find estimates for the frame bounds, which are close to the optimal ones.

In order for the lower condition in (1.3) to be satisfied, it is necessary that $\text{span}\{f_k\}_{k=1}^m = V$. This condition turns out to be sufficient. In fact, every finite sequence is a frame for its span:

Proposition 1.1.2 *Let $\{f_k\}_{k=1}^m$ be a sequence in V. Then $\{f_k\}_{k=1}^m$ is a frame for the vector space $W := \text{span}\{f_k\}_{k=1}^m$.*

Proof. We can assume that not all f_k are zero. As we have seen, the upper frame condition is satisfied with $B = \sum_{k=1}^{m} ||f_k||^2$. Now consider the continuous mapping

$$\phi : W \to \mathbb{R}, \ \phi(f) := \sum_{k=1}^{m} |\langle f, f_k \rangle|^2.$$

The unit ball in W is compact, so we can find $g \in W$ with $||g|| = 1$ such that

$$A := \sum_{k=1}^{m} |\langle g, f_k \rangle|^2 = \inf \left\{ \sum_{k=1}^{m} |\langle f, f_k \rangle|^2 \ : \ f \in W, \ ||f|| = 1 \right\}.$$

It is clear that $A > 0$. Now given $f \in W, f \neq 0$, we have

$$\sum_{k=1}^{m} |\langle f, f_k \rangle|^2 = \sum_{k=1}^{m} |\langle \frac{f}{||f||}, f_k \rangle|^2 \, ||f||^2 \geq A \, ||f||^2. \qquad \square$$

Corollary 1.1.3 *A family of elements $\{f_k\}_{k=1}^{m}$ in V is a frame for V if and only if $\text{span}\{f_k\}_{k=1}^{m} = V$.*

Corollary 1.1.3 shows that a frame might contain more elements than needed to be a basis. In particular, if $\{f_k\}_{k=1}^{m}$ is a frame for V and $\{g_k\}_{k=1}^{n}$ is an arbitrary finite collection of vectors in V, then $\{f_k\}_{k=1}^{m} \cup \{g_k\}_{k=1}^{n}$ is also a frame for V. A frame that is *not* a basis is said to be *overcomplete* or *redundant*.

Consider now a vector space V equipped with a frame $\{f_k\}_{k=1}^{m}$, and define a linear mapping

$$T : \mathbb{C}^m \to V, \ T\{c_k\}_{k=1}^{m} = \sum_{k=1}^{m} c_k f_k. \qquad (1.4)$$

T is usually called the *pre-frame operator*, or the *synthesis operator*. The *adjoint* operator is given by (Exercise 1.1)

$$T^* : V \to \mathbb{C}^m, \ T^* f = \{\langle f, f_k \rangle\}_{k=1}^{m}, \qquad (1.5)$$

and is called the *analysis operator*. Composing T with its adjoint T^*, we obtain the *frame operator*

$$S : V \to V, \ Sf = TT^* f = \sum_{k=1}^{m} \langle f, f_k \rangle f_k. \qquad (1.6)$$

Note that in terms of the frame operator,

$$\langle Sf, f \rangle = \sum_{k=1}^{m} |\langle f, f_k \rangle|^2, \ f \in V; \qquad (1.7)$$

the lower frame bound can thus be considered as some kind of "lower bound" on the frame operator.

A frame $\{f_k\}_{k=1}^m$ is *tight* if we can choose $A = B$ in the definition, i.e., if

$$\sum_{k=1}^m |\langle f, f_k \rangle|^2 = A \, ||f||^2, \ \forall f \in V. \tag{1.8}$$

For a tight frame, the exact value A in (1.8) is simply called the *frame bound*. We note that (1.7) leads to a representation of $f \in V$ in terms of the elements in a frame tight:

Proposition 1.1.4 *Assume that $\{f_k\}_{k=1}^m$ is a tight frame for V with frame bound A. Then $S = AI$ (here I is the identity operator on V), and*

$$f = \frac{1}{A} \sum_{k=1}^m \langle f, f_k \rangle f_k, \ \forall f \in V. \tag{1.9}$$

We ask the reader to prove Proposition 1.1.4 in Exercise 1.2. An interpretation of (1.9) is that if $\{f_k\}_{k=1}^m$ is a tight frame and we want to express $f \in V$ as a linear combination $f = \sum_{k=1}^m c_k f_k$, we can simply define $g_k = \frac{1}{A} f_k$ and take $c_k = \langle f, g_k \rangle$. Formula (1.9) is similar to the representation (1.2) via an orthonormal basis: the only difference is the factor $1/A$ in (1.9). For general frames, we now prove that we still have a representation of each $f \in V$ of the form $f = \sum_{k=1}^m \langle f, g_k \rangle f_k$ for an appropriate choice of $\{g_k\}_{k=1}^m$. The obtained theorem is one of the most important results about frames, and (1.10) below is called the *frame decomposition*:

Theorem 1.1.5 *Let $\{f_k\}_{k=1}^m$ be a frame for V with frame operator S. Then the following holds:*

(i) *S is invertible and self-adjoint.*

(ii) *Every $f \in V$ can be represented as*

$$f = \sum_{k=1}^m \langle f, S^{-1} f_k \rangle f_k = \sum_{k=1}^m \langle f, f_k \rangle S^{-1} f_k. \tag{1.10}$$

(iii) *If $f \in V$ also has the representation $f = \sum_{k=1}^m c_k f_k$ for some scalar coefficients $\{c_k\}_{k=1}^m$, then*

$$\sum_{k=1}^m |c_k|^2 = \sum_{k=1}^m |\langle f, S^{-1} f_k \rangle|^2 + \sum_{k=1}^m |c_k - \langle f, S^{-1} f_k \rangle|^2.$$

Proof. Because $S = TT^*$, it is clear that S is self-adjoint. We now prove that S is injective. Let $f \in V$ and assume that $Sf = 0$. Then

$$0 = \langle Sf, f \rangle = \sum_{k=1}^m |\langle f, f_k \rangle|^2,$$

implying by the frame condition that $f = 0$. That S is injective actually implies that S is surjective, but let us give a direct proof. The frame condition implies by Corollary 1.1.3 that $\text{span}\{f_k\}_{k=1}^m = V$, so the pre-frame operator T is surjective. Given $f \in V$ we can therefore find $g \in V$ such that $Tg = f$; we can choose $g \in \mathcal{N}_T^\perp = \mathcal{R}_{T^*}$, so it follows that $\mathcal{R}_S = \mathcal{R}_{TT^*} = V$. Thus S is surjective, as claimed. Each $f \in V$ has the representation

$$
\begin{aligned}
f &= SS^{-1}f \\
&= TT^*S^{-1}f \\
&= \sum_{k=1}^m \langle S^{-1}f, f_k \rangle f_k;
\end{aligned}
$$

using that S is self-adjoint, we arrive at

$$
f = \sum_{k=1}^m \langle f, S^{-1}f_k \rangle f_k.
$$

The second representation in (1.10) is obtained in the same way, using that $f = S^{-1}Sf$. For the proof of (iii), suppose that $f = \sum_{k=1}^m c_k f_k$. We can write

$$
\{c_k\}_{k=1}^m = \{c_k\}_{k=1}^m - \{\langle f, S^{-1}f_k \rangle\}_{k=1}^m + \{\langle f, S^{-1}f_k \rangle\}_{k=1}^m.
$$

By the choice of $\{c_k\}_{k=1}^m$ we have

$$
\sum_{k=1}^m \left(c_k - \langle f, S^{-1}f_k \rangle \right) f_k = 0,
$$

i.e., $\{c_k\}_{k=1}^m - \{\langle f, S^{-1}f_k \rangle\}_{k=1}^m \in \mathcal{N}_T = \mathcal{R}_{T^*}^\perp$; also, we note that

$$
\{\langle f, S^{-1}f_k \rangle\}_{k=1}^m = \{\langle S^{-1}f, f_k \rangle\}_{k=1}^m \in \mathcal{R}_{T^*}.
$$

Putting all the information together, we obtain that

$$
\begin{aligned}
\sum_{k=1}^m |c_k|^2 &= \left\| \{c_k\}_{k=1}^m - \{\langle f, S^{-1}f_k \rangle\}_{k=1}^m + \{\langle f, S^{-1}f_k \rangle\}_{k=1}^m \right\|^2 \\
&= \sum_{k=1}^m |c_k - \langle f, S^{-1}f_k \rangle|^2 + \sum_{k=1}^m |\langle f, S^{-1}f_k \rangle|^2,
\end{aligned}
$$

which proves (iii).   □

Every frame in a finite-dimensional space contains a subfamily that is a basis (Exercise 1.3). If $\{f_k\}_{k=1}^m$ is a frame but not a basis, there exist non-zero sequences $\{d_k\}_{k=1}^m$ such that $\sum_{k=1}^m d_k f_k = 0$. Therefore, any given element $f \in V$ can be written as

$$
\begin{aligned}
f &= \sum_{k=1}^{m} \langle f, S^{-1} f_k \rangle f_k + \sum_{k=1}^{m} d_k f_k \\
&= \sum_{k=1}^{m} \left(\langle f, S^{-1} f_k \rangle + d_k \right) f_k.
\end{aligned}
$$

This demonstrates that f has many representations as superpositions of the frame elements. Theorem 1.1.5 shows that among all scalar sequences $\{c_k\}_{k=1}^{m}$ for which $f = \sum_{k=1}^{m} c_k f_k$, the coefficients $\{\langle f, S^{-1} f_k \rangle\}_{k=1}^{m}$ have minimal ℓ^2-norm. The numbers

$$
\langle f, S^{-1} f_k \rangle, \quad k = 1, \ldots, m
$$

are called *frame coefficients*. Note that because $S : V \to V$ is bijective, the sequence $\{S^{-1} f_k\}_{k=1}^{m}$ is also a frame by Corollary 1.1.3; it is called the *canonical dual frame* of $\{f_k\}_{k=1}^{m}$.

For frames consisting of only a few elements, the canonical dual frame and the corresponding frame decomposition can be found via elementary calculations:

Example 1.1.6 Let $\{e_k\}_{k=1}^{2}$ be an orthonormal basis for a two-dimensional vector space V with inner product. Let

$$
f_1 = e_1, \quad f_2 = e_1 - e_2, \quad f_3 = e_1 + e_2.
$$

Then $\{f_k\}_{k=1}^{3}$ is a frame for V. Using the definition of the frame operator,

$$
Sf = \sum_{k=1}^{3} \langle f, f_k \rangle f_k,
$$

we obtain that

$$
Se_1 = e_1 + e_1 - e_2 + e_1 + e_2 = 3e_1
$$

and

$$
Se_2 = -(e_1 - e_2) + e_1 + e_2 = 2e_2.
$$

Thus

$$
S^{-1} e_1 = \frac{1}{3} e_1, \quad S^{-1} e_2 = \frac{1}{2} e_2.
$$

By linearity, the canonical dual frame is

$$
\begin{aligned}
\{S^{-1} f_k\}_{k=1}^{3} &= \{S^{-1} e_1, S^{-1} e_1 - S^{-1} e_2, S^{-1} e_1 + S^{-1} e_2\} \\
&= \{\frac{1}{3} e_1, \frac{1}{3} e_1 - \frac{1}{2} e_2, \frac{1}{3} e_1 + \frac{1}{2} e_2\}.
\end{aligned}
$$

Via Theorem 1.1.5, the representation of $f \in V$ in terms of the frame is given by

$$f = \sum_{k=1}^{3} \langle f, S^{-1} f_k \rangle f_k$$

$$= \frac{1}{3} \langle f, e_1 \rangle e_1 + \langle f, \frac{1}{3} e_1 - \frac{1}{2} e_2 \rangle (e_1 - e_2) + \langle f, \frac{1}{3} e_1 + \frac{1}{2} e_2 \rangle (e_1 + e_2). \quad \square$$

Theorem 1.1.5 gives some special information in case $\{f_k\}_{k=1}^{m}$ is a basis:

Corollary 1.1.7 *Assume that $\{f_k\}_{k=1}^{m}$ is a basis for V. Then there exists a unique family $\{g_k\}_{k=1}^{m}$ in V such that*

$$f = \sum_{k=1}^{m} \langle f, g_k \rangle f_k, \quad \forall f \in V. \tag{1.11}$$

In terms of the frame operator, $\{g_k\}_{k=1}^{m} = \{S^{-1} f_k\}_{k=1}^{m}$. Furthermore $\langle f_j, g_k \rangle = \delta_{j,k}$.

Proof. The existence of a family $\{g_k\}_{k=1}^{m}$ satisfying (1.11) follows from Theorem 1.1.5; we leave the proof of the uniqueness to the reader. Applying (1.11) on a fixed element f_j and using that $\{f_k\}_{k=1}^{m}$ is a basis, we obtain that $\langle f_j, g_k \rangle = \delta_{j,k}$ for all $k = 1, 2, \cdots, m$. $\square$

The simplicity of the calculations in Example 1.1.6 is slightly misleading: for a general frame, calculation of the canonical dual frame might be very cumbersome and lengthy if the frame contains many elements. This explains the prominent role of tight frames, for which the complicated representation (1.10) takes the much simpler form (1.9). Another way of obtaining "simple" frame expansions, whose potential has not been completely exploited so far, is to take advantage of the overcompleteness of frames. In fact, if one considers a frame $\{f_k\}_{k=1}^{m}$ that is *not* a basis, one can prove (see Lemma 5.2.3) that there exist frames $\{g_k\}_{k=1}^{m} \neq \{S^{-1} f_k\}_{k=1}^{m}$ such that

$$f = \sum_{k=1}^{m} \langle f, g_k \rangle f_k.$$

Each such frame $\{g_k\}_{k=1}^{m}$ is called a *dual frame*. Thus, rather than restricting attention to tight frames, one could consider frames, for which one can find a dual frame easily (Exercise 1.6). We return to this idea in several of the later chapters, see, e.g., Section 9.4.

If one insists on working with a tight frame, it is worth noticing that every frame can be extended to a tight frame by adding some extra vectors. In the proof of this, we will use the finite-dimensional version of the *spectral theorem*, which is proved in standard textbooks on linear algebra:

Theorem 1.1.8 *If a linear map $U : V \to V$ is self-adjoint, then all eigenvalues are real, and V has an orthonormal basis consisting of eigenvectors for U.*

Proposition 1.1.9 *Let $\{f_k\}_{k=1}^m$ be a frame for a vector space V with dimension n. Then there exist $n-1$ vectors $h_2, \ldots, h_n$ such that the collection $\{f_k\}_{k=1}^m \bigcup \{h_k\}_{k=2}^n$ forms a tight frame for V.*

Proof. Denote the frame operator for $\{f_k\}_{k=1}^m$ by $S : V \to V$. Since S is self-adjoint, Theorem 1.1.8 shows that V has an orthonormal basis consisting of eigenvectors $\{e_k\}_{k=1}^n$ for S. Denote the corresponding eigenvalues by $\{\lambda_k\}_{k=1}^n$. We will assume that the eigenvectors and eigenvalues are ordered such that $\lambda_1 \geq \lambda_2 \geq \cdots \geq \lambda_n$. Now, for $k = 2, \ldots, n$, let $h_k := \sqrt{\lambda_1 - \lambda_k} e_k$. The frame operator $\tilde{S}$ for the family $\{f_k\}_{k=1}^m \bigcup \{h_k\}_{k=2}^n$ is given by

$$\tilde{S} : V \to V, \ \ \tilde{S}f \ = \ Sf + \sum_{k=2}^n \langle f, h_k \rangle h_k. \tag{1.12}$$

Now consider an arbitrary $f \in V$. Using that

$$f = \sum_{k=1}^n \langle f, e_k \rangle e_k,$$

we see that the action of the frame operator S on f is given by

$$Sf = \sum_{k=1}^n \langle f, e_k \rangle S e_k = \sum_{k=1}^n \lambda_k \langle f, e_k \rangle e_k.$$

Inserting this expression and the definition of h_k into (1.12) shows that

$$\begin{aligned}
\tilde{S}f &= \sum_{k=1}^n \lambda_k \langle f, e_k \rangle e_k + \sum_{k=2}^n (\lambda_1 - \lambda_k) \langle f, e_k \rangle e_k \\
&= \lambda_1 \langle f, e_1 \rangle e_1 + \sum_{k=2}^n \lambda_k \langle f, e_k \rangle e_k + \lambda_1 \sum_{k=2}^n \langle f, e_k \rangle e_k - \sum_{k=2}^n \lambda_k \langle f, e_k \rangle e_k \\
&= \lambda_1 \sum_{k=1}^n \langle f, e_k \rangle e_k \\
&= \lambda_1 f.
\end{aligned}$$

This implies that for all $f \in V$,

$$\sum_{k=1}^n |\langle f, f_k \rangle|^2 + \sum_{k=2}^n |\langle f, h_k \rangle|^2 = \langle \tilde{S}f, f \rangle = \lambda_1 \, ||f||^2,$$

i.e., $\{f_k\}_{k=1}^m \bigcup \{h_k\}_{k=2}^n$ is a tight frame with frame bound λ_1.   □

We have already seen that, for given $f \in V$, the frame coefficients $\{\langle f, S^{-1} f_k \rangle\}_{k=1}^m$ have minimal ℓ^2-norm among all sequences $\{c_k\}_{k=1}^m$ for which $f = \sum_{k=1}^{m} c_k f_k$. We can also choose to minimize the norm in other spaces than ℓ^2; we now show the existence of coefficients minimizing the ℓ^1-norm.

Theorem 1.1.10 *Let $\{f_k\}_{k=1}^{m}$ be a frame for a finite-dimensional vector space V. Given $f \in V$, there exist coefficients $\{d_k\}_{k=1}^{m} \in \mathbb{C}^m$ such that $f = \sum_{k=1}^{m} d_k f_k$, and*

$$\sum_{k=1}^{m} |d_k| = \inf \left\{ \sum_{k=1}^{m} |c_k| \; : \; f = \sum_{k=1}^{m} c_k f_k \right\}. \tag{1.13}$$

Proof. Fix $f \in V$, and choose a set of coefficients $\{c_k\}_{k=1}^{m}$ such that $f = \sum_{k=1}^{m} c_k f_k$; let $r := \sum_{k=1}^{m} |c_k|$. Since we want to minimize the ℓ^1-norm of the coefficients, it is clear that we can now restrict our search for a minimizer to sequences $\{d_k\}_{k=1}^{m}$ belonging to the compact set

$$M := \{\{d_k\}_{k=1}^{m} \in \mathbb{C}^m \; : \; |d_k| \leq r, \; k = 1, \ldots, m\}.$$

Now the result follows from the fact that the set

$$\left\{ \{d_k\}_{k=1}^{m} \in M \mid f = \sum_{k=1}^{m} d_k f_k \right\}$$

is compact and that the function

$$\phi : \mathbb{C}^m \to \mathbb{R}, \; \phi\{d_k\}_{k=1}^{m} := \sum_{k=1}^{m} |d_k|$$

is continuous. □

There are some important differences between Theorem 1.1.5 and Theorem 1.1.10. In Theorem 1.1.5, we find the sequence minimizing the ℓ^2-norm of the coefficients in the expansion of f explicitly; it is unique, and it depends linearly on f. On the other hand, Theorem 1.1.10 only gives the existence of an ℓ^1-minimizer, and it might not be unique (Exercise 1.7). Even if the minimizer is unique, it might not depend linearly on f (Exercise 1.8).

As we have seen in Proposition 1.1.2, every finite set of vectors $\{f_k\}_{k=1}^{m}$ is a frame for its span. If $\text{span}\{f_k\}_{k=1}^{m} \neq V$, the frame decomposition associated with $\{f_k\}_{k=1}^{m}$ gives a convenient expression for the orthogonal projection onto $\text{span}\{f_k\}_{k=1}^{m}$. We state it here and ask the reader to provide the proof (Exercise 1.9).

Theorem 1.1.11 *Let $\{f_k\}_{k=1}^{m}$ be a frame for a subspace W of the vector space V. Then the orthogonal projection of V onto W is given by*

$$Pf = \sum_{k=1}^{m} \langle f, S^{-1} f_k \rangle f_k. \tag{1.14}$$

In order to compute the inverse frame operator S^{-1}, it is convenient to consider S as a matrix. The speed of convergence in numerical algorithms involving a strictly positive definite matrix depends heavily on the *condition number* of the matrix, which is defined as the ratio between the largest eigenvalue, $\lambda_{\max}$, and the smallest eigenvalue, $\lambda_{\min}$. In case of the frame operator, these eigenvalues correspond to the optimal frame bounds:

Theorem 1.1.12 *Let $\{f_k\}_{k=1}^m$ be a frame for V. Then the following hold:*

(i) *The optimal lower frame bound is the smallest eigenvalue for S, and the optimal upper frame bound is the largest eigenvalue.*

(ii) *Assume that V has dimension n. Let $\{\lambda_k\}_{k=1}^n$ denote the eigenvalues for S; each eigenvalue appears in the list corresponding to its algebraic multiplicity. Then*

$$\sum_{k=1}^n \lambda_k = \sum_{k=1}^m ||f_k||^2.$$

(iii) *Assume that V has dimension n. If $\{f_k\}_{k=1}^m$ is tight and $||f_k|| = 1$ for all k, then the frame bound is $A = m/n$.*

Proof. Assume that $\{f_k\}_{k=1}^m$ is a frame for V. Since the frame operator $S : V \to V$ is self-adjoint, Theorem 1.1.8 shows that V has an orthonormal basis consisting of eigenvectors $\{e_k\}_{k=1}^n$ for S. Denote the corresponding eigenvalues by $\{\lambda_k\}_{k=1}^n$. Given $f \in V$, we can write

$$f = \sum_{k=1}^n \langle f, e_k \rangle e_k.$$

Then

$$Sf = \sum_{k=1}^n \langle f, e_k \rangle S e_k = \sum_{k=1}^n \lambda_k \langle f, e_k \rangle e_k,$$

and

$$\sum_{k=1}^m |\langle f, f_k \rangle|^2 = \langle Sf, f \rangle = \sum_{k=1}^n \lambda_k |\langle f, e_k \rangle|^2.$$

Therefore

$$\lambda_{\min} ||f||^2 \le \sum_{k=1}^m |\langle f, f_k \rangle|^2 \le \lambda_{\max} ||f||^2.$$

So $\lambda_{\min}$ is a lower frame bound, and $\lambda_{\max}$ is an upper frame bound. That they are the optimal frame bounds follows by taking f to be an eigenvector corresponding to $\lambda_{\min}$ (respectively $\lambda_{\max}$). This proves (i).

For the proof of (ii), we have

$$\sum_{k=1}^{n} \lambda_k = \sum_{k=1}^{n} \lambda_k \, \|e_k\|^2 = \sum_{k=1}^{n} \langle Se_k, e_k \rangle$$

$$= \sum_{k=1}^{n} \sum_{\ell=1}^{m} |\langle e_k, f_\ell \rangle|^2.$$

Interchanging the sums and using that $\{e_k\}_{k=1}^{n}$ is an orthonormal basis for V finally gives (ii). For the proof of (iii), the assumptions imply that the set of eigenvalues $\{\lambda_k\}_{k=1}^{n}$ consists of the frame bound A repeated n times; thus the result follows from (ii). □

Corollary 1.1.13 *Let $\{f_k\}_{k=1}^{m}$ be a frame for V. Then the condition number for the frame operator is equal to the ratio between the optimal upper frame bound and the optimal lower frame bound.*

1.2 Frames in $\mathbb{C}^n$

The natural examples of finite-dimensional vector spaces are

$$\mathbb{R}^n = \{(c_1, c_2, \ldots, c_n) \mid c_i \in \mathbb{R}, \ i = 1, \ldots, n\}$$

and

$$\mathbb{C}^n = \{(c_1, c_2, \ldots, c_n) \mid c_i \in \mathbb{C}, \ i = 1, \ldots, n\};$$

the latter is equipped with the inner product

$$\langle \{c_k\}_{k=1}^{n}, \{d_k\}_{k=1}^{n} \rangle = \sum_{k=1}^{n} c_k \overline{d_k}$$

and the associated norm

$$\|\{c_k\}_{k=1}^{n}\| = \sqrt{\sum_{k=1}^{n} |c_k|^2}.$$

This corresponds to the definitions in $\mathbb{R}^n$, except that complex conjugation and modulus are not needed in the real case. We will describe the theory for bases and frames in $\mathbb{C}^n$, but easy modifications give the corresponding results in $\mathbb{R}^n$. If, for example, $\{f_k\}_{k=1}^{m}$ is a frame for $\mathbb{C}^n$, then the $2m$ vectors consisting of the real parts, respectively the imaginary parts, of the frame vectors will be a frame for $\mathbb{R}^n$ (Exercise 1.11); in particular, if the vectors $\{f_k\}_{k=1}^{m}$ have real coordinates, they constitute a frame for $\mathbb{R}^n$. On the other hand a frame for $\mathbb{R}^n$ is automatically a frame for $\mathbb{C}^n$; we ask the reader to prove this in Exercise 1.12.

The canonical basis for $\mathbb{C}^n$ consists of the vectors $\{\delta_k\}_{k=1}^n$, where δ_k is the vector in $\mathbb{C}^n$ having 1 at the k-th entry and otherwise 0. We will consequently identify vectors in $\mathbb{C}^n$ with their representation in this basis.

From elementary linear algebra, we know many equivalent conditions for a set of vectors to constitute a basis for $\mathbb{C}^n$. Let us list the most important characterizations:

Theorem 1.2.1 *Consider n vectors in $\mathbb{C}^n$ and write them as columns in an $n \times n$ matrix,*

$$
\Lambda = \begin{pmatrix} \lambda_{11} & \lambda_{12} & \cdot & \cdot & \lambda_{1n} \\ \lambda_{21} & \lambda_{22} & \cdot & \cdot & \lambda_{2n} \\ \cdot & \cdot & \cdot & \cdot & \cdot \\ \cdot & \cdot & \cdot & \cdot & \cdot \\ \lambda_{n1} & \lambda_{n2} & \cdot & \cdot & \lambda_{nn} \end{pmatrix}.
$$

Then the following are equivalent:

  (i) *The columns in Λ (i.e., the given vectors) constitute a basis for $\mathbb{C}^n$.*

  (ii) *The rows in Λ constitute a basis for $\mathbb{C}^n$.*

  (iii) *The determinant of Λ is non-zero.*

  (iv) *Λ is invertible.*

  (v) *Λ defines an injective mapping from $\mathbb{C}^n$ into $\mathbb{C}^n$.*

  (vi) *Λ defines a surjective mapping from $\mathbb{C}^n$ onto $\mathbb{C}^n$.*

  (vii) *The columns in Λ are linearly independent.*

(viii) *Λ has rank equal to n.*

Recall that the *rank* of a matrix E is defined as the dimension of its range $\mathcal{R}_E$. We also remind the reader that any basis can be turned into an orthonormal basis by applying the Gram–Schmidt orthogonalization procedure.

We now turn to a discussion of frames for $\mathbb{C}^n$. Note that we *consequently identify operators $V : \mathbb{C}^n \to \mathbb{C}^m$ with their matrix representations with respect to the canonical bases in $\mathbb{C}^n$ and $\mathbb{C}^m$.* Letting $\{e_k\}_{k=1}^n$ denote the canonical orthonormal basis in $\mathbb{C}^n$ and $\{\tilde{e}_k\}_{k=1}^m$ the canonical orthonormal basis in $\mathbb{C}^m$, the matrix representation of V is the $m \times n$ matrix, where the k-th column consists of the coordinates of the image under V of the k-th basis vector in V, in terms of the given basis in W. The jk-th entry in the matrix representation is $\langle Ve_k, \tilde{e}_j \rangle$.

In case $\{f_k\}_{k=1}^m$ is a frame for $\mathbb{C}^n$, the pre-frame operator T defined in (1.4) maps $\mathbb{C}^m$ onto $\mathbb{C}^n$, and its matrix with respect to the canonical bases in $\mathbb{C}^n$ and $\mathbb{C}^m$ is

$$T = \begin{pmatrix} | & | & \cdot & \cdot & | \\ f_1 & f_2 & \cdot & \cdot & f_m \\ | & | & \cdot & \cdot & | \end{pmatrix}, \tag{1.15}$$

i.e., the $n \times m$ matrix having the vectors f_k as columns.

Since m vectors can at most span an m-dimensional space, we necessarily have $m \geq n$ when $\{f_k\}_{k=1}^m$ is a frame for $\mathbb{C}^n$, i.e., the matrix T has at least as many columns as rows.

We now show that frames $\{f_k\}_{k=1}^m$ for $\mathbb{C}^n$ naturally appear by projections of certain bases in $\mathbb{C}^m$ onto $\mathbb{C}^n$, i.e., by removal of some of the coordinates:

Theorem 1.2.2 *Let $\{f_k\}_{k=1}^m$ be a frame for $\mathbb{C}^n$. Then the following holds:*

(i) *The vectors f_k can be considered as the first n coordinates of some vectors g_k in $\mathbb{C}^m$ constituting a basis for $\mathbb{C}^m$.*

(ii) *If $\{f_k\}_{k=1}^m$ is tight, then the vectors f_k are the first n coordinates of some vectors g_k in $\mathbb{C}^m$ constituting an orthogonal basis for $\mathbb{C}^m$.*

Proof. Let $\{f_k\}_{k=1}^m$ be an arbitrary frame for $\mathbb{C}^n$. Then $m \geq n$. Consider the mapping

$$F : \mathbb{C}^n \to \mathbb{C}^m, \quad Fx = \{\langle x, f_k \rangle\}_{k=1}^m.$$

F is the adjoint of the pre-frame operator T. The matrix for F with respect to the canonical bases is the $m \times n$ matrix where the k-th row is the complex conjugate of f_k, i.e.,

$$F = \begin{pmatrix} - & \overline{f_1} & - \\ - & \overline{f_2} & - \\ \cdot & \cdot & \cdot \\ \cdot & \cdot & \cdot \\ - & \overline{f_m} & - \end{pmatrix}.$$

If $Fx = 0$, then $0 = ||Fx||^2 = \sum_{k=1}^m |\langle x, f_k \rangle|^2$. Since $\text{span}\{f_k\}_{k=1}^m = \mathbb{C}^n$, it follows that $x = 0$, so F is an injective mapping. We can therefore extend F to a bijection $\tilde{F}$ of $\mathbb{C}^m$ onto $\mathbb{C}^m$: for example, still letting $\{\delta_k\}_{k=1}^m$ be the canonical basis for $\mathbb{C}^m$, let $\{\phi_k\}_{k=n+1}^m$ be a basis for the orthogonal complement of $\mathcal{R}_F$ in $\mathbb{C}^m$ and extend F by the definition $\tilde{F}\delta_k := \phi_k$, $k = n+1, n+2, \ldots, m$. The matrix for $\tilde{F}$ is an $m \times m$ matrix, whose first n columns are the columns from F:

$$\tilde{F} = \begin{pmatrix} - & \overline{f_1} & - & | & | & \cdot & | \\ \cdot & \cdot & \cdot & | & \phi_{n+1} & \cdot & \phi_m \\ - & \overline{f_m} & - & | & | & \cdot & | \end{pmatrix}.$$

Since $\tilde{F}$ is surjective, the columns span $\mathbb{C}^m$. The rank of the rows equals the rank of the columns, so also the rows in $\tilde{F}$ span $\mathbb{C}^m$, and they are linearly independent. Thus, they constitute a basis for $\mathbb{C}^m$.

If $\{f_k\}_{k=1}^m$ is a tight frame for $\mathbb{C}^n$ with frame bound A and $\{\delta_k\}_{k=1}^n$ still denotes the canonical basis for $\mathbb{C}^n$, Proposition 1.1.4 shows that

$$\langle TT^*\delta_l, \delta_j \rangle \;=\; A\delta_{j,l}, \; j,l = 1,\ldots,n.$$

$\langle TT^*\delta_l, \delta_j \rangle$ is the j,l-th entry in the matrix representation for TT^*, so this calculation shows that the n rows in the matrix representation (1.15) for T are orthogonal, considered as vectors in $\mathbb{C}^m$. By adding $m - n$ rows we can extend the matrix for T to an $m \times m$ matrix in which the rows are orthogonal. Therefore the columns are orthogonal. $\qquad\square$

Geometrically, Theorem 1.2.2 means that if $\{f_k\}_{k=1}^m$ is a frame for $\mathbb{C}^n$, then there exist vectors $\{h_k\}_{k=1}^m$ in $\mathbb{C}^{m-n}$ such that the columns in the $m \times n$ matrix

$$\begin{pmatrix} | & | & \cdot & \cdot & | \\ f_1 & f_2 & \cdot & \cdot & f_m \\ | & | & \cdot & \cdot & | \\ h_1 & h_2 & \cdot & \cdot & h_m \\ | & | & \cdot & \cdot & | \end{pmatrix} \tag{1.16}$$

constitute a basis for $\mathbb{C}^m$.

For a given $m \times n$ matrix Λ, the following proposition gives a condition for the rows constituting a frame for $\mathbb{C}^n$.

Proposition 1.2.3 *For an $m \times n$ matrix*

$$\Lambda = \begin{pmatrix} \lambda_{11} & \lambda_{12} & \cdot & \cdot & \lambda_{1n} \\ \lambda_{21} & \lambda_{22} & \cdot & \cdot & \lambda_{2n} \\ \cdot & \cdot & \cdot & \cdot & \cdot \\ \cdot & \cdot & \cdot & \cdot & \cdot \\ \lambda_{m1} & \lambda_{m2} & \cdot & \cdot & \lambda_{mn} \end{pmatrix},$$

the following are equivalent:

(i) There exists a constant $A > 0$ such that

$$A\sum_{k=1}^n |c_k|^2 \le ||\Lambda\{c_k\}_{k=1}^n||^2, \; \forall\{c_k\}_{k=1}^n \in \mathbb{C}^n.$$

(ii) The columns in Λ constitute a basis for their span in $\mathbb{C}^m$.

(iii) The rows in Λ constitute a frame for $\mathbb{C}^n$.

Proof.   Denote the columns in Λ by $g_1, \ldots, g_n$; they are vectors in $\mathbb{C}^m$. By definition, (i) means that for all $\{c_k\}_{k=1}^n \in \mathbb{C}^n$,

$$A\sum_{k=1}^n |c_k|^2 \le \left|\left|\sum_{k=1}^n c_k g_k\right|\right|^2, \tag{1.17}$$

which is equivalent to $\{g_k\}_{k=1}^n$ being a basis for its span in $\mathbb{C}^m$ (use an argument like in the proof of Proposition 1.1.2). On the other hand, denoting the rows in Λ by $f_1, \ldots, f_m$, (i) can also be written as

$$A \sum_{k=1}^n |c_k|^2 \le \sum_{k=1}^n \left| \left\langle f_k, \begin{pmatrix} \overline{c_1} \\ \overline{c_2} \\ \vdots \\ \overline{c_n} \end{pmatrix} \right\rangle \right|^2 , \ \forall \{c_k\}_{k=1}^n \in \mathbb{C}^n,$$

which is equivalent to (iii). $\qquad \square$

Example 1.2.4 As an illustration of Proposition 1.2.3, consider the matrix

$$\Lambda = \begin{pmatrix} 1 & 0 \\ 0 & 1 \\ 1 & 0 \end{pmatrix}.$$

It is clear that the rows $\begin{pmatrix} 1 \\ 0 \end{pmatrix}, \begin{pmatrix} 0 \\ 1 \end{pmatrix}, \begin{pmatrix} 1 \\ 0 \end{pmatrix}$ constitute a frame for $\mathbb{C}^2$.

The columns $\begin{pmatrix} 1 \\ 0 \\ 1 \end{pmatrix}, \begin{pmatrix} 0 \\ 1 \\ 0 \end{pmatrix}$ constitute a basis for their span in $\mathbb{C}^3$, but the span is only a two-dimensional subspace of $\mathbb{C}^3$. $\qquad \square$

As an immediate consequence of the proof of Proposition 1.2.3, we have the following useful fact:

Corollary 1.2.5 *Let Λ be an $m \times n$ matrix. Denote the columns by $g_1, \ldots, g_n$ and the rows by $f_1, \ldots, f_m$. Given $A, B > 0$, the vectors $\{f_k\}_{k=1}^m$ constitute a frame for $\mathbb{C}^n$ with bounds A, B if and only if*

$$A \sum_{k=1}^n |c_k|^2 \le \left\| \sum_{k=1}^n c_k g_k \right\|^2 \le B \sum_{k=1}^n |c_k|^2, \ \forall \{c_k\}_{k=1}^n \in \mathbb{C}^n.$$

Example 1.2.6 Consider the vectors

$$\begin{pmatrix} 0 \\ \sqrt{\frac{1}{3}} \\ \sqrt{\frac{2}{3}} \end{pmatrix}, \begin{pmatrix} 0 \\ -\sqrt{\frac{1}{3}} \\ \sqrt{\frac{2}{3}} \end{pmatrix}, \begin{pmatrix} 0 \\ 1 \\ 0 \end{pmatrix}, \begin{pmatrix} \sqrt{\frac{5}{6}} \\ 0 \\ \sqrt{\frac{1}{6}} \end{pmatrix}, \begin{pmatrix} -\sqrt{\frac{5}{6}} \\ 0 \\ \sqrt{\frac{1}{6}} \end{pmatrix} \qquad (1.18)$$

in $\mathbb{C}^3$. Corresponding to these vectors, we consider the matrix

$$\Lambda = \begin{pmatrix} 0 & \sqrt{\frac{1}{3}} & \sqrt{\frac{2}{3}} \\ 0 & -\sqrt{\frac{1}{3}} & \sqrt{\frac{2}{3}} \\ 0 & 1 & 0 \\ \sqrt{\frac{5}{6}} & 0 & \sqrt{\frac{1}{6}} \\ -\sqrt{\frac{5}{6}} & 0 & \sqrt{\frac{1}{6}} \end{pmatrix}.$$

The reader can check that the columns $\{g_k\}_{k=1}^3$ are orthogonal in $\mathbb{C}^5$ and all have length $\sqrt{\frac{5}{3}}$. Therefore

$$\left\| \sum_{k=1}^3 c_k g_k \right\|^2 = \frac{5}{3} \sum_{k=1}^3 |c_k|^2$$

for all $c_1, c_2, c_3 \in \mathbb{C}$. By Corollary 1.2.5, we conclude that the vectors defined by (1.18) constitute a tight frame for $\mathbb{C}^3$ with frame bound $\frac{5}{3}$. The frame is normalized. $\qquad\square$

For later use, we state a special case of Corollary 1.2.5; we ask the reader to provide the proof in Exercise 1.13.

Corollary 1.2.7 *Let Λ be an $m \times n$ matrix. Then the following are equivalent:*

(i) *$\Lambda^* \Lambda = I$, the $n \times n$ identity matrix.*

(ii) *The columns $g_1, \ldots, g_n$ in Λ constitute an orthonormal system in $\mathbb{C}^m$.*

(iii) *The rows $f_1, \ldots, f_m$ in Λ constitute a tight frame for $\mathbb{C}^n$ with frame bound equal to 1.*

1.3   The discrete Fourier transform

When working with frames and bases in $\mathbb{C}^n$, one has to be particularly careful with the meaning of the notation. For example, we have used f_k and g_k to denote vectors in $\mathbb{C}^n$, whereas c_k in general is the k-th coordinate of a sequence $\{c_k\}_{k=1}^n \in \mathbb{C}^n$, i.e., c_k is a scalar. In order to avoid confusion, we will change the notation slightly in this section. The key to the new notation is the observation that to have a sequence in $\mathbb{C}^n$ is equivalent to having a function

$$f : \{1, \ldots, n\} \to \mathbb{C};$$

the j-th entry in the sequence corresponds to the j-th function value $f(j)$.

Our purpose is to consider a special orthonormal basis for $\mathbb{C}^n$. Given $f \in \mathbb{C}^n$, we denote the coordinates of f with respect to the canonical orthonormal basis $\{\delta_k\}_{k=1}^n$ by $\{f(j)\}_{j=1}^n$. For $k = 1, \ldots, n$, we define vectors $e_k \in \mathbb{C}^n$ by

$$e_k(j) = \frac{1}{\sqrt{n}} e^{2\pi i (j-1)(k-1)/n}, \quad j = 1, \ldots, n; \tag{1.19}$$

that is

$$e_k = \frac{1}{\sqrt{n}} \begin{pmatrix} 1 \\ e^{2\pi i (k-1)/n} \\ e^{4\pi i (k-1)/n} \\ \cdot \\ \cdot \\ e^{2\pi i (n-1)(k-1)/n} \end{pmatrix}, \quad k = 1, \ldots n. \tag{1.20}$$

Theorem 1.3.1 *The vectors $\{e_k\}_{k=1}^n$ defined by (1.19) constitute an orthonormal basis for $\mathbb{C}^n$.*

Proof. Since $\{e_k\}_{k=1}^n$ are n vectors in an n-dimensional vector space, it is enough to prove that they constitute an orthonormal system. It is clear that $\|e_k\| = 1$ for all k. Now, given $k \neq \ell$,

$$\langle e_k, e_\ell \rangle = \frac{1}{n} \sum_{j=1}^n e^{2\pi i (j-1)(k-1)/n} e^{-2\pi i (j-1)(\ell-1)/n} = \frac{1}{n} \sum_{j=0}^{n-1} e^{2\pi i j(k-\ell)/n}.$$

Using the formula $(1-x)(1+x+\cdots+x^{n-1}) = 1 - x^n$ with $x = e^{2\pi i (k-\ell)/n}$, we get

$$\langle e_k, e_\ell \rangle = \frac{1}{n} \frac{1 - (e^{2\pi i (k-\ell)/n})^n}{1 - e^{2\pi i (k-\ell)/n}} = 0. \qquad \square$$

The basis $\{e_k\}_{k=1}^n$ is called the *discrete Fourier transform basis*. Using this basis, every sequence $f \in \mathbb{C}^n$ has a representation

$$f = \sum_{k=1}^n \langle f, e_k \rangle e_k = \frac{1}{\sqrt{n}} \sum_{k=1}^n \sum_{\ell=1}^n f(\ell) e^{-2\pi i (\ell-1)(k-1)/n} e_k.$$

Written out in coordinates, this means that

$$\begin{aligned} f(j) &= \frac{1}{n} \sum_{k=1}^n \sum_{\ell=1}^n f(\ell) e^{-2\pi i (\ell-1)(k-1)/n} e^{2\pi i (j-1)(k-1)/n} \\ &= \frac{1}{n} \sum_{k=1}^n \sum_{\ell=1}^n f(\ell) e^{2\pi i (j-\ell)(k-1)/n}, \quad j = 1, \ldots, n. \end{aligned}$$

Applications often ask for tight frames because the cumbersome inversion of the frame operator is avoided in this case, see (1.9). It is interesting that overcomplete tight frames can be obtained in $\mathbb{C}^n$ by projecting the discrete Fourier transform basis in any $\mathbb{C}^m$, $m > n$, onto $\mathbb{C}^n$:

Proposition 1.3.2 *Let $m > n$ and define the vectors $\{f_k\}_{k=1}^m$ in $\mathbb{C}^n$ by*

$$f_k = \frac{1}{\sqrt{m}} \begin{pmatrix} 1 \\ e^{2\pi i(k-1)/m} \\ \cdot \\ \cdot \\ e^{2\pi i(n-1)(k-1)/m} \end{pmatrix}, \quad k = 1, 2, \ldots, m.$$

Then $\{f_k\}_{k=1}^m$ is a tight overcomplete frame for $\mathbb{C}^n$ with frame bound equal to one, and $\|f_k\| = \sqrt{\frac{n}{m}}$ for all k.

Proof. Let $\{\delta_j\}_{j=1}^n$ be the canonical basis for $\mathbb{C}^n$, and let $\{e_k\}_{k=1}^m$ be the discrete Fourier transform basis for $\mathbb{C}^m$, i.e.,

$$e_k = \frac{1}{\sqrt{m}} \begin{pmatrix} 1 \\ e^{2\pi i(k-1)/m} \\ \cdot \\ e^{2\pi i(n-1)(k-1)/m} \\ \cdot \\ \cdot \\ e^{2\pi i(m-1)(k-1)/m} \end{pmatrix}.$$

Identifying $\mathbb{C}^n$ with a subspace of $\mathbb{C}^m$, the orthogonal projection of e_k onto $\mathbb{C}^n$ is $Pe_k = f_k$; now the result follows from Exercise 1.14. □

It is important to notice that all the vectors f_k in Proposition 1.3.2 have the same norm. If needed, we can therefore normalize them while keeping a tight frame; we only have to adjust the frame bound accordingly. We formulate the result as an existence result, but it is important to keep in mind that we actually have an explicit construction:

Corollary 1.3.3 *For any $m \geq n$, there exists a tight frame in $\mathbb{C}^n$ consisting of m normalized vectors.*

Example 1.3.4 The discrete Fourier transform basis in $\mathbb{C}^4$ consists of the vectors

$$\frac{1}{2}\begin{pmatrix} 1 \\ 1 \\ 1 \\ 1 \end{pmatrix}, \frac{1}{2}\begin{pmatrix} 1 \\ i \\ -1 \\ -i \end{pmatrix}, \frac{1}{2}\begin{pmatrix} 1 \\ -1 \\ 1 \\ -1 \end{pmatrix}, \frac{1}{2}\begin{pmatrix} 1 \\ -i \\ -1 \\ i \end{pmatrix}.$$

Via Proposition 1.3.2, the vectors

$$\frac{1}{2}\begin{pmatrix}1\\1\end{pmatrix}, \frac{1}{2}\begin{pmatrix}1\\i\end{pmatrix}, \frac{1}{2}\begin{pmatrix}1\\-1\end{pmatrix}, \frac{1}{2}\begin{pmatrix}1\\-i\end{pmatrix},$$

constitute a tight frame in $\mathbb{C}^2$. with frame bound one. The vectors have length $1/\sqrt{2}$. Changing the length of the vectors, i.e., considering the vectors

$$\frac{1}{\sqrt{2}}\begin{pmatrix}1\\1\end{pmatrix}, \frac{1}{\sqrt{2}}\begin{pmatrix}1\\i\end{pmatrix}, \frac{1}{\sqrt{2}}\begin{pmatrix}1\\-1\end{pmatrix}, \frac{1}{\sqrt{2}}\begin{pmatrix}1\\-i\end{pmatrix},$$

we obtain a tight frame with frame bound 2, consisting of normalized vectors. □

1.4 Pseudo-inverses and the singular value decomposition

It is well-known from linear algebra that not all matrices have an inverse. Keeping in mind how useful inverses are, it is natural to search for some types of "generalized inverses" in case no inverse exists; they should capture at least some of the nice properties.

The right definition of a generalized inverse depends on the properties we are interested in, and we shall only define the so-called *pseudo-inverse*. Given an $m \times n$ matrix E, we consider it as a linear mapping of $\mathbb{C}^n$ into $\mathbb{C}^m$. E is not necessarily injective, but by restricting E to the orthogonal complement of the kernel $\mathcal{N}_E$, we obtain an injective linear mapping

$$\tilde{E} : \mathcal{N}_E^\perp \to \mathbb{C}^m.$$

E and $\tilde{E}$ have the same range, $\mathcal{R}_{\tilde{E}} = \mathcal{R}_E$; thus $\tilde{E}$ considered as a mapping from $\mathcal{N}_E^\perp$ to $\mathcal{R}_E$ has an inverse,

$$(\tilde{E})^{-1} : \mathcal{R}_E \to \mathcal{N}_E^\perp.$$

We can extend $(\tilde{E})^{-1}$ to an operator $E^\dagger : \mathbb{C}^m \to \mathbb{C}^n$ by defining

$$E^\dagger(y + z) = (\tilde{E})^{-1}y \text{ if } y \in \mathcal{R}_E, z \in \mathcal{R}_E^\perp. \tag{1.21}$$

With this definition,

$$EE^\dagger x = x, \ \forall x \in \mathcal{R}_E. \tag{1.22}$$

The operator $E^\dagger$ is called the *pseudo-inverse* of E. From the definition, we immediately have that

$$\mathcal{N}_{E^\dagger} = \mathcal{R}_E^\perp = \mathcal{N}_{E^*}, \ \ \mathcal{R}_{E^\dagger} = \mathcal{N}_E^\perp = \mathcal{R}_{E^*}. \tag{1.23}$$

We state two characterizations of the pseudo-inverse:

Proposition 1.4.1 *Let E be an $m \times n$ matrix. Then*

(i) $E^\dagger$ is the unique $n \times m$ matrix for which $EE^\dagger$ is the orthogonal projection onto $\mathcal{R}_E$ and $E^\dagger E$ is the orthogonal projection onto $\mathcal{R}_{E^\dagger}$.

(ii) $E^\dagger$ is the unique $n \times m$ matrix for which $EE^\dagger$ and $E^\dagger E$ are self-adjoint and

$$EE^\dagger E = E, \; E^\dagger E E^\dagger = E^\dagger.$$

Proof. We first prove the equivalence between the conditions stated in (i) and (ii). If a matrix $E^\dagger$ satisfies (i), it immediately follows that (ii) is satisfied. On the other hand, if (ii) is satisfied, then

$$(EE^\dagger)^2 = EE^\dagger E E^\dagger = EE^\dagger.$$

Since $EE^\dagger$ is self-adjoint, it follows that $EE^\dagger$ is the orthogonal projection onto $\mathcal{R}_{EE^\dagger}$. Finally, the identity $EE^\dagger E = E$ shows that $\mathcal{R}_{EE^\dagger} = \mathcal{R}_E$. The proof that $E^\dagger E$ is the orthogonal projection onto $\mathcal{R}_{E^\dagger}$ is similar. Thus (i) is satisfied.

We now prove the equivalence between the properties in Proposition 1.4.1 and the definition (1.21) of the pseudo-inverse. First we note that with our definition of the pseudo-inverse, the conditions in (i) are satisfied; the main ingredients in the following argument are the relations (1.22) and (1.23). In fact, if $y \in \mathcal{R}_E$, then $EE^\dagger y = y$; and if $y \in \mathcal{R}_E^\perp = \mathcal{N}_{E^\dagger}$, then $EE^\dagger y = 0$. This proves that $EE^\dagger$ is the orthogonal projection onto $\mathcal{R}_E$. Also, if $y \in \mathcal{R}_{E^\dagger}^\perp = \mathcal{N}_E$, then $E^\dagger E y = 0$; and if $y \in \mathcal{R}_{E^\dagger}$, $y = E^\dagger x$ for some x, then

$$E^\dagger E y = E^\dagger E E^\dagger x = E^\dagger x - E^\dagger (I - EE^\dagger) x = E^\dagger x = y.$$

Here we used that $I - EE^\dagger$ is the orthogonal projection onto $\mathcal{R}_E^\perp = \mathcal{N}_{E^\dagger}$. We have now proved that $E^\dagger E$ is the orthogonal projection onto $\mathcal{R}_{E^\dagger}$.

To conclude, we only have to prove that if a matrix $E^\dagger$ satisfies (i) and (ii), then it fulfills the requirements in the definition of the pseudo-inverse, i.e., (1.21) is satisfied. First, we note that (ii) implies that

$$E^* = (EE^\dagger E)^* = (E^\dagger E)^* E^* = E^\dagger E E^*;$$

this shows that

$$\mathcal{N}_E^\perp = \mathcal{R}_{E^*} \subseteq \mathcal{R}_{E^\dagger}.$$

Now, if $y \in \mathcal{R}_E$, then we can find $x \in \mathcal{N}_E^\perp$ such that $y = Ex$; thus

$$E^\dagger y = E^\dagger E x = x = (\tilde{E})^{-1} E x = (\tilde{E})^{-1} y.$$

Finally, if $z \in \mathcal{R}_E^\perp = \mathcal{N}_{E^*}$, then by (i), $EE^\dagger z = 0$; using (ii),

$$E^\dagger z = E^\dagger E E^\dagger z = 0. \qquad \square$$

The pseudo-inverse gives the solution to an important minimization problem:

Theorem 1.4.2 *Let E be an $m \times n$ matrix. Given $y \in \mathcal{R}_E$, the equation $Ex = y$ has a unique solution of minimal norm, namely $x = E^\dagger y$.*

Proof. By (1.22), we know that $x := E^\dagger y$ is a solution to the equation $Ex = y$. All solutions have the form $x = E^\dagger y + z$, where $z \in \mathcal{N}_E$. Since $E^\dagger y \in \mathcal{N}_E^\perp$, the norm of the general solution satisfies that

$$||x||^2 = ||E^\dagger y + z||^2 = ||E^\dagger y||^2 + ||z||^2.$$

This expression is minimal when $z = 0$. □

Historically, (i) and (ii) in Proposition 1.4.1 were given as definitions of a "generalized inverse" by Moore, respectively Penrose. For this reason, the pseudo-inverse is frequently called the *Moore–Penrose inverse*.

For computational purposes, it is important to notice that the pseudo-inverse can be found using the singular value decomposition of E. We begin with a lemma.

Lemma 1.4.3 *Let E be an $m \times n$ matrix with rank $r \geq 1$. Then there exist constants $\sigma_1, \ldots, \sigma_r > 0$ and orthonormal bases $\{u_k\}_{k=1}^r$ for $\mathcal{R}_E$ and $\{v_k\}_{k=1}^r$ for $\mathcal{R}_{E^*}$ such that*

$$Ev_k = \sigma_k u_k, \quad k = 1, \ldots, r. \tag{1.24}$$

Proof. Observe that $E^* E$ is a self-adjoint $n \times n$ matrix; by Theorem 1.1.8 this implies that there exists an orthonormal basis $\{v_k\}_{k=1}^n$ for $\mathbb{C}^n$ consisting of eigenvectors for $E^* E$. Let $\{\lambda_k\}_{k=1}^n$ denote the corresponding eigenvalues. Note that for each k,

$$\lambda_k = \lambda_k ||v_k||^2 = \langle E^* E v_k, v_k \rangle = ||Ev_k||^2 \geq 0.$$

The rank of E is given by

$$r = \dim \mathcal{R}_E = \dim \mathcal{R}_{E^*};$$

since $\mathcal{R}_E^\perp = \mathcal{N}_{E^*}$, we have

$$\mathcal{R}_{E^*} = \mathcal{R}_{E^* E} = \text{span}\{E^* E v_k\}_{k=1}^n = \text{span}\{\lambda_k v_k\}_{k=1}^n. \tag{1.25}$$

Thus, the rank is equal to the number of non-zero eigenvalues, counted with multiplicity. We can assume that the eigenvectors $\{v_k\}_{k=1}^n$ are ordered such that $\{v_k\}_{k=1}^r$ corresponds to the non-zero eigenvalues. Then (1.25) shows that $\{v_k\}_{k=1}^r$ is an orthonormal basis for $\mathcal{R}_{E^*}$. Note that for $k > r$, we have $||Ev_k||^2 = \langle E^* E v_k, v_k \rangle = 0$, i.e.,

$$Ev_k = 0, \quad k > r. \tag{1.26}$$

Defining

$$u_k := \frac{1}{\sqrt{\lambda_k}} Ev_k, \quad k = 1, \ldots, r,$$

we therefore obtain that $\{u_k\}_{k=1}^r$ spans $\mathcal{R}_E$; and it is an orthonormal basis for $\mathcal{R}_E$ because for all $k, l = 1, \ldots, r$ we have

$$
\begin{aligned}
\langle u_k, u_l \rangle &= \frac{1}{\sqrt{\lambda_k}} \frac{1}{\sqrt{\lambda_l}} \langle Ev_k, Ev_l \rangle \\
&= \frac{1}{\sqrt{\lambda_k \lambda_l}} \langle E^* Ev_k, v_l \rangle \\
&= \sqrt{\frac{\lambda_k}{\lambda_l}} \langle v_k, v_l \rangle \\
&= \delta_{k,l}.
\end{aligned}
$$

Thus, the conditions in Lemma 1.4.3 are fulfilled with

$$
\sigma_k = \sqrt{\lambda_k}, \quad k = 1, \ldots, r.
$$
□

Lemma 1.4.3 leads to the *singular value decomposition* of E:

Theorem 1.4.4 *Every $m \times n$ matrix E with rank $r \geq 1$ has a decomposition*

$$
E = U \begin{pmatrix} D & 0 \\ 0 & 0 \end{pmatrix} V^*, \tag{1.27}
$$

where U is a unitary $m \times m$ matrix, V is a unitary $n \times n$ matrix, and $\begin{pmatrix} D & 0 \\ 0 & 0 \end{pmatrix}$ is an $m \times n$ block matrix in which D is an $r \times r$ diagonal matrix with positive entries $\sigma_1, \ldots, \sigma_r$ in the diagonal.

Proof. We use the proof of Lemma 1.4.3. Let $\{v_k\}_{k=1}^n$ be the orthonormal basis for $\mathbb{C}^n$ considered there, ordered such that $\{v_k\}_{k=1}^r$ is an orthonormal basis for $\mathcal{R}_{E^*}$. Let V be the $n \times n$ matrix having the vectors $\{v_k\}_{k=1}^n$ as columns. Extend the orthonormal basis $\{u_k\}_{k=1}^r$ for $\mathcal{R}_E$ to an orthonormal basis $\{u_k\}_{k=1}^m$ for $\mathbb{C}^m$ and let U be the $m \times m$ matrix having these vectors as columns. Finally, let D be the $r \times r$ diagonal matrix having $\sigma_1, \ldots, \sigma_r$ in the diagonal. Via (1.24) and (1.26),

$$
\begin{aligned}
EV &= \begin{pmatrix} \sigma_1 u_1 & \cdot & \cdot & \sigma_r u_r & 0 & \cdot & \cdot & 0 \end{pmatrix} \\
&= U \begin{pmatrix} D & 0 \\ 0 & 0 \end{pmatrix}.
\end{aligned}
$$

Multiplying with V^* from the right gives the result. □

The numbers $\sigma_1, \ldots, \sigma_r$ are called *singular values* for E; the proof of Lemma 1.4.3 shows that they are the square roots of the positive eigenvalues for $E^* E$.

Corollary 1.4.5 *With the notation in Theorem 1.4.4, the pseudo-inverse of E is given by*

$$E^\dagger = V \begin{pmatrix} D^{-1} & 0 \\ 0 & 0 \end{pmatrix} U^*, \tag{1.28}$$

where $\begin{pmatrix} D^{-1} & 0 \\ 0 & 0 \end{pmatrix}$ *is an* $n \times m$ *block matrix in which* D^{-1} *is the* $r \times r$ *matrix having* $1/\sigma_1, \ldots, 1/\sigma_r$ *in the diagonal.*

Proof. We check that the matrix $E^\dagger$ defined by (1.28) satisfies the requirements in Proposition 1.4.1(ii). First, via (1.27),

$$
\begin{aligned}
EE^\dagger &= U \begin{pmatrix} D & 0 \\ 0 & 0 \end{pmatrix} V^* V \begin{pmatrix} D^{-1} & 0 \\ 0 & 0 \end{pmatrix} U^* \\
&= U \begin{pmatrix} I & 0 \\ 0 & 0 \end{pmatrix} U^*,
\end{aligned}
$$

which shows that $EE^\dagger$ is self-adjoint. The proof that $E^\dagger E$ is self-adjoint is similar. Furthermore, using the derived expression for $EE^\dagger$,

$$
\begin{aligned}
EE^\dagger E &= U \begin{pmatrix} I & 0 \\ 0 & 0 \end{pmatrix} U^* U \begin{pmatrix} D & 0 \\ 0 & 0 \end{pmatrix} V^* \\
&= E.
\end{aligned}
$$

Similarly, one can verify that $E^\dagger EE^\dagger = E^\dagger$.   □

Let us return to the setting where $\{f_k\}_{k=1}^m$ is a frame for $\mathbb{C}^n$ with pre-frame operator $T : \mathbb{C}^m \to \mathbb{C}^n$. The calculation of the frame coefficients amounts to finding the pseudo-inverse $T^\dagger$:

Theorem 1.4.6 *Let* $\{f_k\}_{k=1}^m$ *be a frame for* $\mathbb{C}^n$, *with pre-frame operator T and frame operator S. Then*

$$T^\dagger f = \{\langle f, S^{-1} f_k \rangle\}_{k=1}^m, \ \forall f \in \mathbb{C}^n. \tag{1.29}$$

Proof. Let $f \in \mathbb{C}^n$. Expressed in terms of the pre-frame operator T, the equation $f = \sum_{k=1}^m c_k f_k$ means that $T\{c_k\}_{k=1}^m = f$. The result now follows by combining Theorem 1.1.5 and Theorem 1.4.2.   □

One interpretation of Theorem 1.4.6 is that when $\{f_k\}_{k=1}^m$ is a frame for $\mathbb{C}^n$, the matrix for $T^\dagger$ is obtained by placing the complex conjugate of the vectors in the canonical dual frame $\{S^{-1} f_k\}_{k=1}^m$ as rows in an $m \times n$ matrix:

$$T^\dagger = \begin{pmatrix} -\overline{S^{-1} f_1}- \\ -\overline{S^{-1} f_2}- \\ \cdot \\ \cdot \\ \cdot \\ -\overline{S^{-1} f_m}- \end{pmatrix}.$$

In operator terms, (1.29) means that

$$T^\dagger = T^*(TT^*)^{-1},$$

a formula that is known to hold generally for the pseudo-inverse of an arbitrary bounded surjective operator T.

The singular value decomposition gives a natural way to obtain coefficients $\{c_k\}_{k=1}^m$ such that $f = \sum_{k=1}^m c_k f_k$. Let $\{f_k\}_{k=1}^m$ be an overcomplete frame for $\mathbb{C}^n$ with pre-frame operator $T : \mathbb{C}^m \to \mathbb{C}^n$. Considered as a matrix, T is an $n \times m$ matrix, and we know that $m > n$. Since T is surjective, its rank equals n, so according to Theorem 1.4.4 its singular value decomposition is

$$T = U \begin{pmatrix} D & 0 \end{pmatrix} V^*.$$

Note that D is now an $n \times n$ matrix; $\begin{pmatrix} D & 0 \end{pmatrix}$ is an $n \times m$ matrix, U is an $n \times n$ matrix, and V is an $m \times m$ matrix. Given any $(m-n) \times n$ matrix F and any $f \in \mathbb{C}^n$, we have that

$$
\begin{aligned}
TV \begin{pmatrix} D^{-1} \\ F \end{pmatrix} U^* f &= U \begin{pmatrix} D & 0 \end{pmatrix} V^* V \begin{pmatrix} D^{-1} \\ F \end{pmatrix} U^* f \\
&= UIU^* f \\
&= f.
\end{aligned}
$$

This means that we can use the coefficients

$$\{c_k\}_{k=1}^m = V \begin{pmatrix} D^{-1} \\ F \end{pmatrix} U^* f$$

for the reconstruction of f, regardless how the entries in the matrix F are chosen. By Corollary 1.4.5, the choice $F = 0$ leads to the pseudo-inverse, which, as noted already in Theorem 1.1.5, is optimal in the sense that the ℓ^2-norm of the coefficients is minimized. However, there are many cases where other properties than minimal ℓ^2-norm are more relevant. The matrix

$$V \begin{pmatrix} D^{-1} \\ F \end{pmatrix} U^*$$

is frequently called a *generalized inverse* of T.

1.5 Applications in signal transmission

Mathematically, the option of having overcompleteness in a frame makes the concept more flexible than that of a basis: we have more freedom, which enhances the chance that we can construct systems having prescribed properties. The overcompleteness is also useful in practice, e.g., in the context of signal transmission. We will explain this in more detail below.

Modern communication networks act by transporting packets of data. Each packet contains the "essential information," i.e., the data we want to transmit, as well as a collection of "control parameters." The purpose of these extra parameters is to check that the data are delivered correctly: in case an error occurs, no packet will be delivered at all. It is clear that if there are no relationships between the various packets, the data belonging to a lost packet cannot be recovered. However, if there is some redundancy built into the system, i.e., relationships between the information in the packets, there is some hope that at least parts of the missing data can be recovered.

Mathematically, one can model the packets to transmit as frame coefficients. Thus, a packet that is not delivered amounts to removal of an element from the frame. If the frame was a basis, it would no longer be a basis after removal of an element; however, if it is overcomplete, it is possible that it remains a frame after deletion of an element.

In practice, one might lose more than one packet, i.e., more than one frame element. Thus, we are facing the question of how to construct frames that are stable toward removal of more than one element.

Let us formulate the relevant mathematical question in terms of frames in $\mathbb{C}^n$. A frame for $\mathbb{C}^n$ needs to contain at least n elements. Thus, how can we construct a frame $\{f_k\}_{k=1}^m$ for $\mathbb{C}^n, m > n$, such that the set remains a frame after removal of $m - n$ *arbitrary elements*? We now show that the frames $\{f_k\}_{k=1}^m$ obtained in Proposition 1.3.2 behave optimally:

Proposition 1.5.1 *Consider the frame* $\{f_k\}_{k=1}^m$ *for* $\mathbb{C}^n$ *defined in Proposition 1.3.2. Any subset containing at least n elements of this frame forms a frame for* $\mathbb{C}^n$.

Proof. Consider an arbitrary subset $\{k_1, k_2, \ldots, k_n\} \subseteq \{1, 2, \ldots, m\}$. Placing the vectors $\{f_{k_i}\}_{i=1}^n$ as rows in an $n \times n$ matrix and letting $z := e^{2\pi i/m}$, we obtain that

$$
\begin{pmatrix} -f_{k_1}- \\ -f_{k_2}- \\ \cdot \\ \cdot \\ -f_{k_n}- \end{pmatrix} = \frac{1}{\sqrt{m}} \begin{pmatrix} 1 & e^{2\pi i(k_1-1)/m} & \cdot & \cdot & e^{2\pi i(k_1-1)(n-1)/m} \\ 1 & e^{2\pi i(k_2-1)/m} & \cdot & \cdot & e^{2\pi i(k_2-1)(n-1)/m} \\ \cdot & \cdot & \cdot & \cdot & \cdot \\ \cdot & \cdot & \cdot & \cdot & \cdot \\ 1 & e^{2\pi i(k_n-1)/m} & \cdot & \cdot & e^{2\pi i(k_n-1)(n-1)/m} \end{pmatrix}
$$

$$
= \frac{1}{\sqrt{m}} \begin{pmatrix} 1 & z^{k_1-1} & \cdot & \cdot & z^{(k_1-1)(n-1)} \\ 1 & z^{k_2-1} & \cdot & \cdot & z^{(k_2-1)(n-1)} \\ \cdot & \cdot & \cdot & \cdot & \cdot \\ \cdot & \cdot & \cdot & \cdot & \cdot \\ 1 & z^{k_n-1} & \cdot & \cdot & z^{(k_n-1)(n-1)} \end{pmatrix};
$$

this is a Vandermonde matrix with determinant

$$\frac{1}{m^{n/2}} \prod_{i,j=1, i \neq j}^{n} (z^{k_i-1} - z^{k_j-1}) \neq 0.$$

Thus, $\{f_{k_i}\}_{i=1}^{n}$ is a basis for $\mathbb{C}^n$ by Theorem 1.2.1. $\qquad\square$

Not all frames behave as well as the one in Proposition 1.3.2: regardless how many elements a frame has, it might happen that the removal of a single particular element destroys the frame property (Exercise 1.17). If we have information on the lower frame bound and the norm of the frame elements, we can provide a criterion for how many elements we can (at least) remove. We ask the reader to provide the proof, see Exercise 1.18.

Proposition 1.5.2 Let $\{f_k\}_{k=1}^{m}$ be a normalized frame for $\mathbb{C}^n$ with lower frame bound $A > 1$. Then, for any index set $I \subset \{1, \ldots, m\}$ with $|I| < A$, the family $\{f_k\}_{k \notin I}$ is a frame for $\mathbb{C}^n$ with lower bound $A - |I|$.

Theorem 1.1.12 shows that if $\{f_k\}_{k=1}^{m}$ is a tight normalized frame for $\mathbb{C}^n$, then Proposition 1.5.2 applies if $|I| < \frac{m}{n}$.

In the context of signal transmission, the overcompleteness of frames has a very useful noise-suppressing effect. We will first give an intuitive explanation and return to a more detailed statistical argument afterwards. Let us assume that we want to transmit the signal f belonging to a vector space V from a transmitter $\mathcal{A}$ to a receiver $\mathcal{R}$. If both $\mathcal{A}$ and $\mathcal{R}$ have knowledge of a frame $\{f_k\}_{k=1}^{m}$ for V, this can be done if $\mathcal{A}$ transmits the coefficients $\{\langle f, f_k \rangle\}_{k=1}^{m}$; based on knowledge of these numbers, the receiver $\mathcal{R}$ can reconstruct the signal f using the frame decomposition

$$f = \sum_{k=1}^{m} \langle f, f_k \rangle S^{-1} f_k.$$

Now assume that $\mathcal{R}$ receives a noisy signal, i.e., a perturbation

$$\{\langle f, f_k \rangle + c_k\}_{k=1}^{m}$$

of the correct coefficients. Based on the received coefficients, $\mathcal{R}$ will claim that the transmitted signal was

$$\sum_{k=1}^{m} (\langle f, f_k \rangle + c_k) S^{-1} f_k = \sum_{k=1}^{m} \langle f, f_k \rangle S^{-1} f_k + S^{-1} \sum_{k=1}^{m} c_k f_k$$

$$= f + S^{-1} \sum_{k=1}^{m} c_k f_k;$$

this differs from the correct signal f by the term $S^{-1} \sum_{k=1}^{m} c_k f_k$. If $\{f_k\}_{k=1}^{m}$ is overcomplete, the pre-frame operator $T\{c_k\}_{k=1}^{m} = \sum_{k=1}^{m} c_k f_k$ has a non-trivial kernel, implying that parts of the noise contribution might add up to zero and cancel. This will *never* happen if $\{f_k\}_{k=1}^{m}$ is an orthonormal basis! In that case

$$\left|\left|S^{-1}\sum_{k=1}^{m}c_kf_k\right|\right| \geq \frac{1}{||S||}\sqrt{\sum_{k=1}^{m}|c_k|^2},$$

so (at least intuitively) each noise contribution will make the reconstruction worse.

The above arguments can be refined using statistical models for noise. Following [36], we will use this to analyze how one should choose the frame $\{f_k\}_{k=1}^{m}$ in order to obtain the maximal noise-suppressing effect. Let us again assume that $\mathcal{A}$ transmits the coefficients $\{\langle f, f_k\rangle\}_{k=1}^{m}$ to the receiver $\mathcal{R}$, and that $\mathcal{R}$ receives a noisy signal $\{\langle f, f_k\rangle + \eta_k\}_{k=1}^{m}$. In contrast with the simplified setting above, we now consider each noise component η_k as a random variable; we will assume that each η_k has mean zero and variance σ^2, and that η_k and η_ℓ are uncorrelated for $k \neq \ell$. Letting E denote the mean, these assumptions can be expressed as

$$E[\eta_k] = 0, \quad E[\eta_k\eta_\ell] = \sigma^2\delta_{k,\ell}, \ k,\ell = 1,\ldots,m. \tag{1.30}$$

As above, based on the coefficients $\{\langle f, f_k\rangle + \eta_k\}_{k=1}^{m}$, the receiver will reconstruct the signal as

$$\tilde{f} = \sum_{k=1}^{m}\left(\langle f, f_k\rangle + \eta_k\right)S^{-1}f_k = f + \sum_{k=1}^{m}\eta_kS^{-1}f_k.$$

Thus, the difference between the reconstructed signal $\tilde{f}$ and the original signal f is

$$\tilde{f} - f = \sum_{k=1}^{m}\eta_kS^{-1}f_k.$$

Remember that $\tilde{f} - f$ is a vector with n coordinates, which depend on the random variables η_k. The associated *mean-square error* MSE is defined by

$$MSE := \frac{1}{n}E||\tilde{f} - f||^2.$$

Now, inserting the expression for $\tilde{f} - f$ shows that

$$\begin{aligned}
MSE &= \frac{1}{n}E(\langle \tilde{f} - f, \tilde{f} - f\rangle) \\
&= \frac{1}{n}E\left[\sum_{k=1}^{m}\sum_{\ell=1}^{m}\eta_k\eta_\ell\langle S^{-1}f_k, S^{-1}f_\ell\rangle\right] \\
&= \frac{1}{n}\sum_{k=1}^{m}\sum_{\ell=1}^{m}E[\eta_k\eta_\ell]\langle S^{-1}f_k, S^{-1}f_\ell\rangle.
\end{aligned}$$

Via the assumptions (1.30), this implies that

$$MSE = \frac{1}{n}\sum_{k=1}^{m}\sum_{\ell=1}^{m}\sigma^2\delta_{k,\ell}\langle S^{-1}f_k, S^{-1}f_\ell\rangle = \frac{1}{n}\sigma^2\sum_{k=1}^{m}||S^{-1}f_k||^2. \quad (1.31)$$

We now show that among all normalized frames containing a fixed number of elements, this expression is minimized for tight frames. We will use the following well-known lemma.

Lemma 1.5.3 *Let $\{a_k\}_{k=1}^{n}$ be a sequence of positive numbers. Then the harmonic mean of the sequence is smaller than or equal to the arithmetic mean, i.e.,*

$$\frac{n}{\sum_{k=1}^{n}\frac{1}{a_k}} \le \frac{1}{n}\sum_{k=1}^{n}a_k.$$

The inequality is an equality if and only if all the a_k are equal.

Theorem 1.5.4 *Consider normalized frames $\{f_k\}_{k=1}^{m}$ for $\mathbb{R}^n$, where $n, m \in \mathbb{N}$ are fixed. Among all such frames $\{f_k\}_{k=1}^{m}$, the MSE is minimal if and only if the frame is tight. The attained minimal value is*

$$MSE = \frac{n}{m}\sigma^2. \quad (1.32)$$

Proof.   Let $\lambda_1, \ldots, \lambda_n$ denote the eigenvalues for the frame operator S associated with $\{f_k\}_{k=1}^{m}$. By Theorem 1.1.12,

$$\sum_{k=1}^{n}\lambda_k = \sum_{k=1}^{m}||f_k||^2 = m. \quad (1.33)$$

The frame $\{S^{-1}f_k\}_{k=1}^{m}$ has S^{-1} as frame operator, and this operator has the eigenvalues $\lambda_1^{-1}, \ldots, \lambda_n^{-1}$. Now, (1.31) together with Theorem 1.1.12 imply that

$$MSE = \frac{1}{n}\sigma^2\sum_{k=1}^{m}||S^{-1}f_k||^2 = \frac{1}{n}\sigma^2\sum_{k=1}^{n}\frac{1}{\lambda_k}. \quad (1.34)$$

Our goal is now to minimize the expression in (1.34) under the constraint (1.33); equivalently, we want to *maximize* the expression

$$\frac{1}{\sum_{k=1}^{n}\frac{1}{\lambda_k}}$$

under the condition that $\sum_{k=1}^{n}\lambda_k = m$. According to Lemma 1.5.3, this happens if and only if all eigenvalues λ_k are equal, i.e., for

$$\lambda_k = \frac{m}{n}, \quad k = 1, \ldots, m.$$

This implies that $\{f_k\}_{k=1}^m$ is a tight frame with frame bound m/n. The attained minimal value of the mean-square error is

$$MSE = \frac{1}{n}\sigma^2 \sum_{k=1}^n \frac{1}{\lambda_k} = \frac{1}{n}\sigma^2 n\frac{n}{m} = \frac{n}{m}\sigma^2. \qquad \square$$

The expression (1.32) shows that for a fixed value of $n \in \mathbb{N}$, the MSE decreases when the number of elements in the frame increases, i.e., for higher redundancy. In this sense, the redundancy in a frame helps to reduce the mean-square error.

1.6   Exercises

1.1 Prove that the adjoint of the pre-frame operator T in (1.4) is given by the expression in (1.5).

1.2 Prove Proposition 1.1.4. (Hint: use that if U is a linear self-adjoint map on V for which $\langle Ux, x \rangle = 0$ for all $x \in V$, then $U = 0$; see Lemma 2.4.3.)

1.3 Show that every frame $\{f_k\}_{k=1}^m$ for a finite-dimensional vector space V contains a subset that is a basis for V.

1.4 Can a frame in a finite-dimensional space contain infinitely many elements?

1.5 Let $\{f_k\}_{k \in I}$ be a frame for a finite-dimensional vector space V and assume that $\|f_k\|$ is bounded below. Prove that I is finite. (w.l.o.g. you may assume that $V = \mathbb{R}^n$ and that $\|f_k\| = 1$, $\forall k$; explain why if you want to use this fact!)

1.6 Find a non-canonical dual frame associated with the frame considered in Example 1.1.6.

1.7 Construct a frame $\{f_k\}_{k=1}^m$ for $\mathbb{C}^2$ for which there exists $f \in \mathbb{C}^2$ such that the coefficients $\{d_k\}_{k=1}^m$ in Theorem 1.1.10 are not unique.

1.8 Let $\{e_1, e_2\}$ be the canonical orthonormal basis for $\mathbb{C}^2$ and consider the frame $\{f_k\}_{k=1}^3 = \{e_1, e_2, e_1 + e_2\}$.

(i) Find the coefficients with minimal ℓ^2-norm among all sequences $\{c_k\}_{k=1}^3$ for which $e_1 = \sum_{k=1}^3 c_k f_k$.

(ii) Find the coefficients $\{c_k^{(1)}\}_{k=1}^3$ and $\{c_k^{(2)}\}_{k=1}^3$ that minimize the ℓ^1-norm in the representation of e_1 and e_2, respectively.

(iii) Clearly, $e_1 + e_2 = \sum_{k=1}^3 (c_k^{(1)} + c_k^{(2)}) f_k$; but is $\{c_k^{(1)} + c_k^{(2)}\}_{k=1}^3$ minimizing the ℓ^1-norm among all sequences representing $e_1 + e_2$?

1.9 Prove Corollary 1.1.11.

1.10 Let $\{e_k\}_{k=1}^n$ be an orthonormal basis for V. Prove that any family $\{g_k\}_{k=1}^n$ of vectors in V for which

$$R := \left(\sum_{k=1}^n \|e_k - g_k\|^2 \right)^{1/2} < 1$$

is a basis for V. (Hint: use

$$\sum_{k=1}^n c_k g_k = \sum_{k=1}^n c_k e_k - \sum_{k=1}^n c_k (e_k - g_k)$$

to prove that $\sum_{k=1}^n c_k g_k \neq 0$ whenever at least one c_k is non-zero.)

1.11 Assume that $\{f_k\}_{k=1}^m$ is a frame for $\mathbb{C}^n$. Prove that the $2m$ vectors consisting of the real parts, respectively the imaginary parts, of the frame vectors constitute a frame for $\mathbb{R}^n$.

1.12 Show that a frame for $\mathbb{R}^n$ is also a frame for $\mathbb{C}^n$.

1.13 Prove Corollary 1.2.7.

1.14 Let $\{f_k\}_{k=1}^m$ be a frame for V with bounds A, B and let P denote the orthogonal projection of V onto a subspace W. Prove that $\{Pf_k\}_{k=1}^m$ is a frame for W with frame bounds A, B.

1.15 Let $\{f_k\}_{k=1}^m$ be a normalized tight frame. Prove that the frame bound A is at least 1, and that $A = 1$ if and only if $\{f_k\}_{k=1}^m$ is an orthonormal basis.

1.16 Let $\{f_k\}_{k=1}^m$ be a frame for an n-dimensional vector space V, and let B denote the optimal upper bound. Prove that

$$B \leq \sum_{k=1}^m \|f_k\|^2 \leq nB.$$

1.17 (i) Find a frame in $\mathbb{C}^3$ with 10 elements, having the property that removal of a single element might destroy the frame property.

(ii) Let $\{f_k\}_{k=1}^m$ be a frame for $\mathbb{C}^n$, $m > n$. Under which condition on f_1 will it happen that removal of f_1 destroys the frame property?

1.18 Prove Proposition 1.5.2.

1.19 Prove that for any $n \in \mathbb{N}$, the polynomials $\{1, x, \ldots, x^n\}$ are linearly independent in the vector space $C(0, 1)$.

1.20 Consider the polynomials $\{1, x, x^2\}$ as functions on the interval $[0, 1]$, and let $V = \mathrm{span}\{1, x, x^2\}$. Equip V with the inner product

$$\langle f, g \rangle = \int_0^1 f(x)\overline{g(x)}dx.$$

Find an orthonormal basis for V.

1.21 Let $\{\lambda_k\}_{k=1}^n$ be a sequence of real numbers. Assume that $\lambda_k \neq \lambda_j$ for $k \neq j$. Let $I \subseteq \mathbb{R}$ be an arbitrary non-empty interval, and consider the complex exponentials $\{e^{i\lambda_k x}\}_{k=1}^n$ as functions on I. Prove that the functions $\{e^{i\lambda_k x}\}_{k=1}^n$ are linearly independent.

1.22 Let $\{\lambda_k\}_{k=1}^n$ be a sequence of real numbers.

(i) Prove that $\{\cos \lambda_k x\}_{k=1}^n$ are linearly independent in $C(-1, 1)$ if and only if $|\lambda_k| \neq |\lambda_j|$ for $k \neq j$.

(ii) Prove that $\{\sin \lambda_k x\}_{k=1}^n$ are linearly independent in $C(-1, 1)$ if and only if all λ_k are non-zero and $|\lambda_k| \neq |\lambda_j|$ for $k \neq j$.

(iii) Under which conditions on sequences $\{\lambda_k\}_{k=1}^n, \{\mu_k\}_{k=1}^m$ are the functions

$$\{\cos \lambda_k x\}_{k=1}^n \cup \{\sin \mu_k x\}_{k=1}^m$$

linearly independent in $C(-1, 1)$?

(iv) Replace the interval $]-1, 1[$ by an arbitrary non-empty interval and generalize (i), (ii), and (iii).

2

Infinite-dimensional Vector Spaces and Sequences

After the introduction to frames in finite-dimensional vector spaces in Chapter 1, the rest of the book will deal with expansions in infinite-dimensional vector spaces. Here great care is needed: we need to replace finite sequences $\{f_k\}_{k=1}^n$ by infinite sequences $\{f_k\}_{k=1}^\infty$, and suddenly the question of convergence properties becomes a central issue. The vector space itself might also cause problems, e.g., in the sense that Cauchy sequences might not be convergent. We expect the reader to have a basic knowledge about these problems and the way to circumvent them, but for completeness we repeat the central themes in Sections 2.1–2.4. Section 2.5 deals with pseudo-inverse operators; this subject is not expected to be known and is treated in more detail. Section 2.6 introduces the so-called moment problems in Hilbert spaces. In Sections 2.7–2.9, we discuss the Hilbert space $L^2(\mathbb{R})$ consisting of the square integrable functions on $\mathbb{R}$ and three classes of operators hereon, as well as the Fourier transform. The material in those sections is not needed for the study of frames and bases on abstract Hilbert spaces in Chapter 3 (except Section 3.5 and Section 3.6) and Chapter 5, but it forms the basis for all the constructions in Chapters 7–11.

2.1 Normed vector spaces and sequences

A central theme in this book is to find conditions on a sequence $\{f_k\}$ in a vector space X such that every $f \in X$ has a representation as a

O. Christensen, *Frames and Bases*. DOI: 10.1007/978-0-8176-4678-3_2,
© Springer Science+Business Media, LLC 2008

superposition of the vectors f_k. In most spaces appearing in functional analysis, this cannot be done with a finite sequence $\{f_k\}$. We are therefore forced to work with infinite sequences, say, $\{f_k\}_{k=1}^{\infty}$, and the representation of f in terms of $\{f_k\}_{k=1}^{\infty}$ will be via an infinite series. For this reason, the starting point must be a discussion of convergence of infinite series. We collect the basic definitions here together with some conventions.

Throughout the section, we let X denote a complex vector space. A *norm* on X is a function $||\cdot|| : X \to [0, \infty[$ satisfying the following three conditions:

(i) $||x|| = 0 \Leftrightarrow x = 0$;

(ii) $||\alpha x|| = |\alpha|\, ||x||, \ \forall x \in X, \ \alpha \in \mathbb{C}$;

(iii) $||x + y|| \leq ||x|| + ||y||, \ \forall x, y \in X.$

In situations where more than one vector space appear, we will frequently denote the norm on X by $||\cdot||_X$. If X is equipped with a norm, we say that X is a *normed vector space*. The *opposite triangle inequality* is satisfied in any normed vector space:

$$||x - y|| \geq |\, ||x|| - ||y||\, |, \ x, y \in X. \tag{2.1}$$

We say that a sequence $\{x_k\}_{k=1}^{\infty}$ in X

(i) converges to $x \in X$ if

$$||x - x_k|| \to 0 \text{ for } k \to \infty;$$

(ii) is a *Cauchy sequence* if for each $\epsilon > 0$ there exists $N \in \mathbb{N}$ such that

$$||x_k - x_l|| \leq \epsilon \text{ whenever } k, l \geq N.$$

A convergent sequence is automatically a Cauchy sequence, but the opposite is not true in general. There are, however, normed vector spaces in which a sequence is convergent if and only if it is a Cauchy sequence; a space X with this property is called a *Banach space*.

Imitating the finite-dimensional setting described in Chapter 1, we want to study sequences $\{f_k\}_{k=1}^{\infty}$ in X with the property that each $f \in X$ has a representation $f = \sum_{k=1}^{\infty} c_k f_k$ for some coefficients $c_k \in \mathbb{C}$. In order to do so, we have to explain exactly what we mean by convergence of an infinite series. There are, in fact, at least three different options; we will now discuss these options.

First, the notation $\{f_k\}_{k=1}^{\infty}$ indicates that we have chosen some ordering of the vectors f_k,

$$f_1, f_2, f_3, \ldots, f_k, f_{k+1}, \cdots .$$

We say that an *infinite series* $\sum_{k=1}^{\infty} c_k f_k$ is convergent with sum $f \in X$ if

$$\left\| f - \sum_{k=1}^{n} c_k f_k \right\| \to 0 \text{ as } n \to \infty.$$

If this condition is satisfied, we write

$$f = \sum_{k=1}^{\infty} c_k f_k. \tag{2.2}$$

Thus, the definition of a convergent infinite series corresponds exactly to our definition of a convergent sequence with $x_n = \sum_{k=1}^{n} c_k f_k$.

Above we insisted on a fixed ordering of the sequence $\{f_k\}_{k=1}^{\infty}$. It is very important to notice that convergence properties of $\sum_{k=1}^{\infty} c_k f_k$ not only depend on the sequence $\{f_k\}_{k=1}^{\infty}$ and the coefficients $\{c_k\}_{k=1}^{\infty}$ but also on the ordering. Even if we consider a sequence in the simplest possible Banach space, i.e., a sequence $\{a_k\}_{k=1}^{\infty}$ in $\mathbb{C}$, it can happen that $\sum_{k=1}^{\infty} a_k$ is convergent but that $\sum_{k=1}^{\infty} a_{\sigma(k)}$ is divergent for a certain permutation σ of the natural numbers (Exercise 2.1). This observation leads to the second definition of convergence. If $\{f_k\}_{k=1}^{\infty}$ is a sequence in X and $\sum_{k=1}^{\infty} f_{\sigma(k)}$ is convergent for all permutations σ, we say that $\sum_{k=1}^{\infty} f_k$ is *unconditionally convergent*. In that case, the limit is the same regardless of the order of summation.

Finally, an infinite series $\sum_{k=1}^{\infty} f_k$ is said to be *absolutely convergent* if

$$\sum_{k=1}^{\infty} ||f_k|| < \infty.$$

In any Banach space, absolute convergence of $\sum_{k=1}^{\infty} f_k$ implies that the series converges unconditionally (Exercise 2.2), but the opposite does not hold in infinite-dimensional spaces. In finite-dimensional spaces, the two types of convergence are identical.

A subset $Z \subseteq X$ (countable or not) is said to be *dense* in X if for each $f \in X$ and each $\epsilon > 0$ there exists $g \in Z$ such that

$$||f - g|| \le \epsilon.$$

In words, this means that elements in X can be approximated arbitrarily well by elements in Z.

For a given sequence $\{f_k\}_{k=1}^{\infty}$ in X, we let span$\{f_k\}_{k=1}^{\infty}$ denote the vector space consisting of all *finite* linear combinations of vectors f_k. The definition of convergence shows that if each $f \in X$ has a representation of the type (2.2), then each $f \in X$ can be approximated arbitrarily well in norm by elements in span$\{f_k\}_{k=1}^{\infty}$, i.e.,

$$\overline{\text{span}}\{f_k\}_{k=1}^{\infty} = X. \tag{2.3}$$

A sequence $\{f_k\}_{k=1}^{\infty}$ having the property (2.3) is said to be *complete* or *total*. We note that there exist normed spaces where no sequence $\{f_k\}_{k=1}^{\infty}$ is complete. A normed vector space, in which a countable and dense family exists, is said to be *separable*.

When we speak about a *finite sequence,* we mean a sequence $\{c_k\}_{k=1}^{\infty}$ where at most finitely many entries c_k are non-zero.

2.2   Operators on Banach spaces

Let X and Y denote Banach spaces. An linear map $U : X \to Y$ is called an *operator*, and U is *bounded* or *continuous* if there exists a constant $K > 0$ such that

$$||Ux||_Y \le K\,||x||_X, \ \forall x \in X. \tag{2.4}$$

Usually, it will be clear from the context which norm we use, so we will write $||\cdot||$ for both $||\cdot||_X$ and $||\cdot||_Y$. The *norm* of the operator U, denoted by $||U||$, is the smallest constant K that can be used in (2.4). Alternatively,

$$||U|| = \sup\{||Ux||\ :\ x \in X, ||x|| = 1\}.$$

If U_1 and U_2 are operators for which the range of U_2 is contained in the domain of U_1, we can consider the composed operator U_1U_2; if U_1 and U_2 are bounded, then also U_1U_2 is bounded, and

$$||U_1U_2|| \le ||U_1||\,||U_2||. \tag{2.5}$$

Now consider a sequence of operators $U_n : X \to Y$, $n \in \mathbb{N}$, which converges pointwise to a mapping $U : X \to Y$, i.e.,

$$U_nx \to Ux, \text{ as } n \to \infty, \ \forall x \in X.$$

We say that U_n converges to U in the *strong operator topology*. The *Banach–Steinhaus Theorem*, also known as the *uniform boundedness principle*, states the following:

Theorem 2.2.1 *Let $U_n : X \to Y$, $n \in \mathbb{N}$, be a sequence of bounded operators, which converges pointwise to a mapping $U : X \to Y$. Then U is linear and bounded. Furthermore, the sequence of norms $||U_n||$ is bounded, and $||U|| \le \liminf ||U_n||$.*

An operator $U : X \to Y$ is *invertible* if U is surjective and injective. For a bounded, invertible operator, the inverse operator is bounded:

Theorem 2.2.2 *A bounded bijective operator between Banach spaces has a bounded inverse.*

In case $X = Y$, it makes sense to speak about the identity operator I on X. The *Neumann Theorem* states that an operator $U : X \to X$ is invertible if it is close enough to the identity operator:

Theorem 2.2.3 *If $U : X \to X$ is bounded and $||I - U|| < 1$, then U is invertible, and*

$$U^{-1} = \sum_{k=0}^{\infty}(I - U)^k. \tag{2.6}$$

Furthermore,

$$||U^{-1}|| \leq \frac{1}{1 - ||I - U||}.$$

Note that (2.6) should be interpreted in the sense of the operator norm, i.e., as

$$\left|\left| U^{-1} - \sum_{k=0}^{N} (I - U)^k \right|\right| \to 0 \text{ as } N \to \infty.$$

2.3   Hilbert spaces

A special class of normed vector spaces is formed by *inner product spaces.* Recall that an inner product on a complex vector space X is a mapping $\langle \cdot, \cdot \rangle : X \times X \to \mathbb{C}$ for which

(i) $\langle \alpha x + \beta y, z \rangle = \alpha \langle x, z \rangle + \beta \langle y, z \rangle, \ \forall x, y, z \in X, \ \alpha, \beta \in \mathbb{C}$;

(ii) $\langle x, y \rangle = \overline{\langle y, x \rangle}, \ \forall x, y \in X$;

(iii) $\langle x, x \rangle \geq 0, \ \forall x \in X$, and $\langle x, x \rangle = 0 \Leftrightarrow x = 0$.

Note that we have chosen to let the inner product be linear in the first entry. It implies that the inner product is conjugated linear in the second entry. Frequently, the opposite convention is used in the literature.

A vector space X with an inner product $\langle \cdot, \cdot \rangle$ can be equipped with the norm

$$||x|| := \sqrt{\langle x, x \rangle}, \ x \in X.$$

If X is a Banach space with respect to this norm, then X is called a *Hilbert space.* We reserve the letter $\mathcal{H}$ for these spaces. We will always assume that $\mathcal{H}$ is *non-trivial,* i.e., that $\mathcal{H} \neq \{0\}$. The standard examples are the spaces $L^2(\mathbb{R})$ and $\ell^2(\mathbb{N})$ discussed in Section 2.7.

In any Hilbert space $\mathcal{H}$ with an inner product $\langle \cdot, \cdot \rangle$, *Cauchy–Schwarz' inequality* holds: it states that

$$|\langle x, y \rangle| \leq ||x|| \, ||y||, \ \ \forall x, y \in \mathcal{H}.$$

Two elements $x, y \in \mathcal{H}$ are *orthogonal* if $\langle x, y \rangle = 0$; and the *orthogonal complement* of a subspace U of $\mathcal{H}$ is

$$U^{\perp} = \{x \in \mathcal{H} : \ \langle x, y \rangle = 0, \ \forall y \in U\}.$$

The above definitions and results are valid whether $\mathcal{H}$ is finite-dimensional or infinite-dimensional. Also note that norms and inner products are defined in a similar way on real vector spaces (just replace the scalars $\mathbb{C}$ by the real scalars $\mathbb{R}$).

We state a few elementary results concerning Hilbert spaces that will be used repeatedly during the book. The proof of the first is left to the reader as Exercise 2.3.

Lemma 2.3.1 *For a sequence $\{x_k\}_{k=1}^{\infty}$ in a Hilbert space $\mathcal{H}$ the following are equivalent:*

(i) $\{x_k\}_{k=1}^{\infty}$ *is complete.*

(ii) *If $\langle x, x_k \rangle = 0$ for all $k \in \mathbb{N}$, then $x = 0$.*

Among the linear operators on a Hilbert space, a special role is played by the continuous linear operators $U : \mathcal{H} \to \mathbb{C}$. They are called *functionals* and are characterized in *Riesz' Representation Theorem*:

Theorem 2.3.2 *Let $U : \mathcal{H} \to \mathbb{C}$ be a continuous linear mapping. Then there exists a unique $y \in \mathcal{H}$ such that $Ux = \langle x, y \rangle$ for all $x \in \mathcal{H}$.*

The uniqueness of the element $y \in \mathcal{H}$ associated with a given functional has the following important consequence.

Corollary 2.3.3 *Let $\mathcal{H}$ be a Hilbert space. Assume that $x, y \in \mathcal{H}$ satisfy that*

$$\langle x, z \rangle = \langle y, z \rangle, \ \forall z \in \mathcal{H}.$$

Then $x = y$.

Finally, we note that the norm of an arbitrary element $x \in \mathcal{H}$ can be recovered based on the inner product between x and the elements in the unit sphere in $\mathcal{H}$:

Lemma 2.3.4 *For any $x \in \mathcal{H}$,*

$$||x|| = \sup_{||y||=1} |\langle x, y \rangle|.$$

2.4 Operators on Hilbert spaces

Let U be a bounded operator from the Hilbert space $(\mathcal{K}, \langle \cdot, \cdot \rangle_{\mathcal{K}})$ into the Hilbert space $(\mathcal{H}, \langle \cdot, \cdot \rangle_{\mathcal{H}})$. The *adjoint* operator is defined as the unique operator $U^* : \mathcal{H} \to \mathcal{K}$ satisfying that

$$\langle x, Uy \rangle_{\mathcal{H}} = \langle U^* x, y \rangle_{\mathcal{K}}, \ \forall x \in \mathcal{H}, y \in \mathcal{K}.$$

Usually, we will write $\langle \cdot, \cdot \rangle$ for both inner products; it will always be clear from the context in which space the inner product is taken.

We collect some relationships between U and U^*; the proofs can be found in, e.g., Theorem 4.14 and Theorem 4.15 in [60].

Lemma 2.4.1 *Let $U : \mathcal{K} \to \mathcal{H}$ be a bounded operator. Then the following holds:*

(i) $\|U\| = \|U^*\|$, *and* $\|UU^*\| = \|U\|^2$.

(ii) $\mathcal{R}_U$ *is closed in $\mathcal{H}$ if and only if $\mathcal{R}_{U^*}$ is closed in $\mathcal{K}$.*

(iii) *U is surjective if and only if there exists a constant $C > 0$ such that*

$$\|U^*y\| \geq C \|y\|, \quad \forall y \in \mathcal{H}.$$

In the rest of this section, we consider the case $\mathcal{K} = \mathcal{H}$. A bounded operator $U : \mathcal{H} \to \mathcal{H}$ is *unitary* if $UU^* = U^*U = I$. If U is unitary, then

$$\langle Ux, Uy \rangle = \langle x, y \rangle, \quad \forall x, y \in \mathcal{H}.$$

A bounded operator $U : \mathcal{H} \to \mathcal{H}$ is *self-adjoint* if $U = U^*$. When U is self-adjoint,

$$\|U\| = \sup_{\|x\|=1} |\langle Ux, x \rangle|. \tag{2.7}$$

For a self-adjoint operator U, the inner product $\langle Ux, x \rangle$ is real for all $x \in \mathcal{H}$. One can introduce a partial order on the set of self-adjoint operators by

$$U_1 \leq U_2 \Leftrightarrow \langle U_1 x, x \rangle \leq \langle U_2 x, x \rangle, \quad \forall x \in \mathcal{H}.$$

Using this order, one can work with self-adjoint operators almost as with real numbers. For example, under certain conditions it is possible to "multiply" an operator inequality with a bounded operator. The precise statement below can be found in [43]:

Theorem 2.4.2 *Let U_1, U_2, U_3 be self-adjoint operators. If $U_1 \leq U_2$, $U_3 \geq 0$, and U_3 commutes with U_1 and U_2, then $U_1 U_3 \leq U_2 U_3$.*

An important class of self-adjoint operators consists of the *orthogonal projections*. Given a closed subspace V of $\mathcal{H}$, the orthogonal projection of $\mathcal{H}$ onto V is the operator $P : \mathcal{H} \to \mathcal{H}$ for which

$$Px = x, \ x \in V, \ Px = 0, \ x \in V^\perp.$$

If $\{e_k\}_{k=1}^\infty$ is an orthonormal basis for V, the operator P is given explicitly by

$$Px = \sum_{k=1}^\infty \langle x, e_k \rangle e_k, \ x \in \mathcal{H}.$$

In case $\mathcal{H}$ is a complex Hilbert space and U is a bounded operator on $\mathcal{H}$, a direct calculation gives that

$$
\begin{aligned}
4\langle Ux, y \rangle = {} & \langle U(x+y), x+y \rangle - \langle U(x-y), x-y \rangle \\
& + i\langle U(x+iy), x+iy \rangle - i\langle U(x-iy), x-iy \rangle. \quad (2.8)
\end{aligned}
$$

In particular, we can recover the inner product in $\mathcal{H}$ from the norm by

$$4\langle x, y \rangle \quad = \quad ||x+y||^2 - ||x-y||^2 + i \, ||x+iy||^2 - i \, ||x-iy||^2, \quad x, y \in \mathcal{H},$$

a result that is known as the *polarization identity*.

Lemma 2.4.3 *Let* $U : \mathcal{H} \to \mathcal{H}$ *be a bounded operator, and assume that* $\langle Ux, x \rangle = 0$ *for all* $x \in \mathcal{H}$. *Then the following holds:*

(i) *If* $\mathcal{H}$ *is a complex Hilbert space, then* $U = 0$.

(ii) *If* $\mathcal{H}$ *is a real Hilbert space and* U *is self-adjoint, then* $U = 0$.

Proof. If $\mathcal{H}$ is complex, we can use (2.8); thus, if $\langle Ux, x \rangle = 0$ for all $x \in \mathcal{H}$, then $\langle Ux, y \rangle = 0$ for all $x, y \in \mathcal{H}$, and therefore $U = 0$.

In case $\mathcal{H}$ is a real Hilbert space, we must use a different approach. Let $\{e_k\}_{k=1}^{\infty}$ be an orthonormal basis for $\mathcal{H}$. Then, for arbitrary $j, k \in \mathbb{N}$,

$$\begin{aligned} 0 \quad &= \quad \langle U(e_k + e_j), e_k + e_j \rangle \\ &= \quad \langle Ue_k, e_k \rangle + \langle Ue_j, e_j \rangle + \langle Ue_k, e_j \rangle + \langle Ue_j, e_k \rangle \\ &= \quad \langle Ue_k, e_j \rangle + \langle e_j, Ue_k \rangle \\ &= \quad 2 \langle Ue_j, e_k \rangle; \end{aligned}$$

therefore $U = 0$.   □

Note that without the assumption $U = U^*$, the second part of the lemma would fail; to see that, let U be a rotation of $90°$ in $\mathbb{R}^2$.

A bounded operator $U : \mathcal{H} \to \mathcal{H}$ is *positive* if $\langle Ux, x \rangle \geq 0$, $\forall x \in \mathcal{H}$. On a complex Hilbert space, every bounded positive operator is self-adjoint. For a positive operator U, we will often use the following result about the existence of a *square root*, i.e., a bounded operator W such that $W^2 = U$:

Lemma 2.4.4 *Every bounded and positive operator* $U : \mathcal{H} \to \mathcal{H}$ *has a unique bounded and positive square root* W. *The operator* W *has the following properties:*

(i) *If* U *is self-adjoint, then* W *is self-adjoint.*

(ii) *If* U *is invertible, then* W *is also invertible.*

(iii) W *can be expressed as a limit (in the strong operator topology) of a sequence of polynomials in* U, *and commutes with* U.

2.5   The pseudo-inverse operator

It is well-known that not all bounded operators U on a Hilbert space $\mathcal{H}$ are invertible: an operator U needs to be injective and surjective in order to be invertible. We will now prove that if an operator U has closed range, there exists a "*right-inverse operator*" $U^{\dagger}$ in the following sense:

Lemma 2.5.1 *Let $\mathcal{H}, \mathcal{K}$ be Hilbert spaces, and suppose that $U : \mathcal{K} \to \mathcal{H}$ is a bounded operator with closed range $\mathcal{R}_U$. Then there exists a bounded operator $U^\dagger : \mathcal{H} \to \mathcal{K}$ for which*

$$UU^\dagger x = x, \ \forall x \in \mathcal{R}_U. \tag{2.9}$$

Proof. Consider the restriction of U to an operator on the orthogonal complement of the kernel of U, i.e., let

$$\tilde{U} := U_{|\mathcal{N}_U^\perp} : \mathcal{N}_U^\perp \to \mathcal{H}.$$

Clearly, $\tilde{U}$ is linear and bounded. $\tilde{U}$ is also injective: if $\tilde{U}x = 0$, it follows that $x \in \mathcal{N}_U^\perp \cap \mathcal{N}_U = \{0\}$. We now prove that the range of $\tilde{U}$ equals the range of U. Given $y \in \mathcal{R}_U$, there exists $x \in \mathcal{K}$ such that $Ux = y$. By writing $x = x_1 + x_2$, where $x_1 \in \mathcal{N}_U^\perp$, $x_2 \in \mathcal{N}_U$, we obtain that

$$\tilde{U}x_1 = Ux_1 = U(x_1 + x_2) = Ux = y.$$

It follows from Theorem 2.2.2 that $\tilde{U}$ has a bounded inverse

$$\tilde{U}^{-1} : \mathcal{R}_U \to \mathcal{N}_U^\perp.$$

Extending $\tilde{U}^{-1}$ by zero on the orthogonal complement of $\mathcal{R}_U$ we obtain a bounded operator $U^\dagger : \mathcal{H} \to \mathcal{K}$ for which $UU^\dagger x = x$ for all $x \in \mathcal{R}_U$. $\square$

The operator $U^\dagger$ constructed in the proof of Lemma 2.5.1 is called the *pseudo-inverse* of U. In the literature, one will often see the pseudo-inverse of an operator U with closed range defined as the unique operator $U^\dagger$ satisfying that

$$\mathcal{N}_{U^\dagger} = \mathcal{R}_U^\perp, \ \ \mathcal{R}_{U^\dagger} = \mathcal{N}_U^\perp, \ \text{and} \ UU^\dagger x = x, x \in \mathcal{R}_U; \tag{2.10}$$

this definition is equivalent to the above construction (Exercise 2.4). We collect some properties of $U^\dagger$ and its relationship to U.

Lemma 2.5.2 *Let $U : \mathcal{K} \to \mathcal{H}$ be a bounded operator with closed range. Then the following holds:*

(i) The orthogonal projection of $\mathcal{H}$ onto $\mathcal{R}_U$ is given by $UU^\dagger$.

(ii) The orthogonal projection of $\mathcal{K}$ onto $\mathcal{R}_{U^\dagger}$ is given by $U^\dagger U$.

(iii) U^* has closed range, and $(U^*)^\dagger = (U^\dagger)^*$.

(iv) On $\mathcal{R}_U$, the operator $U^\dagger$ is given explicitly by

$$U^\dagger = U^*(UU^*)^{-1}. \tag{2.11}$$

Proof. All statements follow from the characterization of $U^\dagger$ in (2.10). For example, it shows that

$$UU^\dagger = I \text{ on } \mathcal{R}_U \text{ and that } UU^\dagger = 0 \text{ on } \mathcal{N}_{U^\dagger} = \mathcal{R}_U^\perp;$$

this gives (i) by the definition of an orthogonal projection. The proof of (ii) is similar. That $\mathcal{R}_{U^*}$ is closed was stated already in Lemma 2.4.1; thus $(U^*)^\dagger$ is well defined. That $(U^*)^\dagger$ equals $(U^\dagger)^*$ follows by verifying that $(U^\dagger)^*$ satisfies (2.10) with U replaced by U^*. Finally, UU^* is invertible as an operator on $\mathcal{R}_U$, and the operator given by

$$U^*(UU^*)^{-1} \text{ on } \mathcal{R}_U, \quad \text{and } 0 \text{ on } \mathcal{R}_U^\perp$$

satisfies the conditions (2.10) characterizing $U^\dagger$.   □

The pseudo-inverse gives the solution to an important optimization problem:

Theorem 2.5.3 *Let $U : \mathcal{K} \to \mathcal{H}$ be a bounded surjective operator. Given $y \in \mathcal{H}$, the equation $Ux = y$ has a unique solution of minimal norm, namely $x = U^\dagger y$.*

The proof is identical with the proof of Theorem 1.4.2.

2.6   A moment problem

Before we leave the discussion of abstract Hilbert spaces, we mention a special class of equations, known as moment problems. For the purpose of the current book, they are only needed in Section 7.4.

The general version of a *moment problem* is as follows: given a collection of elements $\{x_k\}_{k=1}^\infty$ in a Hilbert space $\mathcal{H}$ and a sequence $\{a_k\}_{k=1}^\infty$ of complex numbers, can we find an element $x \in \mathcal{H}$ such that

$$\langle x, x_k \rangle = a_k, \text{ for all } k \in \mathbb{N}?$$

We will only need a special moment problem:

Lemma 2.6.1 *Let $\{x_k\}_{k=1}^N$ be a collection of vectors in $\mathcal{H}$ and consider the moment problem*

$$\langle x, x_k \rangle = \begin{cases} 1 \text{ if } k = 1, \\ 0 \text{ if } k = 2, \dots, N. \end{cases} \tag{2.12}$$

Then the following are equivalent:

(i) The moment problem (2.12) has a solution x.

(ii) If $\sum_{k=1}^N c_k x_k = 0$ for some scalar coefficients c_k, then $c_1 = 0$.

(iii) $x_1 \notin span\{x_k\}_{k=2}^N$.

In case the moment problem (2.12) has a solution, it can be chosen of the form $x = \sum_{k=1}^N d_k x_k$ for some scalar coefficients d_k.

Proof. Assume first that (i) is satisfied, i.e., (2.12) has a solution x. Then, if $\sum_{k=1}^{N} c_k x_k = 0$ for some coefficients $\{c_k\}_{k=1}^{N}$, we have that

$$0 = \langle x, \sum_{k=1}^{N} c_k x_k \rangle = \sum_{k=1}^{N} c_k \langle x, x_k \rangle = c_1,$$

i.e., (ii) holds. Now assume that (ii) is satisfied. Then $x_1 \notin \mathrm{span}\{x_k\}_{k=2}^{N}$. Let P denote the orthogonal projection of $\mathcal{H}$ onto $\mathrm{span}\{x_k\}_{k=2}^{N}$, and put $\varphi = x_1 - P x_1$. Then

$$\langle \varphi, x_1 \rangle = \langle x_1 - P x_1, x_1 - P x_1 \rangle + \langle x_1 - P x_1, P x_1 \rangle = ||x_1 - P x_1||^2 \neq 0,$$

and $\langle \varphi, x_k \rangle = 0$ for $k = 2, \ldots, N$. Thus, the element

$$x := \frac{\varphi}{||x_1 - P x_1||^2} \tag{2.13}$$

solves the moment problem (2.12), i.e., (i) is satisfied. The equivalence of (ii) and (iii) is clear. In case the equivalent conditions are satisfied, the construction of x in (2.13) shows that $x \in \mathrm{span}\{x_k\}_{k=1}^{N}$. $\square$

2.7   The spaces $L^p(\mathbb{R})$, $L^2(\mathbb{R})$, and $\ell^2(\mathbb{N})$

The most important class of Banach spaces is formed by the L^p-spaces, $1 \le p \le \infty$. Before we define these spaces, we will remind the reader about some basic facts from the theory of integration. The proofs and further results can be found in any standard book on the subject, e.g., [59].

We begin with *Fatou's Lemma*. For our purpose, it is enough to consider the case of the Lebesgue measure on the real axis $\mathbb{R}$, equipped with the (Borel-) measurable sets:

Lemma 2.7.1 *Let $f_n : \mathbb{R} \to [0, \infty]$, $n \in \mathbb{N}$ be a sequence of measurable functions. Then the function $\liminf_{n \to \infty} f_n$ is measurable, and*

$$\int_{-\infty}^{\infty} \liminf_{n \to \infty} f_n(x)\, dx \le \liminf_{n \to \infty} \int_{-\infty}^{\infty} f_n(x)\, dx.$$

Lebesgue's Dominated Convergence Theorem is the main tool to interchange limits and integrals:

Theorem 2.7.2 *Suppose that $f_n : \mathbb{R} \to \mathbb{C}$, $n \in \mathbb{N}$ is a sequence of measurable functions, that $f_n(x) \to f(x)$ pointwise as $n \to \infty$, and that there exists a positive, measurable function g such that $|f_n| \le g$ for all $n \in \mathbb{N}$ and $\int_{-\infty}^{\infty} g(x)\, dx < \infty$. Then f is integrable, and*

$$\lim_{n \to \infty} \int_{-\infty}^{\infty} f_n(x)\, dx = \int_{-\infty}^{\infty} f(x)\, dx.$$

A *null set* is a measurable set with measure zero. A condition holds *almost everywhere* (abbreviated a.e.) if it holds except on a null set.

We are now ready to define the Banach spaces $L^p(\mathbb{R})$ for $1 \le p \le \infty$. First, we define $L^\infty(\mathbb{R})$ as the space of essentially bounded measurable functions $f : \mathbb{R} \to \mathbb{C}$, equipped with the essential supremums-norm. For $1 \le p < \infty$, $L^p(\mathbb{R})$ is the space of functions f for which $|f|^p$ is integrable with respect to the Lebesgue measure:

$$L^p(\mathbb{R}) := \left\{ f : \mathbb{R} \to \mathbb{C} \mid f \text{ is measurable and } \int_{-\infty}^{\infty} |f(x)|^p \, dx < \infty \right\}.$$

The norm on $L^p(\mathbb{R})$ is

$$\|f\| = \left(\int_{-\infty}^{\infty} |f(x)|^p \, dx \right)^{1/p}.$$

To be more precise, $L^p(\mathbb{R})$ consists of equivalence classes of functions that are equal almost everywhere, and for which a representative (and hence all) for the equivalence class satisfies the integrability condition. In order not to be too tedious, we adopt the standard terminology and speak about functions in $L^p(\mathbb{R})$ rather than equivalence classes.

The case $p = 2$ plays a special role: in fact, the space

$$L^2(\mathbb{R}) = \left\{ f : \mathbb{R} \to \mathbb{C} \mid f \text{ is measurable and } \int_{-\infty}^{\infty} |f(x)|^2 \, dx < \infty \right\}$$

is the only one of the $L^p(\mathbb{R})$-spaces that can be equipped with an inner product. Actually, $L^2(\mathbb{R})$ is a Hilbert space with respect to the inner product

$$\langle f, g \rangle = \int_{-\infty}^{\infty} f(x)\overline{g(x)} \, dx, \ \ f, g \in L^2(\mathbb{R}).$$

In $L^2(\mathbb{R})$, Cauchy–Schwarz' inequality states that for all $f, g \in L^2(\mathbb{R})$,

$$\left| \int_{-\infty}^{\infty} f(x)g(x) \, dx \right| \le \left(\int_{-\infty}^{\infty} |f(x)|^2 \, dx \right)^{1/2} \left(\int_{-\infty}^{\infty} |g(x)|^2 \, dx \right)^{1/2}.$$

The spaces $L^2(\Omega)$, where Ω is an open subset of $\mathbb{R}$, are defined similarly. According to the general definition, a sequence of functions $\{g_k\}_{k=1}^{\infty}$ in $L^2(\Omega)$ converges to $g \in L^2(\Omega)$ if

$$\|g - g_k\| = \left(\int_{\Omega} |g(x) - g_k(x)|^2 \, dx \right)^{1/2} \to 0 \text{ as } k \to \infty.$$

Convergence in L^2 is very different from pointwise convergence. As a positive result, we have *Riesz' Subsequence Theorem:*

Theorem 2.7.3 *Let $\Omega \subseteq \mathbb{R}$ be an open set, and let $\{g_k\}$ be a sequence in $L^2(\Omega)$ that converges to $g \in L^2(\Omega)$. Then $\{g_k\}$ has a subsequence $\{g_{n_k}\}_{k=1}^{\infty}$ such that*

$$g(x) = \lim_{k \to \infty} g_{n_k}(x)$$

for a.e. $x \in \Omega$.

The result holds no matter how we choose the representatives for the equivalence classes. This is typical for this book, where we rarely deal with a specific representative for a given class. There are, however, a few important exceptions. When we speak about a continuous function, it is clear that we have chosen a specific representative, and the same is the case when we discuss *Lebesgue points*. By definition, a point $y \in \mathbb{R}$ is a Lebesgue point for a function f if

$$\lim_{\epsilon \to 0} \frac{1}{\epsilon} \int_{y - \frac{1}{2}\epsilon}^{y + \frac{1}{2}\epsilon} |f(y) - f(x)| \, dx = 0.$$

If f is continuous in y, then y is a Lebesgue point (Exercise 2.5). More generally, one can prove that if $f \in L^1(\mathbb{R})$, then almost every $y \in \mathbb{R}$ is a Lebesgue point.

It is clear from the definition that different representatives for the same equivalence class will have different Lebesgue points. For example, every $y \in \mathbb{R}$ is a Lebesgue point for the function $f = 0$; changing the definition of f in a single point y will not change the equivalence class, but y will no longer be a Lebesgue point. See Exercise 2.5 for some related observations.

The discrete analogue of $L^2(\mathbb{R})$ is $\ell^2(I)$, the space of square sumable scalar sequences with a countable index set I:

$$\ell^2(I) := \left\{ \{x_k\}_{k \in I} \mid x_k \in \mathbb{C}, \ \sum_{k \in I} |x_k|^2 < \infty \right\}.$$

The definition of the space $\ell^2(I)$ corresponds to our definition of $L^2(\mathbb{R})$ with the set $\mathbb{R}$ replaced by I and the Lebesgue measure replaced by the counting measure. $\ell^2(I)$ is a Hilbert space with respect to the inner product

$$\langle \{x_k\}, \{y_k\} \rangle = \sum_{k \in I} x_k \overline{y_k};$$

in this case, Cauchy–Schwarz' inequality gives that

$$\left| \sum_{k \in I} x_k \overline{y_k} \right|^2 \le \sum_{k \in I} |x_k|^2 \sum_{k \in I} |y_k|^2, \ \{x_k\}_{k \in I}, \{y_k\}_{k \in I} \in \ell^2(I).$$

We will frequently use the discrete version of Fatou's lemma:

Lemma 2.7.4 *Let I be a countable index set and $f_n : I \to [0, \infty]$, $n \in \mathbb{N}$, a sequence of functions. Then*

$$\sum_{k \in I} \liminf_{n \to \infty} f_n(k) \leq \liminf_{n \to \infty} \sum_{k \in I} f_n(k).$$

2.8   The Fourier transform and convolution

For $f \in L^1(\mathbb{R})$, the *Fourier transform* $\hat{f} : \mathbb{R} \to \mathbb{C}$ is defined by

$$\hat{f}(\gamma) := \int_{-\infty}^{\infty} f(x)e^{-2\pi i x \gamma} \, dx, \ \gamma \in \mathbb{R}.$$

Frequently, we will also denote the Fourier transform of f by $\mathcal{F}f$.

If $(L^1 \cap L^2)(\mathbb{R})$ is equipped with the $L^2(\mathbb{R})$-norm, the Fourier transform is an isometry from $(L^1 \cap L^2)(\mathbb{R})$ into $L^2(\mathbb{R})$. If $f \in L^2(\mathbb{R})$ and $\{f_k\}_{k=1}^{\infty}$ is a sequence of functions in $(L^1 \cap L^2)(\mathbb{R})$ that converges to f in L^2-sense, then the sequence $\{\hat{f}_k\}_{k=1}^{\infty}$ is also convergent in $L^2(\mathbb{R})$, with a limit that is independent of the choice of $\{f_k\}_{k=1}^{\infty}$. Defining

$$\hat{f} := \lim_{k \to \infty} \hat{f}_k$$

we can extend the Fourier transform to a unitary mapping of $L^2(\mathbb{R})$ onto $L^2(\mathbb{R})$. We will use the same notation to denote this extension. In particular, we have *Plancherel's equation:*

$$\langle \hat{f}, \hat{g} \rangle = \langle f, g \rangle, \ \forall f, g \in L^2(\mathbb{R}), \ \text{and} \ ||\hat{f}|| = ||f||. \tag{2.14}$$

If $f \in L^1(\mathbb{R})$, then $\hat{f}$ is continuous. If the function f as well as $\hat{f}$ belong to $L^1(\mathbb{R})$, the *inversion formula* describes how to come back to f from the function values $\hat{f}(\gamma)$, see [2]:

Theorem 2.8.1 *Assume that $f, \hat{f} \in L^1(\mathbb{R})$. Then*

$$f(x) = \int_{-\infty}^{\infty} \hat{f}(\gamma)e^{2\pi i x \gamma} d\gamma, \ a.e. \ x \in \mathbb{R}. \tag{2.15}$$

If f is continuous, the pointwise formula (2.15) holds for all $x \in \mathbb{R}$. In general, it holds at least for all Lebesgue points for f.

Given two functions $f, g \in L^1(\mathbb{R})$, the *convolution* $f * g : \mathbb{R} \to \mathbb{C}$ is defined by

$$f * g(y) = \int_{-\infty}^{\infty} f(y - x)g(x) \, dx, \ y \in \mathbb{R}.$$

The function $f * g$ is well defined for all $y \in \mathbb{R}$ and belongs to $L^1(\mathbb{R})$. If $f \in L^1(\mathbb{R})$ and $g \in L^2(\mathbb{R})$, the convolution $f * g(y)$ is well defined for a.e. $y \in \mathbb{R}$ and defines a function in $L^2(\mathbb{R})$.

The Fourier transform and convolution are related by the following important result.

Theorem 2.8.2 *If $f, g \in L^1(\mathbb{R})$, then $\widehat{f * g}(\gamma) = \hat{f}(\gamma)\hat{g}(\gamma)$ for all $\gamma \in \mathbb{R}$; if $f \in L^1(\mathbb{R})$ and $g \in L^2(\mathbb{R})$, the formula holds for a.e. $\gamma \in \mathbb{R}$.*

2.9 Operators on $L^2(\mathbb{R})$

In this section, we consider three classes of operators on $L^2(\mathbb{R})$ that will play a key role in our analysis of Gabor frames and wavelets. Their definitions are as follows:

$$\text{Translation by } a \in \mathbb{R}, \ T_a : L^2(\mathbb{R}) \to L^2(\mathbb{R}), \ (T_a f)(x) = f(x - a); \quad (2.16)$$

$$\text{Modulation by } b \in \mathbb{R}, \ E_b : L^2(\mathbb{R}) \to L^2(\mathbb{R}), \ (E_b f)(x) = e^{2\pi i b x} f(x); \quad (2.17)$$

$$\text{Dilation by } a \neq 0, \ D_a : L^2(\mathbb{R}) \to L^2(\mathbb{R}), \ (D_a f)(x) = \frac{1}{\sqrt{|a|}} f(\frac{x}{a}). \quad (2.18)$$

A comment about notation: we will usually skip the parentheses and simply write $T_a f(x)$, and similarly for the other operators. Frequently, we will also let E_b denote the function $x \mapsto e^{2\pi i b x}$. We collect some of the most important properties for the operators in (2.16)–(2.18):

Lemma 2.9.1 *The translation operators satisfy the following:*

(i) T_a is unitary for all $a \in \mathbb{R}$.

(ii) For each $f \in L^2(\mathbb{R})$, the mapping $y \mapsto T_y f$ is continuous from $\mathbb{R}$ to $L^2(\mathbb{R})$.

Similar statements hold for $E_b, b \in \mathbb{R}$, and $D_a, a \neq 0$.

Proof. Let us prove that the operators T_a are unitary. Since

$$\langle T_a f, g \rangle = \int_{-\infty}^{\infty} f(x - a)\overline{g(x)} \, dx \ = \ \int_{-\infty}^{\infty} f(x)\overline{g(x + a)} \, dx$$

$$= \ \langle f, T_{-a} g \rangle, \ \forall f, g \in L^2(\mathbb{R}),$$

we see that $T_a^* = T_{-a}$. On the other hand, T_a is clearly an invertible operator with $T_a^{-1} = T_{-a}$, so we conclude that $T_a^{-1} = T_a^*$.

To prove the continuity of the mapping $y \mapsto T_y f$, we first assume that the function f is continuous and has compact support, say, contained in the bounded interval $[c, d]$. For notational convenience, we prove the continuity in $y_0 = 0$. First, for $y \in] - \frac{1}{2}, \frac{1}{2} [$ the function

$$\phi(x) = T_y f(x) - T_{y_0} f(x) = f(x - y) - f(x)$$

has support in the interval $[-\frac{1}{2}+c, d+\frac{1}{2}]$. Since f is uniformly continuous, we can for any given $\epsilon > 0$ find $\delta > 0$ such that

$$|f(x-y) - f(x)| \le \epsilon \text{ for all } x \in \mathbb{R} \text{ whenever } |y| \le \delta;$$

with this choice of δ, we thus obtain that

$$
\begin{aligned}
||T_y f - T_{y_0} f|| &= \left(\int_{-\frac{1}{2}+c}^{\frac{1}{2}+d} |f(x-y) - f(x)|^2 \, dx \right)^{1/2} \\
&\le \epsilon \sqrt{d - c + 1}.
\end{aligned}
$$

This proves the continuity in the considered special case. The case of an arbitrary function $f \in L^2(\mathbb{R})$ follows by an approximation argument, using that the continuous functions with compact support are dense in $L^2(\mathbb{R})$ (Exercise 2.6). The proofs of the statements for E_b and D_a are left to the reader (Exercise 2.7).   □

Chapters 9–11 will deal with Gabor systems and wavelet systems in $L^2(\mathbb{R})$; both classes consist of functions in $L^2(\mathbb{R})$ that are defined by compositions of some of the operators $T_a, E_b,$ and D_a. For this reason, the following *commutator relations* are important:

$$
\begin{aligned}
T_a E_b f(x) &= e^{-2\pi i b a} E_b T_a f(x) = e^{2\pi i b (x-a)} f(x-a), & (2.19) \\
T_b D_a f(x) &= D_a T_{b/a} f(x) = \frac{1}{\sqrt{|a|}} f\left(\frac{x}{a} - \frac{b}{a}\right), & (2.20) \\
D_a E_b f(x) &= \frac{1}{\sqrt{|a|}} e^{2\pi i x b / a} f\left(\frac{x}{a}\right) = E_{\frac{b}{a}} D_a f(x). & (2.21)
\end{aligned}
$$

In wavelet analysis, the dilation operator $D_{1/2}$ plays a special role, and we simply write

$$Df(x) := 2^{1/2} f(2x).$$

With this notation, the commutator relation (2.20) in particular implies that

$$T_k D^j = D^j T_{2^j k} \text{ and } D^j T_k = T_{2^{-j} k} D^j, \ j, k \in \mathbb{Z}. \tag{2.22}$$

We will often use the Fourier transformation in connection with Gabor systems and wavelet systems. In this context, we need the commutator relations

$$\mathcal{F} T_a = E_{-a} \mathcal{F}, \quad \mathcal{F} E_a = T_a \mathcal{F}, \quad \mathcal{F} D_a = D_{1/a} \mathcal{F}, \quad \mathcal{F} D = D^{-1} \mathcal{F}. \tag{2.23}$$

2.10   Exercises

2.1 Find a sequence $\{a_k\}_{k=1}^{\infty}$ of real numbers for which $\sum_{k=1}^{\infty} a_k$ is convergent but not unconditionally convergent.

2.2 Let $\{f_k\}_{k=1}^{\infty}$ be a sequence in a Banach space. Prove that absolute convergence of $\sum_{k=1}^{\infty} f_k$ implies unconditional convergence.

2.3 Prove Lemma 2.3.1.

2.4 Prove that the conditions in (2.10) are equivalent to the construction of the pseudo-inverse in Lemma 2.5.1.

2.5 Here we ask the reader to prove some results concerning Lebesgue points.

(i) Assume that $f : \mathbb{R} \to \mathbb{C}$ is continuous. Prove that every $y \in \mathbb{R}$ is a Lebesgue point.

(ii) Prove that $x = 0$ is not a Lebesgue point for the function $\chi_{[0,1]}$.

(iii) Let $f = \chi_{\mathbb{Q}}$. Prove that every $y \notin \mathbb{Q}$ is a Lebesgue point and that the rational numbers are not Lebesgue points.

2.6 Complete the proof of Lemma 2.9.1 by showing the continuity of the mapping $y \mapsto T_y f$ for $f \in L^2(\mathbb{R})$.

2.7 Prove the statements about E_b and D_a in Lemma 2.9.1.

2.8 Prove the commutator relations (2.23).

3
Bases

Bases play a prominent role in the analysis of vector spaces, as well in the finite-dimensional as in the infinite-dimensional case. The idea is the same in both cases, namely to consider a family of elements in the considered space such that all vectors can be expressed in a unique way as a linear combination of these elements. In the infinite-dimensional case, the situation is complicated: we are forced to work with infinite series, and different concepts of a basis are possible, depending on how we want the series to converge. For example, are we asking for the series to converge with respect to a fixed order of the elements (conditional convergence) or do we want it to converge regardless of how the elements are ordered (unconditional convergence)? In Hilbert spaces, unconditional convergence can be obtained by considering Bessel sequences, so in Section 3.1 we analyze such sequences in detail. In Section 3.2, we discuss the most important properties of orthonormal bases in Hilbert spaces; we expect the reader to have some basic knowledge about this subject. A slight (but useful) modification leads to the definition of Riesz bases, which are treated in detail in Section 3.3. In Section 3.4, Riesz bases and Bessel sequences are described in terms of the so-called Gram matrix. Concrete examples of bases in function spaces are given in Sections 3.5 and 3.6, where the basic theory for Fourier series is revisited (again this subject is expected to be known), and Gabor bases as well as wavelet bases for $L^2(\mathbb{R})$ are introduced. These sections form the background for Chapters 7–11. Section 3.7 gives a short introduction to Schauder bases in Banach spaces. Finally, Section 3.8 presents the sampling problem and relates it to bases in a particular Hilbert space.

O. Christensen, *Frames and Bases*. DOI: 10.1007/978-0-8176-4678-3_3,
© Springer Science+Business Media, LLC 2008

3.1 Bessel sequences in Hilbert spaces

The rest of the book will deal with infinite-dimensional vector spaces; thus, we need to consider expansions in terms of infinite series. The purpose of this section is to introduce a condition which ensures that the relevant infinite series actually converge.

Let $\mathcal{H}$ be a separable Hilbert space, with the inner product $\langle \cdot, \cdot \rangle$ chosen to be linear in the first entry. Recall from Section 2.1 that when speaking about a *sequence* $\{f_k\}_{k=1}^\infty$ in $\mathcal{H}$, we mean an *ordered* set, i.e.,

$$\{f_k\}_{k=1}^\infty = \{f_1, f_2, \dots\}.$$

That we have chosen to index the sequence by the natural numbers is just for convenience: soon, we will see that all results hold with arbitrary countable index sets and the elements f_k ordered in an arbitrary way.

Lemma 3.1.1 *Let $\{f_k\}_{k=1}^\infty$ be a sequence in $\mathcal{H}$, and suppose that $\sum_{k=1}^\infty c_k f_k$ is convergent for all $\{c_k\}_{k=1}^\infty \in \ell^2(\mathbb{N})$. Then*

$$T : \ell^2(\mathbb{N}) \to \mathcal{H}, \ T\{c_k\}_{k=1}^\infty := \sum_{k=1}^\infty c_k f_k \tag{3.1}$$

defines a bounded linear operator. The adjoint operator is given by

$$T^* : \mathcal{H} \to \ell^2(\mathbb{N}), \quad T^* f = \{\langle f, f_k \rangle\}_{k=1}^\infty. \tag{3.2}$$

Furthermore,

$$\sum_{k=1}^\infty |\langle f, f_k \rangle|^2 \leq ||T||^2 \, ||f||^2, \ \forall f \in \mathcal{H}. \tag{3.3}$$

Proof. Consider the sequence of bounded linear operators

$$T_n : \ell^2(\mathbb{N}) \to \mathcal{H}, \ T_n\{c_k\}_{k=1}^\infty := \sum_{k=1}^n c_k f_k.$$

Clearly, $T_n \to T$ pointwise as $n \to \infty$, so by Theorem 2.2.1 the map T defines a bounded linear operator. In order to find the expression for T^*, let $f \in \mathcal{H}$ and $\{c_k\}_{k=1}^\infty \in \ell^2(\mathbb{N})$. Then

$$\langle f, T\{c_k\}_{k=1}^\infty \rangle_\mathcal{H} = \langle f, \sum_{k=1}^\infty c_k f_k \rangle_\mathcal{H} = \sum_{k=1}^\infty \langle f, f_k \rangle \overline{c_k}. \tag{3.4}$$

We mention two ways to find $T^* f$ from here.

(1) The convergence of the series $\sum_{k=1}^\infty \langle f, f_k \rangle \overline{c_k}$ for all $\{c_k\}_{k=1}^\infty \in \ell^2(\mathbb{N})$ implies that $\{\langle f, f_k \rangle\}_{k=1}^\infty \in \ell^2(\mathbb{N})$; see for example [43], page 145. Thus we can write

$$\langle f, T\{c_k\}_{k=1}^\infty \rangle_\mathcal{H} \ = \ \langle \{\langle f, f_k \rangle\}, \{c_k\} \rangle_{\ell^2(\mathbb{N})}$$

and conclude that

$$T^*f = \{\langle f, f_k \rangle\}_{k=1}^{\infty}.$$

(2) Alternatively, when $T : \ell^2(\mathbb{N}) \to \mathcal{H}$ is bounded, we already know that T^* is a bounded operator from $\mathcal{H}$ to $\ell^2(\mathbb{N})$. Therefore, the k-th coordinate function is bounded from $\mathcal{H}$ to $\mathbb{C}$; by Riesz' representation theorem, T^* therefore has the form

$$T^*f = \{\langle f, g_k \rangle\}_{k=1}^{\infty}$$

for some $\{g_k\}_{k=1}^{\infty}$ in $\mathcal{H}$. By definition of T^*, (3.4) now shows that

$$\sum_{k=1}^{\infty} \langle f, g_k \rangle \overline{c_k} = \sum_{k=1}^{\infty} \langle f, f_k \rangle \overline{c_k}, \ \forall \{c_k\}_{k=1}^{\infty} \in \ell^2(\mathbb{N}), \ f \in \mathcal{H}.$$

It follows from here that $g_k = f_k$.

The adjoint of a bounded operator T is itself bounded, and $||T|| = ||T^*||$. Under the assumption in Lemma 3.1.1, we therefore have

$$||T^*f||^2 \leq ||T||^2 \, ||f||^2, \ \forall f \in \mathcal{H},$$

which leads to (3.3). $\qquad\qquad\qquad\qquad\qquad\qquad\qquad\qquad\qquad\square$

Sequences $\{f_k\}_{k=1}^{\infty}$ for which an inequality of the type (3.3) holds will play a crucial role in the sequel.

Definition 3.1.2 *A sequence $\{f_k\}_{k=1}^{\infty}$ in $\mathcal{H}$ is called a Bessel sequence if there exists a constant $B > 0$ such that*

$$\sum_{k=1}^{\infty} |\langle f, f_k \rangle|^2 \leq B \, ||f||^2, \ \forall f \in \mathcal{H}. \qquad\qquad (3.5)$$

Any number B satisfying (3.5) is called a *Bessel bound* for $\{f_k\}_{k=1}^{\infty}$. The *optimal bound* for a given Bessel sequence $\{f_k\}_{k=1}^{\infty}$ is the smallest possible value of $B > 0$ satisfying (3.5). Except for the case $f_k = 0, \ \forall k \in \mathbb{N}$, the optimal bound always exists.

Theorem 3.1.3 *Let $\{f_k\}_{k=1}^{\infty}$ be a sequence in $\mathcal{H}$ and $B > 0$ be given. Then $\{f_k\}_{k=1}^{\infty}$ is a Bessel sequence with Bessel bound B if and only if*

$$T : \{c_k\}_{k=1}^{\infty} \to \sum_{k=1}^{\infty} c_k f_k$$

defines a bounded operator from $\ell^2(\mathbb{N})$ into $\mathcal{H}$ and $||T|| \leq \sqrt{B}$.

Proof. First assume that $\{f_k\}_{k=1}^{\infty}$ is a Bessel sequence with Bessel bound B. Let $\{c_k\}_{k=1}^{\infty} \in \ell^2(\mathbb{N})$. First we want to show that $T\{c_k\}_{k=1}^{\infty}$ is well-defined, i.e., that $\sum_{k=1}^{\infty} c_k f_k$ is convergent. Consider $n, m \in \mathbb{N}, n > m$.

Then

$$\left\|\sum_{k=1}^{n} c_k f_k - \sum_{k=1}^{m} c_k f_k\right\| = \left\|\sum_{k=m+1}^{n} c_k f_k\right\|.$$

Using Lemma 2.3.4 and Cauchy–Schwarz' inequality, it follows that

$$\left\|\sum_{k=1}^{n} c_k f_k - \sum_{k=1}^{m} c_k f_k\right\| = \sup_{||g||=1} \left|\left\langle \sum_{k=m+1}^{n} c_k f_k, g\right\rangle\right|$$

$$\leq \sup_{||g||=1} \sum_{k=m+1}^{n} |c_k \langle f_k, g\rangle|$$

$$\leq \left(\sum_{k=m+1}^{n} |c_k|^2\right)^{1/2} \sup_{||g||=1} \left(\sum_{k=m+1}^{n} |\langle f_k, g\rangle|^2\right)^{1/2}$$

$$\leq \sqrt{B} \left(\sum_{k=m+1}^{n} |c_k|^2\right)^{1/2}.$$

Since $\{c_k\}_{k=1}^{\infty} \in \ell^2(\mathbb{N})$, we know that $\{\sum_{k=1}^{n} |c_k|^2\}_{n=1}^{\infty}$ is a Cauchy sequence in $\mathbb{C}$. The above calculation now shows that $\{\sum_{k=1}^{n} c_k f_k\}_{n=1}^{\infty}$ is a Cauchy sequence in $\mathcal{H}$ and therefore convergent. Thus $T\{c_k\}_{k=1}^{\infty}$ is well-defined. Clearly T is linear; since $||T\{c_k\}_{k=1}^{\infty}|| = \sup_{||g||=1} |\langle T\{c_k\}_{k=1}^{\infty}, g\rangle|$, a calculation as above shows that T is bounded and that $||T|| \leq \sqrt{B}$.

For the opposite implication, suppose that T defines a bounded operator with $||T|| \leq \sqrt{B}$. Then Lemma 3.1.1 shows that $\{f_k\}_{k=1}^{\infty}$ is a Bessel sequence with Bessel bound B. □

It is a consequence of Lemma 3.1.1 that if we only need to know that $\{f_k\}_{k=1}^{\infty}$ is a Bessel sequence and the value for the Bessel bound is irrelevant, we can just check that the operator T is well defined:

Corollary 3.1.4 *If $\{f_k\}_{k=1}^{\infty}$ is a sequence in $\mathcal{H}$ and $\sum_{k=1}^{\infty} c_k f_k$ is convergent for all $\{c_k\}_{k=1}^{\infty} \in \ell^2(\mathbb{N})$, then $\{f_k\}_{k=1}^{\infty}$ is a Bessel sequence.*

The Bessel condition (3.5) remains the same, regardless of how the elements $\{f_k\}_{k=1}^{\infty}$ are numbered. This leads to a very important consequence of Theorem 3.1.3:

Corollary 3.1.5 *If $\{f_k\}_{k=1}^{\infty}$ is a Bessel sequence in $\mathcal{H}$, then $\sum_{k=1}^{\infty} c_k f_k$ converges unconditionally for all $\{c_k\}_{k=1}^{\infty} \in \ell^2(\mathbb{N})$.*

Thus a reordering of the elements in $\{f_k\}_{k=1}^{\infty}$ will not affect the series $\sum_{k=1}^{\infty} c_k f_k$ when $\{c_k\}_{k=1}^{\infty}$ is reordered the same way: the series will converge toward the same element as before, see Exercise 3.2. For this reason we can choose an arbitrary indexing of the elements in the Bessel sequence; in particular, it is not a restriction that we present all results with the natural numbers as index set. As we will see in the sequel, all orthonormal bases, Riesz bases, and frames are Bessel sequences.

It is enough to check the Bessel condition (3.5) on a dense subset of $\mathcal{H}$:

Lemma 3.1.6 *Suppose that $\{f_k\}_{k=1}^{\infty}$ is a sequence of elements in $\mathcal{H}$ and that there exists a constant $B > 0$ such that*

$$\sum_{k=1}^{\infty} |\langle f, f_k \rangle|^2 \leq B \, ||f||^2$$

for all f in a dense subset V of $\mathcal{H}$. Then $\{f_k\}_{k=1}^{\infty}$ is a Bessel sequence with bound B.

We leave the proof to the reader (Exercise 3.3). See Exercise 3.4 for a result in the same spirit.

3.2   General bases and orthonormal bases

We are now ready to introduce one of the central themes, namely, bases in Hilbert spaces. In particular, we will discuss orthonormal bases, which are the infinite-dimensional counterparts of the canonical bases in $\mathbb{C}^n$. Orthonormal bases are widely used in mathematics as well as physics, signal processing, and many other areas where one needs to represent functions in terms of "elementary building blocks."

Definition 3.2.1 *Consider a sequence $\{e_k\}_{k=1}^{\infty}$ of vectors in $\mathcal{H}$.*

(i) *The sequence $\{e_k\}_{k=1}^{\infty}$ is a (Schauder) basis for $\mathcal{H}$ if for each $f \in \mathcal{H}$ there exist unique scalar coefficients $\{c_k(f)\}_{k=1}^{\infty}$ such that*

$$f = \sum_{k=1}^{\infty} c_k(f) e_k. \tag{3.6}$$

(ii) *A basis $\{e_k\}_{k=1}^{\infty}$ is an unconditional basis if the series (3.6) converges unconditionally for each $f \in \mathcal{H}$.*

(iii) *A basis $\{e_k\}_{k=1}^{\infty}$ is an orthonormal basis if $\{e_k\}_{k=1}^{\infty}$ is an orthonormal system, i.e., if*

$$\langle e_k, e_j \rangle = \delta_{k,j} = \begin{cases} 1 & \text{if } k = j, \\ 0 & \text{if } k \neq j. \end{cases}$$

Note that an orthonormal system $\{e_k\}_{k=1}^{\infty}$ is a Bessel sequence. In fact, if $\{c_k\}_{k=1}^{\infty} \in \ell^2(\mathbb{N})$ and $m, n \in \mathbb{N}, n > m$, then

$$\left\| \sum_{k=1}^{n} c_k e_k - \sum_{k=1}^{m} c_k e_k \right\|^2 = \left\| \sum_{k=m+1}^{n} c_k e_k \right\|^2 = \sum_{k=m+1}^{n} |c_k|^2;$$

as in the proof of Theorem 3.1.3 this implies that $\sum_{k=1}^{\infty} c_k e_k$ is convergent, and that

$$\left\| \sum_{k=1}^{\infty} c_k e_k \right\|^2 = \sum_{k=1}^{\infty} |c_k|^2.$$

By Lemma 3.1.1, this shows that $\{e_k\}_{k=1}^{\infty}$ is a Bessel sequence.

The next theorem gives equivalent conditions for an orthonormal system $\{e_k\}_{k=1}^{\infty}$ to be an orthonormal basis.

Theorem 3.2.2 *For an orthonormal system $\{e_k\}_{k=1}^{\infty}$, the following are equivalent:*

(i) $\{e_k\}_{k=1}^{\infty}$ *is an orthonormal basis.*

(ii) $f = \sum_{k=1}^{\infty} \langle f, e_k \rangle e_k, \ \forall f \in \mathcal{H}$.

(iii) $\langle f, g \rangle = \sum_{k=1}^{\infty} \langle f, e_k \rangle \langle e_k, g \rangle, \ \forall f, g \in \mathcal{H}.$

(iv) $\sum_{k=1}^{\infty} |\langle f, e_k \rangle|^2 = ||f||^2, \ \forall f \in \mathcal{H}.$

(v) $\overline{span}\{e_k\}_{k=1}^{\infty} = \mathcal{H}.$

(vi) *If* $\langle f, e_k \rangle = 0, \ \forall k \in \mathbb{N},$ *then* $f = 0.$

Proof. For the proof of (i) $\Rightarrow$ (ii), let $f \in \mathcal{H}$. If $\{e_k\}_{k=1}^{\infty}$ is an orthonormal basis, there exist coefficients $\{c_k\}_{k=1}^{\infty}$ such that $f = \sum_{k=1}^{\infty} c_k e_k$. Given any $j \in \mathbb{N}$, we have

$$\langle f, e_j \rangle = \langle \sum_{k=1}^{\infty} c_k e_k, e_j \rangle = \sum_{k=1}^{\infty} c_k \delta_{k,j} = c_j,$$

and (ii) follows. (iii) is an obvious consequence of (ii), and (iv) is a special case of (iii). The implication (iv) $\Rightarrow$ (v) follows from Lemma 2.3.1; in fact, if $f \in \mathcal{H}$ is perpendicular to all $e_k, k \in \mathbb{N}$, then (iv) shows that $f = 0$. The implication (v) $\Rightarrow$ (vi) also follows from Lemma 2.3.1. For the proof of (vi) $\Rightarrow$ (i), let $f \in \mathcal{H}$. Since $\{e_k\}_{k=1}^{\infty}$ is a Bessel sequence, we know that $g := \sum_{k=1}^{\infty} \langle f, e_k \rangle e_k$ is well defined; furthermore, $\langle f - g, e_j \rangle = 0$ for all $j \in \mathbb{N}$, so by (vi), $f = g = \sum_{k=1}^{\infty} \langle f, e_k \rangle e_k$. To prove that $\{e_k\}_{k=1}^{\infty}$ is a basis, we only need to show that no other linear combination of $\{e_k\}_{k=1}^{\infty}$ can be equal to f, and this follows by the argument we used to prove that (ii) follows from (i). $\qquad \square$

The equality in (iv) is called *Parseval's equation*. Via Corollary 3.1.5, we obtain the following important consequence of Theorem 3.2.2:

Corollary 3.2.3 *If $\{e_k\}_{k=1}^{\infty}$ is an orthonormal basis, then each $f \in \mathcal{H}$ has an unconditionally convergent expansion*

$$f = \sum_{k=1}^{\infty} \langle f, e_k \rangle e_k. \tag{3.7}$$

The expansion property (3.7) is the main reason for considering orthonormal bases. Fortunately, they exist in all separable Hilbert spaces:

Theorem 3.2.4 *Every separable Hilbert space $\mathcal{H}$ has an orthonormal basis.*

Proof. Since $\mathcal{H}$ is assumed separable, we can choose a sequence $\{f_k\}_{k=1}^{\infty}$ in $\mathcal{H}$ such that $\overline{\text{span}}\{f_k\}_{k=1}^{\infty} = \mathcal{H}$. By extracting a subsequence if necessary, we can assume that for each $n \in \mathbb{N}, f_{n+1} \notin \text{span}\{f_k\}_{k=1}^{n}$. By applying the Gram–Schmidt process to $\{f_k\}_{k=1}^{\infty}$, we obtain an orthonormal system $\{e_k\}_{k=1}^{\infty}$ in $\mathcal{H}$ for which $\overline{\text{span}}\{e_k\}_{k=1}^{\infty} = \overline{\text{span}}\{f_k\}_{k=1}^{\infty} = \mathcal{H}$.   □

Usually, the mere existence of orthonormal bases is not enough: we need to be able to construct them. In the Hilbert space $\ell^2(\mathbb{N})$, we have an orthonormal basis given by a particularly simple expression:

Example 3.2.5 For $k \in \mathbb{N}$, let e_k be the sequence in $\ell^2(\mathbb{N})$ whose k-th entry is 1, and all other entries are zero. Then $\{e_k\}_{k=1}^{\infty}$ is an orthonormal basis for $\ell^2(\mathbb{N})$; it is called the *canonical orthonormal basis*. We will often denote this special basis by $\{\delta_k\}_{k=1}^{\infty}$.   □

In practice, orthonormal bases are certainly the most convenient bases to use: we will later see that, for other types of bases, the representation (3.7) has to be replaced by a more complicated expression. Unfortunately, the conditions for $\{e_k\}_{k=1}^{\infty}$ being an orthonormal basis are strong, and often it is impossible to construct orthonormal bases satisfying extra conditions. We discuss this in more detail in Chapter 4. Note also that it is not always a good idea to use the Gram–Schmidt orthonormalization procedure to construct an orthonormal basis from a given basis: it might destroy special properties of the basis at hand. For example, the special structure of Gabor bases and wavelet bases (to be discussed later) will get lost.

Based on Theorem 3.2.4, we can prove that every separable Hilbert space can be identified with $\ell^2(\mathbb{N})$:

Theorem 3.2.6 *Every separable infinite-dimensional Hilbert space $\mathcal{H}$ is isometrically isomorphic to $\ell^2(\mathbb{N})$.*

Proof. Let $\{e_k\}_{k=1}^{\infty}$ be an orthonormal basis for $\mathcal{H}$. We have already observed that $\sum_{k=1}^{\infty} c_k e_k$ is convergent for all $\{c_k\}_{k=1}^{\infty} \in \ell^2(\mathbb{N})$. Furthermore, each $f \in \mathcal{H}$ has a unique expansion with ℓ^2-coefficients, namely $f = \sum \langle f, e_k \rangle e_k$. Letting $\{\delta_k\}_{k=1}^{\infty}$ be the canonical orthonormal basis for $\ell^2(\mathbb{N})$, we can define the operator

$$U : \mathcal{H} \to \ell^2(\mathbb{N}), \ U\left(\sum c_k e_k\right) = \sum c_k \delta_k, \ \{c_k\}_{k=1}^{\infty} \in \ell^2(\mathbb{N}).$$

Then U maps $\mathcal{H}$ bijectively onto $\ell^2(\mathbb{N})$. For $f \in \mathcal{H}, f = \sum \langle f, e_k \rangle e_k$, we have

$$
\begin{aligned}
\|Uf\|^2 &= \|\sum \langle f, e_k \rangle \delta_k \|^2 \\
&= \sum |\langle f, e_k \rangle|^2 \\
&= \|f\|^2;
\end{aligned}
$$

thus U is an isometry.   □

The following theorem characterizes all orthonormal bases for $\mathcal{H}$ starting with one arbitrary orthonormal basis.

Theorem 3.2.7 Let $\{e_k\}_{k=1}^{\infty}$ be an orthonormal basis for $\mathcal{H}$. Then the orthonormal bases for $\mathcal{H}$ are precisely the sets $\{Ue_k\}_{k=1}^{\infty}$, where $U : \mathcal{H} \to \mathcal{H}$ is a unitary operator.

Proof. Let $\{f_k\}_{k=1}^{\infty}$ be an orthonormal basis for $\mathcal{H}$. Define the operator

$$
U : \mathcal{H} \to \mathcal{H}, \ U\left(\sum c_k e_k\right) = \sum c_k f_k, \ \{c_k\}_{k=1}^{\infty} \in \ell^2(\mathbb{N}).
$$

Then U maps $\mathcal{H}$ boundedly and bijectively onto $\mathcal{H}$, and $f_k = Ue_k$. For $f, g \in \mathcal{H}$, write $f = \sum \langle f, e_k \rangle e_k$ and $g = \sum \langle g, e_k \rangle e_k$; then, via the definition of U and Theorem 3.2.2,

$$
\begin{aligned}
\langle U^*Uf, g \rangle &= \langle Uf, Ug \rangle \\
&= \left\langle \sum \langle f, e_k \rangle f_k, \sum \langle g, e_k \rangle f_k \right\rangle \\
&= \sum \langle f, e_k \rangle \overline{\langle g, e_k \rangle} = \langle f, g \rangle.
\end{aligned}
$$

This implies that $U^*U = I$. Since U is surjective, it follows that U is unitary. On the other hand, if U is a given unitary operator, then

$$
\langle Ue_k, Ue_j \rangle = \langle U^*Ue_k, e_j \rangle = \langle e_k, e_j \rangle = \delta_{k,j},
$$

i.e., $\{Ue_k\}_{k=1}^{\infty}$ is an orthonormal system. That it is a basis follows from Theorem 3.2.2 and the fact that U is surjective.   □

Condition (iv) in Theorem 3.2.2 has an interpretation in terms of frames, see Definition 5.1.3. Without assuming that $\{e_k\}_{k=1}^{\infty}$ is an orthonormal system, it implies that $\{e_k\}_{k=1}^{\infty}$ is an orthonormal basis if the vectors are normalized (we ask the reader to provide the proof, see Exercise 3.5):

Proposition 3.2.8 Assume that $\{e_k\}_{k=1}^{\infty}$ is a sequence of normalized vectors in $\mathcal{H}$ and that

$$
\sum_{k=1}^{\infty} |\langle f, e_k \rangle|^2 = \|f\|^2, \ \forall f \in \mathcal{H}.
$$

Then $\{e_k\}_{k=1}^{\infty}$ is an orthonormal basis for $\mathcal{H}$.

3.3   Riesz bases

In Theorem 3.2.7, we characterized all orthonormal bases in terms of unitary operators acting on a single orthonormal basis. Formally, the definition of a *Riesz basis* appears by weakening the condition on the operator:

Definition 3.3.1 *A Riesz basis for $\mathcal{H}$ is a family of the form $\{Ue_k\}_{k=1}^{\infty}$, where $\{e_k\}_{k=1}^{\infty}$ is an orthonormal basis for $\mathcal{H}$ and $U : \mathcal{H} \to \mathcal{H}$ is a bounded bijective operator.*

A Riesz basis $\{f_k\}_{k=1}^{\infty}$ is actually a basis; this follows from the proof of Theorem 3.3.2, which we state now. Note that the expansion (3.8) of elements $f \in \mathcal{H}$ in terms of a Riesz basis is more involved than the expression (3.7) we obtained via orthonormal bases:

Theorem 3.3.2 *If $\{f_k\}_{k=1}^{\infty}$ is a Riesz basis for $\mathcal{H}$, then $\{f_k\}_{k=1}^{\infty}$ is a Bessel sequence. Furthermore, there exists a unique sequence $\{g_k\}_{k=1}^{\infty}$ in $\mathcal{H}$ such that*

$$f = \sum_{k=1}^{\infty} \langle f, g_k \rangle f_k, \ \forall f \in \mathcal{H}. \tag{3.8}$$

The sequence $\{g_k\}_{k=1}^{\infty}$ is also a Riesz basis, and the series (3.8) converges unconditionally for all $f \in \mathcal{H}$.

Proof.   According to the definition, we can write $\{f_k\}_{k=1}^{\infty} = \{Ue_k\}_{k=1}^{\infty}$, where U is a bounded bijective operator and $\{e_k\}_{k=1}^{\infty}$ is an orthonormal basis. Let now $f \in \mathcal{H}$. By expanding $U^{-1}f$ in the orthonormal basis $\{e_k\}_{k=1}^{\infty}$, we have

$$U^{-1}f = \sum_{k=1}^{\infty} \langle U^{-1}f, e_k \rangle e_k = \sum_{k=1}^{\infty} \langle f, (U^{-1})^* e_k \rangle e_k.$$

Therefore, with $g_k := (U^{-1})^* e_k$,

$$f = UU^{-1}f = \sum_{k=1}^{\infty} \langle f, (U^{-1})^* e_k \rangle Ue_k$$

$$= \sum_{k=1}^{\infty} \langle f, g_k \rangle f_k.$$

Since the operator $(U^{-1})^*$ is bounded and bijective, $\{g_k\}_{k=1}^{\infty}$ is a Riesz basis by definition.

For $f \in \mathcal{H}$,

$$\sum_{k=1}^{\infty} |\langle f, f_k \rangle|^2 = \sum_{k=1}^{\infty} |\langle f, Ue_k \rangle|^2 \quad = \quad ||U^* f||^2 \tag{3.9}$$

$$\leq \quad ||U^*||^2 \, ||f||^2$$

$$= \quad ||U||^2 \, ||f||^2. \tag{3.10}$$

This proves that a Riesz basis is a Bessel sequence. Thus, the series (3.8) converges unconditionally by Corollary 3.1.5. We complete the proof by showing that the sequence $\{g_k\}_{k=1}^{\infty}$ constructed in the proof is the only one that satisfies (3.8). For that purpose, we first note that if

$$f = \sum_{k=1}^{\infty} c_k(f) f_k = \sum_{k=1}^{\infty} d_k(f) f_k \tag{3.11}$$

for some coefficients $c_k(f)$ and $d_k(f)$, then necessarily $c_k(f) = d_k(f)$ for all $k \in \mathbb{N}$; this follows by applying the operator U^{-1} on both sides of the equality and using that $\{e_k\}_{k=1}^{\infty}$ is known to be a basis. This argument shows that a Riesz basis actually is a basis. Now we only have to show that if $\{g_k\}_{k=1}^{\infty}$ and $\{h_k\}_{k=1}^{\infty}$ are sequences in $\mathcal{H}$ such that

$$f = \sum_{k=1}^{\infty} \langle f, g_k \rangle f_k = \sum_{k=1}^{\infty} \langle f, h_k \rangle f_k, \ \forall f \in \mathcal{H}, \tag{3.12}$$

then $g_k = h_k$ for all $k \in \mathbb{N}$. However, due to the argument above, (3.12) implies that for all $k \in \mathbb{N}$,

$$\langle f, g_k \rangle = \langle f, h_k \rangle, \ \forall f \in \mathcal{H};$$

the desired result now follows from Lemma 2.3.3.   □

The unique sequence $\{g_k\}_{k=1}^{\infty}$ satisfying (3.8) is called the *dual Riesz basis* of $\{f_k\}_{k=1}^{\infty}$. Let us find the dual of $\{g_k\}_{k=1}^{\infty}$. In the notation used in the proof of Theorem 3.3.2, we have that the dual of $\{f_k\}_{k=1}^{\infty} = \{Ue_k\}_{k=1}^{\infty}$ is given by $\{g_k\}_{k=1}^{\infty} = \{(U^{-1})^* e_k\}_{k=1}^{\infty}$; thus, the dual of $\{g_k\}_{k=1}^{\infty}$ is

$$\left\{ \left(\left((U^{-1})^* \right)^{-1} \right)^* e_k \right\}_{k=1}^{\infty} = \{Ue_k\}_{k=1}^{\infty} = \{f_k\}_{k=1}^{\infty}.$$

That is, $\{f_k\}_{k=1}^{\infty}$ and $\{g_k\}_{k=1}^{\infty}$ are duals of each other. For this reason, we frequently speak about a *pair of dual Riesz bases*. In particular, this implies a symmetric version of (3.8), see (3.13). Furthermore, a Riesz basis and its dual Riesz basis satisfy an important orthogonality relationship. We state the result right after the following definition.

Definition 3.3.3 *Two sequences* $\{f_k\}_{k=1}^{\infty}$ *and* $\{g_k\}_{k=1}^{\infty}$ *in a Hilbert space are biorthogonal if*

$$\langle f_k, g_j \rangle = \delta_{k,j}.$$

The proof of the next result is left to the reader (Exercise 3.6).

Corollary 3.3.4 *For a pair of dual Riesz bases $\{f_k\}_{k=1}^{\infty}$ and $\{g_k\}_{k=1}^{\infty}$ the following holds:*

(i) $\{f_k\}_{k=1}^{\infty}$ and $\{g_k\}_{k=1}^{\infty}$ are biorthogonal.

(ii) For all $f \in \mathcal{H}$,

$$f = \sum_{k=1}^{\infty} \langle f, g_k \rangle f_k = \sum_{k=1}^{\infty} \langle f, f_k \rangle g_k. \tag{3.13}$$

Already in the proof of Theorem 3.8, we saw that a Riesz basis is a Bessel sequence. For later use, we now show that it also satisfies some kind of "opposite inequality":

Proposition 3.3.5 *If $\{f_k\}_{k=1}^{\infty} = \{Ue_k\}_{k=1}^{\infty}$ is a Riesz basis for $\mathcal{H}$, there exist constants $A, B > 0$ such that*

$$A \, ||f||^2 \le \sum_{k=1}^{\infty} |\langle f, f_k \rangle|^2 \le B \, ||f||^2, \ \forall f \in \mathcal{H}. \tag{3.14}$$

The largest possible value for the constant A is $\frac{1}{||U^{-1}||^2}$, and the smallest possible value for B is $||U||^2$.

Proof. That a Riesz basis $\{Ue_k\}_{k=1}^{\infty}$ is a Bessel sequence with optimal upper bound $||U||^2$ follows already from the estimate in (3.10). The result about the lower bound is a consequence of (3.9) and the following estimate:

$$||f|| = ||(U^*)^{-1}U^*f|| \le ||(U^*)^{-1}|| \, ||U^*f|| = ||U^{-1}|| \, ||U^*f||. \qquad \square$$

We now aim at an equivalent characterization of Riesz bases. For this purpose, we need the lemma below.

Lemma 3.3.6 *Let $\mathcal{H}, \mathcal{K}$ be Hilbert spaces, and let $\{h_k\}_{k=1}^{\infty}$ be a sequence in $\mathcal{H}, \{g_k\}_{k=1}^{\infty}$ a sequence in $\mathcal{K}$. Assume that $\{g_k\}_{k=1}^{\infty}$ is a Bessel sequence with bound B, that $\{h_k\}_{k=1}^{\infty}$ is complete in $\mathcal{H}$, and that there exists a constant $A > 0$ such that*

$$A \sum |c_k|^2 \le \left\| \sum c_k h_k \right\|^2 \tag{3.15}$$

for all finite scalar sequences $\{c_k\}$. Then

$$U\left(\sum c_k h_k\right) := \sum c_k g_k \quad (\{c_k\} \text{ finite})$$

defines a linear bounded operator from $span\{h_k\}_{k=1}^{\infty}$ into $span\{g_k\}_{k=1}^{\infty}$ and U has a unique extension to a bounded operator from $\mathcal{H}$ into $\mathcal{K}$; the norm of U as well as its extension is at most $\sqrt{\frac{B}{A}}$.

Proof. By the assumption (3.15), every $h \in \text{span}\{h_k\}_{k=1}^{\infty}$ has a unique representation $h = \sum c_k h_k$ with $\{c_k\}$ finite; it follows that U is well defined and linear. Given a finite sequence $\{c_k\}$,

$$
\begin{aligned}
\left\| U\left(\sum c_k h_k\right)\right\|^2 &= \left\|\sum c_k g_k\right\|^2 \\
&\leq B \sum |c_k|^2 \\
&\leq \frac{B}{A}\left\|\sum c_k h_k\right\|^2.
\end{aligned}
$$

Thus U is bounded. Because $\{h_k\}_{k=1}^{\infty}$ is assumed to be complete, U has an extension to a bounded operator on $\mathcal{H}$. The rest is standard. $\qquad\square$

The next theorem gives an equivalent condition for $\{f_k\}_{k=1}^{\infty}$ being a Riesz basis. Condition (ii) will be used throughout the book and is, in fact, by several authors used as the definition of a Riesz basis.

Theorem 3.3.7 *For a sequence $\{f_k\}_{k=1}^{\infty}$ in $\mathcal{H}$, the following conditions are equivalent:*

(i) $\{f_k\}_{k=1}^{\infty}$ *is a Riesz basis for $\mathcal{H}$.*

(ii) $\{f_k\}_{k=1}^{\infty}$ *is complete in $\mathcal{H}$, and there exist constants $A, B > 0$ such that for every finite scalar sequence $\{c_k\}$, one has*

$$
A \sum |c_k|^2 \leq \left\|\sum c_k f_k\right\|^2 \leq B \sum |c_k|^2. \tag{3.16}
$$

Proof. (i)$\Rightarrow$(ii). Assume that $\{f_k\}_{k=1}^{\infty}$ is a Riesz basis, and write it in the form $\{U e_k\}_{k=1}^{\infty}$ as in the definition. Note that as a consequence of Theorem 3.3.2, $\{f_k\}_{k=1}^{\infty}$ is complete. Given any finite scalar sequence $\{c_k\}$,

$$
\left\|\sum c_k f_k\right\|^2 = \left\| U\left(\sum c_k e_k\right)\right\|^2 \leq \|U\|^2 \left\|\sum c_k e_k\right\|^2 = \|U\|^2 \sum |c_k|^2
$$

and

$$
\left\|\sum c_k e_k\right\|^2 = \left\| U^{-1} U\left(\sum c_k e_k\right)\right\|^2 \leq \|U^{-1}\|^2 \left\|\sum c_k f_k\right\|^2,
$$

from which we deduce that

$$
\frac{1}{\|U^{-1}\|^2} \sum |c_k|^2 \leq \left\|\sum c_k f_k\right\|^2 \leq \|U\|^2 \sum |c_k|^2.
$$

(ii)$\Rightarrow$(i). The right-hand inequality in (3.16) implies that $\{f_k\}_{k=1}^{\infty}$ is a Bessel sequence with bound B (Exercise 3.8). Choose an orthonormal basis $\{e_k\}_{k=1}^{\infty}$ for $\mathcal{H}$, and extend by Lemma 3.3.6 the mapping $U e_k := f_k$ to a bounded operator on $\mathcal{H}$. In the same way, extend $V f_k := e_k$ to a bounded operator on $\mathcal{H}$. Then $VU = UV = I$, so U is invertible; thus $\{f_k\}_{k=1}^{\infty}$ is a Riesz basis. $\qquad\square$

A sequence $\{f_k\}_{k=1}^{\infty}$ satisfying (3.16) for all finite sequences $\{c_k\}_{k=1}^{\infty}$ is called a *Riesz sequence*. By Theorem 3.3.7, a Riesz sequence $\{f_k\}_{k=1}^{\infty}$ is a Riesz basis for $\overline{\text{span}}\{f_k\}_{k=1}^{\infty}$, which might just be a subspace of $\mathcal{H}$. Note that if the condition (3.16) is satisfied for a family $\{f_k\}_{k=1}^{\infty}$, then it is clearly satisfied for any subsequence of $\{f_k\}_{k=1}^{\infty}$. This leads to the following important consequence of Theorem 3.3.7.

Corollary 3.3.8 *Every subfamily of a Riesz sequence is a Riesz sequence.*

If (3.16) holds for all finite scalar sequences $\{c_k\}$, then it automatically holds for all $\{c_k\}_{k=1}^{\infty} \in \ell^2(\mathbb{N})$ (Exercise 3.8). If $\{f_k\}_{k=1}^{\infty}$ is a Riesz basis, numbers $A, B > 0$ that satisfy (3.16) are called *lower Riesz bounds*, respectively, *upper Riesz bounds*. They are clearly not unique, and we define the *optimal Riesz bounds* as the largest possible value for A and the smallest possible value for B.

If (3.16) holds with $A = B = 1$, the sequence $\{f_k\}_{k=1}^{\infty}$ is orthonormal:

Proposition 3.3.9 *Assume that* $\overline{\text{span}}\{f_k\}_{k=1}^{\infty} = \mathcal{H}$ *and that*

$$\left\|\sum c_k f_k\right\|^2 = \sum |c_k|^2 \tag{3.17}$$

for all finite scalar sequences $\{c_k\}$. *Then* $\{f_k\}_{k=1}^{\infty}$ *is an orthonormal basis for* $\mathcal{H}$.

Proof. The assumptions imply by Theorem 3.3.7 that $\{f_k\}_{k=1}^{\infty}$ is a Riesz basis for $\mathcal{H}$, so by letting $\{e_k\}_{k=1}^{\infty}$ be an orthonormal basis for $\mathcal{H}$, we can write $\{f_k\}_{k=1}^{\infty} = \{Ue_k\}_{k=1}^{\infty}$ for an appropriate bounded invertible operator U. Then, for all $\{c_k\}_{k=1}^{\infty} \in \ell^2(\mathbb{N})$,

$$\sum_{k=1}^{\infty} |c_k|^2 = \left\|\sum_{k=1}^{\infty} c_k f_k\right\|^2 = \left\|U\left(\sum_{k=1}^{\infty} c_k e_k\right)\right\|^2.$$

It follows from here that $||U|| = ||U^{-1}|| = 1$; by Proposition 3.3.5, we conclude that

$$\sum_{k=1}^{\infty} |\langle f, f_k \rangle|^2 = ||f||^2, \ \forall f \in \mathcal{H}.$$

The assumption (3.17) implies that $||f_k|| = 1$ for all $k \in \mathbb{N}$; we now obtain the result via Proposition 3.2.8. $\qquad\square$

So far, we have focused on theoretical properties of Riesz bases in general Hilbert spaces; we will return to more concrete results about Riesz bases in subspaces of $L^2(\mathbb{R})$ in Chapter 7.

3.4 The Gram matrix

The conditions for a sequence $\{f_k\}_{k=1}^{\infty}$ being a Bessel sequence or a Riesz basis can conveniently be expressed in terms of the so-called Gram matrix. In this section, we introduce this matrix and prove some of its main properties.

If $\{f_k\}_{k=1}^{\infty}$ is a Bessel sequence, we can compose the pre-frame operator T and its adjoint T^*; hereby we obtain the bounded operator

$$T^*T : \ell^2(\mathbb{N}) \to \ell^2(\mathbb{N}), \ T^*T\{c_k\}_{k=1}^{\infty} = \left\{\left\langle \sum_{\ell=1}^{\infty} c_\ell f_\ell, f_k \right\rangle\right\}_{k=1}^{\infty}.$$

Letting $\{e_k\}_{k=1}^{\infty}$ be the canonical orthonormal basis for $\ell^2(\mathbb{N})$, the jk-th entry in the matrix representation for T^*T is

$$\langle T^*Te_k, e_j \rangle = \langle Te_k, Te_j \rangle = \langle f_k, f_j \rangle.$$

Identifying T^*T with its matrix representation, we write

$$T^*T = \{\langle f_k, f_j \rangle\}_{j,k=1}^{\infty}.$$

The matrix $\{\langle f_k, f_j \rangle\}_{j,k=1}^{\infty}$ is called the *Gram matrix* associated with $\{f_k\}_{k=1}^{\infty}$, and the above argument shows that it defines a bounded operator on $\ell^2(\mathbb{N})$ when $\{f_k\}_{k=1}^{\infty}$ is a Bessel sequence. In principle, one can consider the Gram matrix associated with any sequence $\{f_k\}_{k=1}^{\infty}$ in $\mathcal{H}$, but if we want it to define a bounded operator on $\ell^2(\mathbb{N})$, we cannot avoid the Bessel condition:

Lemma 3.4.1 *For a sequence $\{f_k\}_{k=1}^{\infty}$ in $\mathcal{H}$, the following are equivalent:*

(i) $\{f_k\}_{k=1}^{\infty}$ *is a Bessel sequence with bound B.*

(ii) *The Gram matrix associated with $\{f_k\}_{k=1}^{\infty}$ defines a bounded operator on $\ell^2(\mathbb{N})$, with norm at most B.*

Proof. The implication (i) $\Rightarrow$ (ii) follows from the arguments above together with the norm estimate $\|T\| \leq \sqrt{B}$ in Theorem 3.1.3. Now assume that (ii) is satisfied, and let $\{c_k\}_{k=1}^{\infty} \in \ell^2(\mathbb{N})$. Then

$$\sum_{j=1}^{\infty} \left| \sum_{k=1}^{\infty} \langle f_k, f_j \rangle c_k \right|^2 \leq B^2 \sum_{k=1}^{\infty} |c_k|^2. \tag{3.18}$$

Given arbitrary $n, m \in \mathbb{N}, n > m$,

$$\left\| \sum_{k=1}^{n} c_k f_k - \sum_{k=1}^{m} c_k f_k \right\|^4 = \left\| \sum_{k=m+1}^{n} c_k f_k \right\|^4$$

$$= \left| \left\langle \sum_{k-m+1}^{n} c_k f_k, \sum_{j=m+1}^{n} c_j f_j \right\rangle \right|^2$$

$$= \left| \sum_{j=m+1}^{n} \overline{c_j} \sum_{k=m+1}^{n} c_k \langle f_k, f_j \rangle \right|^2$$

$$\leq \left(\sum_{j=m+1}^{n} |c_j|^2 \right) \left(\sum_{j=m+1}^{n} \left| \sum_{k=m+1}^{n} c_k \langle f_k, f_j \rangle \right|^2 \right),$$

where Cauchy–Schwarz' inequality was used on the sum over j in the last step. Via (3.18) applied to the finite sequence

$$(\cdots, 0, 0, c_{m+1}, c_{m+2}, \cdots, c_n, 0, 0, \cdots),$$

$$\sum_{j=m+1}^{n} \left| \sum_{k=m+1}^{n} c_k \langle f_k, f_j \rangle \right|^2 \leq \sum_{j=1}^{\infty} \left| \sum_{k=m+1}^{n} c_k \langle f_k, f_j \rangle \right|^2$$

$$\leq B^2 \sum_{k=m+1}^{n} |c_k|^2.$$

Altogether we arrive at

$$\left\| \sum_{k=1}^{n} c_k f_k - \sum_{k=1}^{m} c_k f_k \right\|^4 \leq B^2 \left(\sum_{j=m+1}^{\infty} |c_j|^2 \right)^2.$$

It follows that $\sum_{k=1}^{\infty} c_k f_k$ is convergent and, by repeating the argument,

$$\left\| \sum_{k=1}^{\infty} c_k f_k \right\| \leq \sqrt{B} \left(\sum_{j=1}^{\infty} |c_j|^2 \right)^{1/2}.$$

By Theorem 3.1.3, we now conclude that $\{f_k\}_{k=1}^{\infty}$ is a Bessel sequence with bound B. $\qquad \square$

Proposition 3.4.3 will give a sufficient condition for $\{f_k\}_{k=1}^{\infty}$ being a Bessel sequence. The proof uses *Schur's Lemma*:

Lemma 3.4.2 *Let $M = \{M_{j,k}\}_{j,k=1}^{\infty}$ be a matrix for which $M_{j,k} = \overline{M_{k,j}}$ for all $j, k \in \mathbb{N}$, and for which there exists a constant $B > 0$ such that*

$$\sum_{k=1}^{\infty} |M_{j,k}| \leq B, \ \forall j \in \mathbb{N}.$$

Then M defines a bounded operator on $\ell^2(\mathbb{N})$ of norm at most B.

Proof.   Let $\{c_k\}_{k=1}^{\infty} \in \ell^2(\mathbb{N})$. The assumptions imply that $M\{c_k\}_{k=1}^{\infty}$ is well defined as a sequence indexed by $\mathbb{N}$, whose j-th coordinate is $\sum_{k=1}^{\infty} M_{j,k}c_k$. It is, however, not immediately clear that this sequence belongs to $\ell^2(\mathbb{N})$. Abusing the notation, it is enough to show that the map

$$\{d_k\}_{k=1}^{\infty} \to \langle\{d_k\}_{k=1}^{\infty}, M\{c_k\}_{k=1}^{\infty}\rangle_{\ell^2(\mathbb{N})} \tag{3.19}$$

is a continuous linear functional on $\ell^2(\mathbb{N})$. In fact, this implies that $M\{c_k\}_{k=1}^{\infty}$ belongs to the dual of $\ell^2(\mathbb{N})$, which is $\ell^2(\mathbb{N})$ itself. Now, for $\{d_k\}_{k=1}^{\infty} \in \ell^2(\mathbb{N})$,

$$\sum_{j=1}^{\infty} \left| \sum_{k=1}^{\infty} \overline{M_{j,k}c_k}d_j \right| \ \leq \ \sum_{j=1}^{\infty}\sum_{k=1}^{\infty} |M_{j,k}c_kd_j|$$

$$= \ \sum_{j=1}^{\infty}\sum_{k=1}^{\infty} \left(|M_{j,k}|^{1/2}|c_k|\right)\left(|M_{j,k}|^{1/2}|d_j|\right) = (*).$$

Using Cauchy–Schwarz' inequality,

$$(*) \ \leq \ \left(\sum_{j=1}^{\infty}\sum_{k=1}^{\infty} |M_{j,k}|\,|c_k|^2\right)^{1/2} \left(\sum_{j=1}^{\infty}\sum_{k=1}^{\infty} |M_{j,k}|\,|d_j|^2\right)^{1/2}$$

$$\leq \ B\left(\sum_{k=1}^{\infty} |c_k|^2\right)^{1/2} \left(\sum_{j=1}^{\infty} |d_j|^2\right)^{1/2}.$$

This shows that (3.19) indeed defines a continuous linear functional on $\ell^2(\mathbb{N})$, so M maps $\ell^2(\mathbb{N})$ into $\ell^2(\mathbb{N})$. Also,

$$\|M\{c_k\}_{k=1}^{\infty}\| \ = \ \sup_{\|\{d_k\}\|=1} \left|\langle\{d_k\}_{k=1}^{\infty}, M\{c_k\}_{k=1}^{\infty}\rangle_{\ell^2(\mathbb{N})}\right|$$

$$\leq \ B\left(\sum_{k=1}^{\infty} |c_k|^2\right)^{1/2},$$

which shows that M is bounded with norm at most B, as desired (see Exercise 3.9 for a question about the proof). $\qquad\square$

An application of Schur's lemma gives a sufficient condition for the Gram matrix defining a bounded operator on $\ell^2(\mathbb{N})$, and thus for $\{f_k\}_{k=1}^{\infty}$ being a Bessel sequence. For the proof, we just have to refer to Lemma 3.4.1:

Proposition 3.4.3 *Let* $\{f_k\}_{k=1}^{\infty}$ *be a sequence in* $\mathcal{H}$ *and assume that there exists a constant* $B > 0$ *such that*

$$\sum_{k=1}^{\infty} |\langle f_j, f_k \rangle| \leq B, \ \forall j \in \mathbb{N}.$$

Then $\{f_k\}_{k=1}^{\infty}$ *is a Bessel sequence with bound* B.

Compared with the Bessel condition (3.5), Proposition 3.4.3 has the advantage that it only involves inner products between the elements in $\{f_k\}_{k=1}^{\infty}$; that is, only a countable number of conditions must be verified, whereas the Bessel condition has to be checked for all $f \in \mathcal{H}$.

We will now present a further equivalent condition for a sequence $\{f_k\}_{k=1}^{\infty}$ being a Riesz basis, expressed in terms of the Gram matrix.

Theorem 3.4.4 *For a sequence* $\{f_k\}_{k=1}^{\infty}$ *in* $\mathcal{H}$, *the following conditions are equivalent:*

(i) $\{f_k\}_{k=1}^{\infty}$ *is a Riesz basis for* $\mathcal{H}$.

(ii) $\{f_k\}_{k=1}^{\infty}$ *is complete, and its Gram matrix* $\{\langle f_k, f_j \rangle\}_{j,k=1}^{\infty}$ *defines a bounded, invertible operator on* $\ell^2(\mathbb{N})$.

(iii) $\{f_k\}_{k=1}^{\infty}$ *is a complete Bessel sequence, and it has a complete biorthogonal sequence* $\{g_k\}_{k=1}^{\infty}$ *that is also a Bessel sequence.*

Proof. (i)$\Rightarrow$(ii). Write again $\{f_k\}_{k=1}^{\infty} = \{Ue_k\}_{k=1}^{\infty}$ as in the definition of a Riesz basis. For any $k, j \in \mathbb{N}$,

$$\langle f_k, f_j \rangle = \langle Ue_k, Ue_j \rangle = \langle U^*Ue_k, e_j \rangle$$

i.e., the Gram matrix is the matrix representing the bounded invertible operator U^*U in the basis $\{e_k\}_{k=1}^{\infty}$.

(ii)$\Rightarrow$(i). Assume that (ii) is satisfied. Then Lemma 3.4.1 together with Theorem 3.1.3 shows that the upper condition in (3.16) is satisfied. Let G denote the operator on $\ell^2(\mathbb{N})$ given by the Gram matrix $\{\langle f_k, f_j \rangle\}_{j,k=1}^{\infty}$. Given a sequence $\{c_k\}_{k=1}^{\infty} \in \ell^2(\mathbb{N})$, the j-th element in the image sequence $G\{c_k\}_{k=1}^{\infty}$ is $\sum_{k=1}^{\infty} \langle f_k, f_j \rangle c_k$. Thus

$$\langle G\{c_k\}_{k=1}^{\infty}, \{c_k\}_{k=1}^{\infty} \rangle = \sum_{j=1}^{\infty} \sum_{k=1}^{\infty} \langle f_k, f_j \rangle c_k \overline{c_j}$$

$$= \left\| \sum_{k=1}^{\infty} c_k f_k \right\|^2.$$

Thus G is positive, and a similar calculation shows that G is self-adjoint. Let V denote the positive square-root of G, cf. Lemma 2.4.4. Then the above calculation gives that

$$\left\|\sum_{k=1}^{\infty} c_k f_k\right\|^2 = \|V\{c_k\}_{k=1}^{\infty}\|^2 \geq \frac{1}{\|V^{-1}\|^2} \sum_{k=1}^{\infty} |c_k|^2.$$

Now the result follows from Theorem 3.3.7.

(i) $\Rightarrow$ (iii). A Riesz basis is clearly complete. Now, Corollary 3.3.4 shows that the Riesz basis and its dual Riesz basis form a biorthogonal system; and by Proposition 3.3.5, both are Bessel sequences.

(iii) $\Rightarrow$ (i). Every $f \in \text{span}\{f_k\}_{k=1}^{\infty}$ has a representation $f = \sum c_k f_k$ for a finite sequence $\{c_k\}$, and under the assumptions in (iv) it is unique: if $f = \sum c_k f_k$, then $c_k = \langle f, g_k \rangle$. Letting $\{e_k\}_{k=1}^{\infty}$ be an orthonormal basis for $\mathcal{H}$, we can therefore define an operator

$$V : \text{span}\{f_k\}_{k=1}^{\infty} \to \mathcal{H}, \ V \sum c_k f_k = \sum c_k e_k.$$

Writing $f \in \text{span}\{f_k\}_{k=1}^{\infty}$ as $f = \sum \langle f, g_k \rangle f_k$, and letting C denote a Bessel bound for $\{g_k\}_{k=1}^{\infty}$, we have

$$\begin{aligned}
\|Vf\|^2 &= \left\|\sum \langle f, g_k \rangle e_k\right\|^2 \\
&= \sum |\langle f, g_k \rangle|^2 \\
&\leq C \|f\|^2.
\end{aligned}$$

By completeness of $\{f_k\}_{k=1}^{\infty}$, V has an extension to a bounded operator on $\mathcal{H}$. Since the assumptions in (iv) are symmetric in f_k and g_k, we can also extend the linear map

$$W : \text{span}\{g_k\}_{k=1}^{\infty} \to \mathcal{H}, \ W \sum c_k g_k = \sum c_k e_k$$

to a bounded operator on $\mathcal{H}$.

Consider finite linear combinations of $\{f_k\}_{k=1}^{\infty}$ and $\{g_k\}_{k=1}^{\infty}$, say,

$$f = \sum c_k f_k, \ g = \sum d_k g_k.$$

Because $\{f_k\}_{k=1}^{\infty}$ and $\{g_k\}_{k=1}^{\infty}$ are biorthogonal, we have

$$\langle Vf, Wg \rangle = \left\langle \sum c_k e_k, \sum d_k e_k \right\rangle = \sum c_k \overline{d_k} = \langle f, g \rangle;$$

by continuity and completeness we therefore have $\langle Vf, Wg \rangle = \langle f, g \rangle$ for all $f, g \in \mathcal{H}$. Thus, for any $h \in \mathcal{H}$,

$$\|h\|^2 = \langle h, h \rangle = \langle Vh, Wh \rangle \leq \|Vh\| \, \|W\| \, \|h\|.$$

It follows that V is injective. V is also surjective: in fact, given $g \in \mathcal{H}$, we can write $g = \sum_{k=1}^{\infty} \langle g, e_k \rangle e_k = V \left(\sum_{k=1}^{\infty} \langle g, e_k \rangle f_k \right)$. Since $f_k = V^{-1} e_k$, we conclude that $\{f_k\}_{k=1}^{\infty}$ is a Riesz basis. $\square$

The optimal Riesz bounds can be characterized in terms of the operators appearing in the proof of Theorem 3.4.4:

Proposition 3.4.5 *Let* $\{f_k\}_{k=1}^\infty = \{Ue_k\}_{k=1}^\infty$ *be a Riesz basis for* $\mathcal{H}$, *and let* $G : \ell^2(\mathbb{N}) \to \ell^2(\mathbb{N})$ *be the Gram matrix. Then the optimal Riesz bounds are*

$$A = \frac{1}{||U^{-1}||^2} = \frac{1}{||G^{-1}||} \quad and \quad B = ||U||^2 = ||G||.$$

Proof. The bounds involving U follow directly from the proof of Theorem 3.3.7. Also, by Lemma 2.4.1,

$$||G|| = ||U^*U|| = ||U||^2 \text{ and } ||G^{-1}|| = ||(U^*U)^{-1}|| = ||U^{-1}||^2.$$

That the optimal upper Riesz bound equals $||G||$ was also proved in Lemma 3.4.1. $\qquad\qquad\qquad\qquad\qquad\qquad\qquad\qquad\qquad\qquad\qquad\square$

Note that the same optimal bounds involving U were obtained in the inequalities in Proposition 3.3.5.

3.5   Fourier series and trigonometric polynomials

Let us now consider some concrete orthonormal bases for the function spaces $L^2(\mathbb{R})$ and $L^2(I)$ for a given interval $I \subset \mathbb{R}$. Here we will need to use other index sets than the natural numbers; as we have seen in Corollary 3.1.5, Bessel sequences can be ordered any way we want without affecting the convergence of the relevant series expansions, so we can apply all results presented so far without problems.

The starting point is *Fourier series*. We expect the reader to be familiar with the basic theory, so we only give a short overview.

Fourier series can be associated with functions in any space $L^2(I)$, where I is a bounded interval in $\mathbb{R}$. For our purpose, it will be convenient to consider functions in $L^2(0, 1/b)$, where $b > 0$. Since the functions

$$e_k(x) := b^{1/2}e^{2\pi ikbx}, \ k \in \mathbb{Z} \tag{3.20}$$

constitute an orthonormal basis for $L^2(0, 1/b)$, every $f \in L^2(0, 1/b)$ has an expansion

$$f = \sum_{k\in\mathbb{Z}} \langle f, e_k \rangle e_k. \tag{3.21}$$

We will usually expand the functions f directly in terms of the functions $\{e^{2\pi ikbx}\}_{k\in\mathbb{Z}}$ rather than $\{e_k\}_{k\in\mathbb{Z}}$. Thus, we arrive at

$$f(\cdot) = \sum_{k\in\mathbb{Z}} c_k e^{2\pi ikb(\cdot)}, \tag{3.22}$$

where

$$c_k = b^{1/2}\langle f, e_k \rangle = b \int_0^{1/b} f(x)e^{-2\pi ikbx}dx. \tag{3.23}$$

The expansion (3.22) is called the *Fourier series* of f, and the numbers $\{c_k\}_{k\in\mathbb{Z}}$ are the *Fourier coefficients*.

The exact meaning of the Fourier expansion (3.22) is that

$$\left\| f - \sum_{k=-n}^{n} c_k e^{2\pi i k b(\cdot)} \right\|_{L^2(0,1/b)} = \left(\int_0^{1/b} \left| f(x) - \sum_{k=-n}^{n} c_k e^{2\pi i k b x} \right|^2 dx \right)^{1/2}$$
$$\to 0 \text{ as } n \to \infty.$$

Convergence in $L^2(0,1/b)$-sense is different from pointwise convergence, so we *cannot* claim that (3.22) holds for a given $x \in [0,1/b]$ without extra assumptions. For an arbitrary function in $L^2(0,1/b)$, the Fourier series converges pointwise almost everywhere; conditions implying convergence for all x are presented in the following well-known result.

Theorem 3.5.1 *Assume that $f \in L^2(0,1/b)$ is continuous, periodic with period $1/b$, and that the Fourier coefficients $\{c_k\}_{k\in\mathbb{Z}} \in \ell^1(\mathbb{Z})$. Then*

$$f(x) = \sum_{k\in\mathbb{Z}} c_k e^{2\pi i k b x},$$

pointwise for all $x \in \mathbb{R}$.

Parseval's equation, see Theorem 3.2.2, gives us an important relationship between a given function $f \in L^2(0,1/b)$ and its Fourier coefficients $\{c_k\}_{k\in\mathbb{Z}}$:

$$b \int_0^{1/b} |f(x)|^2 dx = \sum_{k\in\mathbb{Z}} |c_k|^2. \tag{3.24}$$

We now state a lemma, which is an immediate consequence of the functions $\{e_k\}_{k=1}^{\infty}$ in (3.20) being an orthonormal basis for $L^2(0,1/b)$.

Lemma 3.5.2 *Let $f,g \in L^2(0,1/b)$ for some $b > 0$, and consider two series expansions*

$$f = \sum_{k\in\mathbb{Z}} a_k e_k, \quad g = \sum_{k\in\mathbb{Z}} b_k e_k,$$

with e_k given by (3.20) and $\{a_k\}_{k\in\mathbb{Z}}, \{b_k\}_{k\in\mathbb{Z}} \in \ell^2(\mathbb{Z})$. Then

$$\langle f, g \rangle = \sum_{k\in\mathbb{Z}} a_k \overline{b_k}.$$

A $\frac{1}{b}$-periodic function $f : \mathbb{R} \to \mathbb{C}$ can equally well be considered as a function in $L^2(0,1/b)$ as in $L^2(-\frac{1}{2b}, \frac{1}{2b})$; the latter choice will sometimes be more convenient, e.g., in our discussion of sampling problems in Section 3.8. For reasons of periodicity, all definitions and results in this section still apply if we consider our functions as members in $L^2(-\frac{1}{2b}, \frac{1}{2b})$ instead of

$L^2(0, 1/b)$, but we can also choose to exchange all the integrals over $]0, 1/b[$ with integrals over $] - \frac{1}{2b}, \frac{1}{2b}[$; for example, the expression for the Fourier coefficients in (3.23) takes the form

$$c_k = b \int_{-\frac{1}{2b}}^{\frac{1}{2b}} f(x) e^{-2\pi i k b x} dx. \tag{3.25}$$

In the following example, we show how to construct an orthonormal basis for $L^2(\mathbb{R})$ based on the orthonormal basis $\{e^{2\pi i k x}\}_{k \in \mathbb{Z}}$ for $L^2(0, 1)$. The example gives an introduction to a special system of functions, Gabor systems, which will be discussed in detail in Chapter 9.

Example 3.5.3 Let $\chi_{[0,1]}$ denote the characteristic function for the interval $[0, 1]$. Then $\{e^{2\pi i k x} \chi_{[0,1]}(x)\}_{k \in \mathbb{Z}}$ is an orthonormal basis for $L^2(0, 1)$; by translation, we see that for each $n \in \mathbb{Z}$, the space $L^2(n, n+1)$ has the orthonormal basis

$$\{e^{2\pi i k (x-n)} \chi_{[0,1]}(x - n)\}_{k \in \mathbb{Z}} = \{e^{2\pi i k x} \chi_{[0,1]}(x - n)\}_{k \in \mathbb{Z}}.$$

Putting these bases together, we conclude that $L^2(\mathbb{R})$ has the orthonormal basis

$$\{e^{2\pi i k x} \chi_{[0,1]}(x - n)\}_{k,n \in \mathbb{Z}}.$$

Note that all elements in the basis consist of translated versions of $\chi_{[0,1]}$ that have been *modulated*, i.e., multiplied with a complex exponential function; using the operators introduced in Section 2.9, we can write the basis as $\{E_k T_n g\}_{k,n \in \mathbb{Z}}$, where $g = \chi_{[0,1]}$. Bases of the form $\{E_k T_n g\}_{k,n \in \mathbb{Z}}$ are called *Gabor bases*. Calculations with Gabor bases are convenient because of their *coherent structure*, i.e., the fact that the elements in the basis appear by the action of a family of operators, namely $E_k T_n, k, n \in \mathbb{Z}$, on the single function g. We will consider some of the limitations on such bases in Chapter 4 and extensions to frames in Chapter 9. $\quad\square$

In concrete applications, a Fourier expansion will always need to be truncated to a finite sum. A function f that is a *finite* linear combination of the type

$$f(x) = \sum_{k=N_1}^{N_2} c_k e^{2\pi i k x} \quad \text{for some } c_k \in \mathbb{C}, \ N_1, N_2 \in \mathbb{Z}, N_2 \geq N_1 \tag{3.26}$$

is called a *trigonometric polynomial*. A trigonometric polynomial f can also be written as a linear combination of functions $\sin(2\pi k x), \cos(2\pi k x)$, in general with complex coefficients. It will be useful later to note that if the function f in (3.26) is real-valued and the coefficients c_k are real, then f is a linear combination of functions $\cos(2\pi k x)$ alone:

Lemma 3.5.4 *Assume that the trigonometric polynomial f in (3.26) is real-valued and that the coefficients $c_k \in \mathbb{R}$. Then*

$$f(x) = \sum_{k=N_1}^{N_2} c_k \cos(2\pi kx). \tag{3.27}$$

We leave the short proof to the reader. Note that we need the assumption that $c_k \in \mathbb{R}$: for example, the function

$$f(x) = \frac{1}{2i} e^{2\pi ix} - \frac{1}{2i} e^{-2\pi ix} = \sin(2\pi x),$$

is real-valued but does not have the form (3.27).

For later use, we also mention that a positive-valued trigonometric polynomial with real coefficients has a square root (in the sense of (3.30) below), which again is a trigonometric polynomial. For convenience, we formulate the result for a slight rewriting of the series (3.27):

Lemma 3.5.5 *Let f be a positive-valued trigonometric polynomial of the form*

$$f(x) = \sum_{k=0}^{N} c_k \cos(2\pi kx), \quad c_k \in \mathbb{R}. \tag{3.28}$$

Then there exists a trigonometric polynomial

$$g(x) = \sum_{k=0}^{N} d_k e^{2\pi ikx} \text{ with } d_k \in \mathbb{R}, \tag{3.29}$$

such that

$$|g(x)|^2 = f(x), \quad \forall x \in \mathbb{R}. \tag{3.30}$$

A constructive proof can be found in [26]. Note that by definition, the function g in (3.29) is complex-valued, unless f is constant; actually, despite the fact that f is assumed to be positive, there might not exist a *positive* trigonometric polynomial g satisfying (3.30). See Exercise 3.10.

3.6   Wavelet bases

Wavelet bases constitute another important class of bases. We will only give a short overview and refer to the many excellent wavelet books for more information (see for example [64] for an elementary treatment or [26], [65] for more advanced presentations).

Given a function $\psi \in L^2(\mathbb{R})$ and $j, k \in \mathbb{Z}$, let

$$\psi_{j,k}(x) := 2^{j/2} \psi(2^j x - k), \quad x \in \mathbb{R}. \tag{3.31}$$

In terms of the translation operators T_k and the dilation operator D introduced in Section 2.9, we have that

$$\psi_{j,k} = D^j T_k \psi, \quad j, k \in \mathbb{Z}.$$

If $\{\psi_{j,k}\}_{j,k \in \mathbb{Z}}$ is an orthonormal basis for $L^2(\mathbb{R})$, the function ψ is called a *wavelet*. The first example of such a function appeared a long time before the systematic study of wavelet bases began around 1985:

Example 3.6.1 The *Haar function* is defined by

$$\psi(x) = \begin{cases} 1 & \text{if } 0 \le x < \frac{1}{2}, \\ -1 & \text{if } \frac{1}{2} \le x < 1, \\ 0 & \text{otherwise.} \end{cases} \tag{3.32}$$

Already in 1910, it was proved by Haar that the functions $\{\psi_{j,k}\}_{j,k \in \mathbb{Z}}$ constitute an orthonormal basis for $L^2(\mathbb{R})$ for this choice of ψ. For the orthonormality, one can argue as follows. If we first consider $\psi_{j,k}$ and $\psi_{j,k'}$, i.e., elements with the same dilation parameter, then

$$\langle \psi_{j,k}, \psi_{j,k'} \rangle = \langle D^j T_k \psi, D^j T_{k'} \psi \rangle = \langle T_k \psi, T_{k'} \psi \rangle = \delta_{k,k'}.$$

Now assume that $j' \ne j$, say, $j' > j$. The commutator relations (2.22) give that

$$\begin{aligned} \langle \psi_{j,k}, \psi_{j',k'} \rangle &= \langle D^j T_k \psi, D^{j'} T_{k'} \psi \rangle \\ &= \langle T_{-k'} D^{j-j'} T_k \psi, \psi \rangle \\ &= \langle D^{j-j'} T_{-k'2^{j-j'}+k} \psi, \psi \rangle. \end{aligned}$$

The function $D^{j-j'} T_{-k'2^{j-j'}+k} \psi$ has support in the interval

$$\begin{aligned} I: &= [2^{j'-j}(-k'2^{j-j'} + k), 2^{j'-j}(-k'2^{j-j'} + k + 1)[\\ &= [-k' + 2^{j'-j}k, -k' + 2^{j'-j}(k + 1)[. \end{aligned}$$

The length of I is $2^{j'-j}$, which can take the values $2, 4, 8, \ldots$ Now, the support of ψ has length 1 and is contained in an interval on which $D^{j-j'} T_{-k'2^{j-j'}+k} \psi$ is constant (make a picture!); it follows that

$$\langle \psi_{j',k'}, \psi_{j,k} \rangle = \int_{-\infty}^{\infty} \left(D^{j-j'} T_{-k'2^{j-j'}+k} \psi \right)(x) \psi(x)\, dx = 0.$$

For the proof of the basis property, we refer to [26] or [41].   □

In 1986, Mallat and Meyer introduced *multiresolution analysis* as a general tool to construct wavelet orthonormal bases:

Definition 3.6.2 *A multiresolution analysis for $L^2(\mathbb{R})$ consists of a sequence of closed subspaces $\{V_j\}_{j\in\mathbb{Z}}$ of $L^2(\mathbb{R})$ and a function $\phi \in V_0$, such that the following conditions hold:*

(i) $\cdots V_{-1} \subset V_0 \subset V_1 \cdots$.

(ii) $\overline{\cup_j V_j} = L^2(\mathbb{R})$ *and* $\cap_j V_j = \{0\}$.

(iii) $f \in V_j \Leftrightarrow [x \to f(2x)] \in V_{j+1}$.

(iv) $f \in V_0 \Rightarrow T_k f \in V_0, \ \forall k \in \mathbb{Z}$.

(v) $\{T_k\phi\}_{k\in\mathbb{Z}}$ *is an orthonormal basis for* V_0.

When (i) is satisfied, we say that the spaces V_j are *nested*. This is a very convenient property in for example approximation theory, especially when there is an easy recipe for moving around between the spaces V_j. The latter property is also guaranteed by Definition 3.6.2 because (iii) implies that $V_j = D^j V_0$, i.e., that all of the spaces V_j are scaled versions of V_0.

If we want to approximate a function $f \in L^2(\mathbb{R})$ via a multiresolution analysis, the natural starting point is to search for an approximation within a certain V_j-space. In case no element in this space approximates f well enough, we choose a larger j-value; then we obtain a better approximation, and it is taken from a space that is just a scaled version of the previous space.

We will now describe how a multiresolution analysis can be used to construct an orthonormal basis for $L^2(\mathbb{R})$. Assume that the conditions in Definition 3.6.2 are satisfied. For $j \in \mathbb{Z}$, we let W_j denote the orthogonal complement of V_j in V_{j+1}. Letting Q_j denote the orthogonal projection onto W_j, it follows from (i) and (ii) that each $f \in L^2(\mathbb{R})$ has a representation $f = \sum_{j\in\mathbb{Z}} Q_j f$, where $Q_j f \perp Q_{j'} f$ for $j \neq j'$; that is,

$$L^2(\mathbb{R}) = \bigoplus_{j\in\mathbb{Z}} W_j. \tag{3.33}$$

The spaces W_j satisfy the same dilation relationship as V_j, i.e.,

$$\psi \in W_0 \Leftrightarrow [x \to \psi(2^j x)] \in W_j. \tag{3.34}$$

In order to obtain an orthonormal basis $\{\psi_{j,k}\}_{j,k\in\mathbb{Z}}$ for $L^2(\mathbb{R})$, it is now enough to find $\psi \in W_0$ such that $\{\psi(\cdot - k)\}_{k\in\mathbb{Z}}$ is an orthonormal basis for W_0; via the dilation property (3.34) and (3.33), this implies that $\{\psi_{j,k}\}_{j,k\in\mathbb{Z}}$ is an orthonormal basis for $L^2(\mathbb{R})$. We will now explain how to find a suitable function ψ. First, the condition $\phi \in V_0 \subset V_1$ implies by (iii) that

$$\frac{1}{\sqrt{2}} D^{-1}\phi \in V_0.$$

Because $\{T_{-k}\phi\}_{k\in\mathbb{Z}}$ is an orthonormal basis for V_0, there exist coefficients $\{c_k\}_{k\in\mathbb{Z}} \in \ell^2(\mathbb{Z})$ such that

$$\frac{1}{\sqrt{2}}D^{-1}\phi = \sum_{k\in\mathbb{Z}} c_k T_{-k}\phi. \tag{3.35}$$

We will now rewrite (3.35) in terms of the Fourier transform of ϕ. Let

$$E_k(x) = e^{2\pi ikx}, \ x\in\mathbb{R},$$

as in Section 2.9. Now, applying the Fourier transform on both sides of (3.35), the commutator relations in (2.23) imply that

$$\frac{1}{\sqrt{2}}D\hat{\phi} = \sum_{k\in\mathbb{Z}} c_k E_k \hat{\phi}.$$

Defining the 1-periodic function $H_0 := \sum_{k\in\mathbb{Z}} c_k E_k$, this can be written as

$$\hat{\phi}(2\gamma) = H_0(\gamma)\hat{\phi}(\gamma), \quad a.e. \ \gamma\in\mathbb{R}. \tag{3.36}$$

The equation (3.36) is called a *scaling equation* or *refinement equation*. Now, it turns out that with a certain choice of a 1-periodic function H_1, the function ψ defined via

$$\hat{\psi}(2\gamma) = H_1(\gamma)\hat{\phi}(\gamma) \tag{3.37}$$

generates a wavelet orthonormal basis $\{D^j T_k\psi\}_{j,k\in\mathbb{Z}}$. One choice of H_1 is to take

$$H_1(\gamma) = \overline{H_0(\gamma+\frac{1}{2})}e^{-2\pi i\gamma}.$$

Note that (3.37) leads to an explicit expression of the function ψ in terms of the given function ϕ:

Lemma 3.6.3 *Assume that* (3.37) *holds for a 1-periodic and bounded function H_1 with Fourier expansion $H_1 = \sum_{k\in\mathbb{Z}} c_k E_k$. Then*

$$\psi(x) = \sqrt{2}\sum_{k\in\mathbb{Z}} c_k DT_{-k}\phi(x) = 2\sum_{k\in\mathbb{Z}} c_k\phi(2x+k), \ a.e. \ x\in\mathbb{R}. \tag{3.38}$$

In particular, if H_1 is a trigonometric polynomial, $H_1(x) = \sum_{k=N_1}^{N_2} c_k e^{2\pi ikx}$, then

$$\psi(x) = \sqrt{2}\sum_{k=N_1}^{N_2} c_k DT_{-k}\phi(x) = 2\sum_{k=N_1}^{N_2} c_k\phi(2x+k), \ \forall x\in\mathbb{R}. \tag{3.39}$$

Proof. We can rewrite (3.37) as

$$\hat{\psi}(\gamma) = H_1(\gamma/2)\hat{\phi}(\gamma/2);$$

formulated in terms of the Fourier series for H_1 and the dilation operator D, this means that

$$
\begin{aligned}
\mathcal{F}\psi &= \sqrt{2}\sum_{k\in\mathbb{Z}} c_k E_{k/2} D^{-1}\mathcal{F}\phi \\
&= \sqrt{2}\sum_{k\in\mathbb{Z}} c_k E_{k/2}\mathcal{F}D\phi.
\end{aligned}
$$

Now, using the commutator relations in Section 2.9,

$$
\begin{aligned}
\mathcal{F}\psi &= \sqrt{2}\mathcal{F}\sum_{k\in\mathbb{Z}} c_k T_{-k/2}D\phi \\
&= \sqrt{2}\mathcal{F}\sum_{k\in\mathbb{Z}} c_k DT_{-k}\phi.
\end{aligned}
$$

$\square$

The Haar basis considered in Example 3.6.1 can be constructed via the multiresolution analysis defined by

$$
\begin{cases}
\phi = \chi_{[0,1[}; \\
V_j = \{f \in L^2(\mathbb{R}) : f \text{ is constant a.e. on } [2^{-j}k, 2^{-j}(k+1)[, \ \forall k \in \mathbb{Z}\}.
\end{cases}
$$

In terms of the function ϕ, the Haar function in (3.32) is

$$
\psi = \frac{1}{\sqrt{2}}\phi_{1,0} - \frac{1}{\sqrt{2}}\phi_{1,1}. \tag{3.40}
$$

The Haar function is a special case of a *spline wavelet*. In fact, one can consider higher-order splines N_n (see Section 6.1 for the definition) and define associated multiresolution analyses, which leads to wavelets of the type

$$
\psi(x) = \sum_{k\in\mathbb{Z}} c_k N_n(2x - k). \tag{3.41}
$$

These wavelets are called *Battle–Lemarié wavelets*. The coefficients $\{c_k\}_{k\in\mathbb{Z}}$ are calculated in, e.g., [26]; except for the case $n = 1$, all coefficients c_k are non-zero, which implies that the wavelet ψ has support equal to $\mathbb{R}$. However, the wavelets have exponential decay.

The *Daubechies wavelets* are the most well-known constructions based on the multiresolution analysis setup. For each $N \in \mathbb{N}$, the construction yields a wavelet ψ with N vanishing moments, having compact support on the interval $[0, 2N - 1]$. One can prove that with the given support size, no more vanishing moments can be obtained for any wavelet coming from a multiresolution analysis. The smoothness of these wavelets increases with N. These wavelets are not given by an explicit formula; see [25] or [64] for a more detailed description.

However, not all wavelets can be constructed via multiresolution analysis. It is shown, e.g., in [41], that a function $\psi \in L^2(\mathbb{R})$ is a wavelet if and only if $||\psi|| = 1$ and the equations

$$\begin{cases} \displaystyle\sum_{j \in \mathbb{Z}} |\hat{\psi}(2^j \gamma)|^2 = 1, \\ \displaystyle\sum_{j=0}^{\infty} \hat{\psi}(2^j \gamma)\overline{\hat{\psi}(2^j(\gamma + q))} = 0 \;\; \text{for all odd integers } q \end{cases} \tag{3.42}$$

hold for a.e. $\gamma \in \mathbb{R}$. Among all wavelets, the wavelets generated from a multiresolution analysis are characterized by the equation

$$\sum_{j=1}^{\infty}\sum_{k \in \mathbb{Z}} |\hat{\psi}(2^j(\gamma + k)|^2 = 1, \; a.e. \; \gamma \in \mathbb{R},$$

a result that is also proved in [41].

Let us now return to the multiresolution analysis setup. As we have seen, the conditions in Definition 3.6.2 determine the spaces V_j uniquely; in fact, $V_0 = \overline{\text{span}}\{T_k\phi\}_{k \in \mathbb{Z}}$, and, via the condition (iii),

$$V_j = \overline{\text{span}}\{D^j T_k \phi\}_{k \in \mathbb{Z}}. \tag{3.43}$$

On the other hand, assuming that ϕ is a given function such that $\{T_k\phi\}_{k \in \mathbb{Z}}$ forms an orthonormal basis for its closed linear span, we only have to verify the conditions (i) and (ii) in order to show that ϕ and the spaces V_j in (3.43) form a multiresolution analysis. It turns out that these conditions are satisfied under very weak assumptions. Let us state a general result obtained by de Boor, DeVore, and Ron [5]:

Lemma 3.6.4 *Let $\phi \in L^2(\mathbb{R})$ and define the spaces V_j by (3.43). Then the following holds:*

(i) $\cap_j V_j = \{0\}$.

(ii) Assume that the spaces V_j in (3.43) are nested. If

$$|\hat{\phi}| > 0 \tag{3.44}$$

on a neighborhood of 0, then $\cup_j V_j$ is dense in $L^2(\mathbb{R})$.

Thus, if (3.44) is satisfied, all what we need is a condition ensuring that the spaces V_j are nested. But under a weak condition, this also follows from the assumption that $\{T_k\phi\}_{k \in \mathbb{Z}}$ forms an orthonormal basis for its closed linear span. In fact, it is enough to assume that $\{T_k\phi\}_{k \in \mathbb{Z}}$ forms a Bessel sequence; this extension will play a role in Chapter 11.

Lemma 3.6.5 *Assume that $\phi \in L^2(\mathbb{R})$ and that $\{T_k\phi\}_{k\in\mathbb{Z}}$ is a Bessel sequence. Define the spaces V_j by (3.43). Then the following holds:*

(i) *If $\psi \in L^2(\mathbb{R})$ and there exists a bounded 1-periodic function H_1 such that $\hat{\psi}(2\gamma) = H_1(\gamma)\hat{\phi}(\gamma)$, then $\psi \in V_1$.*

(ii) *If there exists a bounded 1-periodic function H_0 such that*

$$\hat{\phi}(2\gamma) = H_0(\gamma)\hat{\phi}(\gamma), \tag{3.45}$$

then $V_j \subseteq V_{j+1}$ for all $j \in \mathbb{Z}$.

Proof. If the conditions in (i) are satisfied, the expression for the function ψ in Lemma 3.6.3 shows that $\psi \in V_1$. This proves (i). For the proof of (ii), we note that, via (i), $\phi \in V_1$; since V_1 is closed and invariant under integer-translations, it follows that $V_0 \subseteq V_1$. A scaling now implies that $V_j \subseteq V_{j+1}$ for all $j \in \mathbb{Z}$. $\qquad\square$

Via Lemma 3.6.4 and Lemma 3.6.5, we obtain the following:

Theorem 3.6.6 *Let $\phi \in L^2(\mathbb{R})$, and assume that $|\hat{\phi}| > 0$ on a neighborhood of 0. Assume further that (3.45) is satisfied for a bounded 1-periodic function H_0. Define the spaces V_j by (3.43). Then the following holds:*

(i) *If $\{T_k\phi\}_{k\in\mathbb{Z}}$ is an orthonormal system, then ϕ and the spaces V_j form a multiresolution analysis.*

(i) *If $\{T_k\phi\}_{k\in\mathbb{Z}}$ is a Bessel sequence, then the spaces V_j satisfy the conditions (i)–(iv) in Definition 3.6.2.*

The instrumental condition (3.45) is satisfied, e.g., if ϕ is a B-spline. We introduce the B-splines in Chapter 6, and ask the reader to verify the scaling equation for these functions in Exercise 6.4.

3.7 Bases in Banach spaces

We now give a short introduction to bases in Banach spaces, as defined by Schauder in 1927.

Definition 3.7.1 *Let X be a separable Banach space. A sequence of vectors $\{e_k\}_{k=1}^{\infty}$ belonging to X is a (Schauder) basis for X if, for each $f \in X$, there exist unique scalar coefficients $\{c_k(f)\}_{k=1}^{\infty}$ such that*

$$f = \sum_{k=1}^{\infty} c_k(f)e_k. \tag{3.46}$$

If the series (3.46) converges unconditionally for each $f \in X$, we say that $\{e_k\}_{k=1}^{\infty}$ is an *unconditional basis*.

A Banach space having a basis is necessarily separable. Most of the known separable Banach spaces have a basis; the first example of a separable Banach space not having a basis was constructed by Enflo in 1972.

It is clear that a basis for X is complete and consists of non-zero vectors. Adding an extra condition leads to a characterization of bases:

Theorem 3.7.2 *Let $\{e_k\}_{k=1}^{\infty}$ be a complete family of non-zero vectors in X. Then the following are equivalent.*

(i) $\{e_k\}_{k=1}^{\infty}$ is a basis for X.

(ii) There exists a constant K such that for all $m, n \in \mathbb{N}$ with $m \leq n$,

$$\left\| \sum_{k=1}^{m} c_k e_k \right\| \leq K \left\| \sum_{k=1}^{n} c_k e_k \right\| \tag{3.47}$$

for all scalar-valued sequences $\{c_k\}_{k=1}^{\infty}$.

Proof. Suppose that $\{e_k\}_{k=1}^{\infty}$ is a basis. Then each $f \in X$ has a unique expansion $f = \sum_{k=1}^{\infty} c_k e_k$, and

$$|||f||| := \sup_{m \in \mathbb{N}} \left\| \sum_{k=1}^{m} c_k e_k \right\| < \infty. \tag{3.48}$$

Note that if $|||f||| = 0$, then $\|\sum_{k=1}^{m} c_k e_k\| = 0$ for all $m \in \mathbb{N}$; it follows that $c_k = 0$ for all $k \in \mathbb{N}$, and $f = 0$. One can check (Exercise 3.12) that $||| \cdot |||$ satisfies the other conditions for a norm on X, and that X is a Banach space with respect to this norm. By definition of $||| \cdot |||$, we have $\|f\| \leq |||f|||$, $\forall f \in X$, meaning that the identity operator is a continuous and injective mapping of $(X, ||| \cdot |||)$ onto $(X, \| \cdot \|)$. By Theorem 2.2.2, it follows that this operator has a continuous inverse, i.e., that there exists a constant $K > 0$ such that $|||f||| \leq K \|f\|$ for all $f \in X$. In particular, fixing an arbitrary $n \in \mathbb{N}$ and considering $f = \sum_{k=1}^{n} c_k e_k$, we obtain (3.47).

For the implication (ii)$\Rightarrow$(i), assume that a complete family $\{e_k\}_{k=1}^{\infty}$ of non-zero vectors satisfies (3.47). We begin by showing an inequality that will be used in the proof. Consider any $f \in X$ with an expansion $f = \sum_{k=1}^{\infty} c_k e_k$; then, for any choice of $i \in \mathbb{N}$ and $m \geq i$, (3.47) shows that

$$|c_i| \, \|e_i\| = \left\| \sum_{k=1}^{i} c_k e_k - \sum_{k=1}^{i-1} c_k e_k \right\| \leq \left\| \sum_{k=1}^{i} c_k e_k \right\| + \left\| \sum_{k=1}^{i-1} c_k e_k \right\|$$

$$\leq K \left\| \sum_{k=1}^{m} c_k e_k \right\| + K \left\| \sum_{k=1}^{m} c_k e_k \right\|$$

$$= 2K \left\| \sum_{k=1}^{m} c_k e_k \right\|. \tag{3.49}$$

Now let $\mathcal{A}$ denote the vector space consisting of all $f \in X$, which can be expanded as $f = \sum_{k=1}^{\infty} c_k e_k$ for some coefficients $\{c_k\}_{k=1}^{\infty}$. We will prove that $\mathcal{A} = X$; because $\{e_k\}_{k=1}^{\infty}$ is assumed to be complete, we know that $\mathcal{A}$ is dense in X, so it is enough to prove that $\mathcal{A}$ is closed. Let $f \in X$, and choose a sequence $\{f_j\}_{j=1}^{\infty} \subset \mathcal{A}$ such that $f_j \to f$ as $j \to \infty$. Write $f_j = \sum_{k=1}^{\infty} c_k^{(j)} e_k$ for appropriate coefficients $\{c_k^{(j)}\}_{k=1}^{\infty}$. By (3.49), for each $i \in \mathbb{N}$ and all $n \geq m \geq i$, we have for all $j, \ell \in \mathbb{N}$ that

$$|c_i^{(j)} - c_i^{(\ell)}| \, \|e_i\| \leq 2K \left\| \sum_{k=1}^{m} \left(c_k^{(j)} - c_k^{(\ell)} \right) e_k \right\| \tag{3.50}$$

$$\leq 2K^2 \left\| \sum_{k=1}^{n} \left(c_k^{(j)} - c_k^{(\ell)} \right) e_k \right\|$$

$$\leq 2K^2 \left(\left\| \sum_{k=1}^{n} c_k^{(j)} e_k - f_j \right\| + \|f_j - f\| \right)$$

$$+ 2K^2 \left(\|f - f_\ell\| + \left\| f_\ell - \sum_{k=1}^{n} c_k^{(\ell)} e_k \right\| \right).$$

Given $\epsilon > 0$, choose $N \in \mathbb{N}$ such that

$$\|f - f_j\| \leq \frac{\epsilon}{4K^2} \text{ for } j \geq N.$$

Letting $n \to \infty$, it follows from the above estimate that

$$|c_i^{(j)} - c_i^{(\ell)}| \, \|e_i\| \leq \epsilon \text{ for all } i \in \mathbb{N}, \ j, \ell \geq N, \tag{3.51}$$

and, via the intermediate step (3.50),

$$2K \left\| \sum_{k=1}^{m} \left(c_k^{(j)} - c_k^{(\ell)} \right) e_k \right\| \leq \epsilon \text{ for all } m \in \mathbb{N}, \ j, \ell \geq N. \tag{3.52}$$

For each $i \in \mathbb{N}$, the sequence $\{c_i^{(\ell)}\}_{\ell=1}^{\infty}$ is convergent by (3.51), say, $c_i^{(\ell)} \to c_i$ as $\ell \to \infty$. Letting $\ell \to \infty$ in (3.51) and (3.52), we obtain that

$$|c_i^{(j)} - c_i| \, \|e_i\| \leq \epsilon \text{ for all } i \in \mathbb{N}, \ j \geq N, \tag{3.53}$$

and

$$2K \left\| \sum_{k=1}^{m} \left(c_k^{(j)} - c_k \right) e_k \right\| \leq \epsilon \text{ for all } m \in \mathbb{N}, \ j \geq N. \tag{3.54}$$

Now, for given $m \in \mathbb{N}$ and all $j \in \mathbb{N}$,

$$\left\| f - \sum_{k=1}^{m} c_k e_k \right\| \leq \|f - f_j\|$$

$$+ \left\| f_j - \sum_{k=1}^{m} c_k^{(j)} e_k \right\| + \left\| \sum_{k=1}^{m} \left(c_k^{(j)} - c_k \right) e_k \right\|;$$

We will now show that $\sum_{k=1}^{\infty} c_k e_k$ converges to f. In fact, for a given $\epsilon > 0$, we can choose $N \in \mathbb{N}$ so that (3.54) holds. By fixing a sufficiently large value for $j > N$, we obtain that $||f - f_j|| \leq \epsilon$; after that, we can obtain that $\left|\left| f_j - \sum_{k=1}^{m} c_k^{(j)} e_k \right|\right| \leq \epsilon$ by choosing $m \in \mathbb{N}$ sufficiently large. Thus

$$||f - \sum_{k=1}^{m} c_k e_k|| \leq 2\epsilon + \frac{\epsilon}{2K} = \epsilon \left(2 + \frac{1}{2K} \right)$$

for m sufficiently large. We conclude that $f = \sum_{k=1}^{\infty} c_k e_k$, i.e., $f \in \mathcal{A}$ as desired. To prove that $\{e_k\}_{k=1}^{\infty}$ is a basis, we only need to show that if $\sum_{k=1}^{\infty} c_k e_k = 0$, then $c_k = 0$ for all $k \in \mathbb{N}$. This again follows from (3.49). In fact, if $\sum_{k=1}^{\infty} c_k e_k = 0$, then for each $i \in \mathbb{N}$ and all $n \geq i$,

$$|c_i| \, ||e_i|| \leq 2K \left|\left| \sum_{k=1}^{n} c_k e_k \right|\right|;$$

from here we obtain that $c_i = 0$ by letting $n \to \infty$.   □

Given a basis $\{e_k\}_{k=1}^{\infty}$, it is clear that the coefficients $\{c_k(f)\}_{k=1}^{\infty}$ in (3.46) depend linearly on f. The mappings $f \to c_k(f)$ are called *coefficient functionals*. As a consequence of Theorem 3.7.2, they are continuous:

Corollary 3.7.3 *The coefficient functionals $\{c_k\}_{k=1}^{\infty}$ associated with a basis $\{e_k\}_{k=1}^{\infty}$ for X are continuous and are thus elements in the dual X^*. If there exists a constant $C > 0$ such that $||e_k|| \geq C$ for all $k \in \mathbb{N}$, then the norms of $\{c_k\}_{k=1}^{\infty}$ are uniformly bounded.*

Proof. We will use (3.49) from the proof of Theorem 3.7.2. Given $f \in X$, write $f = \sum_{k=1}^{\infty} c_k(f) e_k$. Then, for any $i \in \mathbb{N}$ and all $n \geq i$,

$$|c_i(f)| \, ||e_i|| \leq 2K \left|\left| \sum_{k=1}^{n} c_k(f) e_k \right|\right|.$$

Letting $n \to \infty$, we obtain that

$$|c_i(f)| \leq \frac{2K}{||e_i||} \, ||f||.$$ □

The concept of biorthogonal sequences in Hilbert spaces has a natural extension to Banach spaces. In fact, a sequence $\{f_k\}_{k=1}^{\infty}$ in X and a sequence $\{g_k\}_{k=1}^{\infty}$ in X^* are said to be *biorthogonal* if

$$g_k(f_j) = \delta_{k,j} = \begin{cases} 1 & \text{if } k = j, \\ 0 & \text{if } k \neq j. \end{cases} \tag{3.55}$$

This leads to an extension of Corollary 3.3.4; we leave the proof to the reader (Exercise 3.13).

Corollary 3.7.4 *Suppose that $\{e_k\}_{k=1}^{\infty}$ is a basis for X. Then $\{e_k\}_{k=1}^{\infty}$ and the coefficient functionals $\{c_k\}_{k=1}^{\infty}$ constitute a biorthogonal system.*

Let us consider the (Schauder) bases in the special case where the underlying space is a Hilbert space. It turns out that Theorem 3.3.2 generalizes to such bases, except that we do not automatically obtain unconditional convergence:

Theorem 3.7.5 *Assume that $\{e_k\}_{k=1}^{\infty}$ is a basis for the Hilbert space $\mathcal{H}$. Then there exists a unique family $\{g_k\}_{k=1}^{\infty}$ in $\mathcal{H}$ for which*

$$f = \sum_{k=1}^{\infty} \langle f, g_k \rangle e_k, \ \forall f \in \mathcal{H}. \tag{3.56}$$

$\{g_k\}_{k=1}^{\infty}$ is a basis for $\mathcal{H}$, and $\{e_k\}_{k=1}^{\infty}$ and $\{g_k\}_{k=1}^{\infty}$ are biorthogonal.

Proof. By Corollary 3.7.3, the coefficient functionals $\{c_k\}_{k=1}^{\infty}$ associated with $\{e_k\}_{k=1}^{\infty}$ are continuous; using Riesz' representation theorem, Theorem 2.3.2, there exists a unique family $\{g_k\}_{k=1}^{\infty}$ in $\mathcal{H}$ such that

$$c_k(f) = \langle f, g_k \rangle, \ \forall f \in \mathcal{H}.$$

Therefore

$$f = \sum_{k=1}^{\infty} \langle f, g_k \rangle e_k, \ \forall f \in \mathcal{H}.$$

We leave it to the reader to verify that no other family $\{g_k\}_{k=1}^{\infty}$ can satisfy (3.56) and that $\{e_k\}_{k=1}^{\infty}$ and $\{g_k\}_{k=1}^{\infty}$ are biorthogonal. The fact that $\{g_k\}_{k=1}^{\infty}$ is a basis for $\mathcal{H}$ follows from the fact that a Hilbert space is reflexive; see, e.g., Section 1.7 in [62]. $\square$

The basis $\{g_k\}_{k=1}^{\infty}$ satisfying (3.56) is called the *dual basis*, or the *biorthogonal basis*, associated with $\{e_k\}_{k=1}^{\infty}$.

We mention that one can characterize Riesz bases in terms of bases satisfying extra conditions:

Lemma 3.7.6 *A sequence $\{f_k\}_{k=1}^{\infty}$ is a Riesz basis for $\mathcal{H}$ if and only if it is an unconditional basis for $\mathcal{H}$ and*

$$0 < \inf_k \|f_k\| \le \sup_k \|f_k\| < \infty.$$

Lemma 3.7.6 was originally proved by Köthe and Lorch and has been rediscovered many times. We refer to [52] for a proof.

3.8   Sampling and analog–digital conversion

A short and not yet precise formulation of the *sampling problem* is: How can we recover a function $f : \mathbb{R} \to \mathbb{C}$ if we only know a countable set of function values $\{f(\lambda_k)\}_{k\in I}$? Formulated this way the problem is ill-posed: there are infinitely many functions that take the same prescribed values on a given countable set, so we need to impose some condition on the function f for the problem to make sense. Traditionally, this is done by requiring f to belong to a certain function space. A classical example is to consider a space of *band-limited* functions, i.e., functions for which the Fourier transform has compact support. Let us consider the *Paley–Wiener space PW*, defined by

$$PW := \left\{ f \in L^2(\mathbb{R}) : \ \mathrm{supp}\, \hat{f} \subseteq [-\frac{1}{2}, \frac{1}{2}] \right\}. \tag{3.57}$$

As always when dealing with L^2-functions, the Paley–Wiener space really consists of equivalence classes of functions; however, due to the fact that the Fourier transform of these functions has compact support, each of these equivalence classes contains a continuous function.

We will now show that the Paley–Wiener space has an orthonormal basis consisting of translates of a single function. Define the *sinc-function* by

$$\mathrm{sinc}(x) = \begin{cases} \frac{\sin(\pi x)}{\pi x} & \text{if } x \neq 0, \\ 1 & \text{if } x = 0. \end{cases}$$

Shannon's Sampling Theorem states that any continuous function in the Paley–Wiener space can be fully recovered from its samples at the integers:

Theorem 3.8.1 *The functions $\{\mathrm{sinc}(\cdot - k)\}_{k\in\mathbb{Z}}$ form an orthonormal basis for PW. If $f \in PW$ is continuous, then*

$$f(x) = \sum_{k\in\mathbb{Z}} f(k)\,\mathrm{sinc}(x - k),$$

with convergence of the symmetric partial sums in $L^2(\mathbb{R})$ and pointwise for all $x \in \mathbb{R}$.

Proof.   The proof is based on classical Fourier analysis as described in Section 3.5. Because of our definition of the Paley–Wiener space, it will be convenient to work with Fourier series in the space $L^2(-1/2, 1/2)$ rather than $L^2(0, 1)$.

We first show that the functions $\{\mathrm{sinc}(\cdot - k)\}_{k\in\mathbb{Z}}$ form an orthonormal sequence in $L^2(\mathbb{R})$. We know that the functions $\{e^{2\pi i k(\cdot)}\chi_{]-1/2,1/2[}(\cdot)\}_{k\in\mathbb{Z}}$ form an orthononormal sequence in $L^2(\mathbb{R})$; taking the Fourier transform of these functions, we arrive at

$$
\begin{aligned}
\mathcal{F}\left(e^{2\pi i k(\cdot)}\chi_{]-1/2,1/2[}(\cdot)\right)(\gamma) &= \int_{-1/2}^{1/2} e^{2\pi i k x} e^{-2\pi i x \gamma} dx \\
&= \int_{-1/2}^{1/2} e^{-2\pi i (\gamma-k)x} dx \\
&= \left[\frac{1}{-2\pi i(\gamma-k)} e^{-2\pi i(\gamma-k)x}\right]_{x=-1/2}^{1/2} \\
&= \operatorname{sinc}(\gamma-k).
\end{aligned}
$$

Because the Fourier transform is unitary, this implies that the functions $\{\operatorname{sinc}(\cdot - k)\}_{k\in\mathbb{Z}}$ are orthonormal as well.

Suppose that f is continuous and that $f \in L^1(\mathbb{R}) \cap PW$. On the interval $]-1/2, 1/2[$ we can expand $\hat{f}$ in a Fourier series,

$$
\hat{f}(\cdot) = \sum_{k\in\mathbb{Z}} c_k e^{2\pi i k(\cdot)},
$$

where

$$
c_k = \int_{-1/2}^{1/2} \hat{f}(\gamma) e^{-2\pi i k \gamma} d\gamma. \tag{3.58}
$$

Recall that the partial sums of the Fourier series converge in the norm of $L^2(-1/2, 1/2)$, i.e.,

$$
\int_{-1/2}^{1/2}\left|\hat{f}(\gamma) - \sum_{n=-N}^{N} c_k e^{2\pi i k \gamma}\right|^2 d\gamma \to 0 \text{ as } N \to \infty.
$$

Note that because we are dealing with a finite interval, convergence in L^2 implies convergence in L^1, so

$$
\int_{-1/2}^{1/2}\left|\hat{f}(\gamma) - \sum_{n=-N}^{N} c_k e^{2\pi i k \gamma}\right| d\gamma \to 0 \text{ as } N \to \infty. \tag{3.59}
$$

Because supp $\hat{f} \subseteq [-\frac{1}{2}, \frac{1}{2}]$, the expression for c_k in (3.58) implies by Theorem 2.8.1 that $c_k = f(-k)$. Using Theorem 2.8.1 once more, we arrive at the following formula, valid pointwise for all $x \in \mathbb{R}$:

$$
f(x) = \int_{-\infty}^{\infty} \hat{f}(\gamma) e^{2\pi i x \gamma} d\gamma = \int_{-1/2}^{1/2}\left(\sum_{k\in\mathbb{Z}} f(-k) e^{2\pi i k \gamma}\right) e^{2\pi i x \gamma} d\gamma.
$$

Because of (3.59), we can interchange the order of summation and integration; thus, for all $x \in \mathbb{R}$,

$$
\begin{aligned}
f(x) &= \sum_{k \in \mathbb{Z}} f(-k) \int_{-1/2}^{1/2} e^{2\pi i(x+k)\gamma} d\gamma \\
&= \sum_{k \in \mathbb{Z}} f(-k) \left[\frac{1}{2\pi i(x+k)} e^{2\pi i(x+k)\gamma} \right]_{\gamma=-1/2}^{1/2} \\
&= \sum_{k \in \mathbb{Z}} f(-k) \operatorname{sinc}(x+k) \\
&= \sum_{k \in \mathbb{Z}} f(k) \operatorname{sinc}(x-k).
\end{aligned}
$$

The series converges in $L^2(\mathbb{R})$ as well: in fact, since $\{\operatorname{sinc}(\cdot - k)\}_{k \in \mathbb{Z}}$ is an orthonormal system,

$$
\left\| f - \sum_{n=-N}^{N} f(k) \operatorname{sinc}(\cdot - k) \right\| = \left\| \sum_{|n|>N} f(k) \operatorname{sinc}(\cdot - k) \right\| = \sqrt{\sum_{|n|>N} |f(k)|^2},
$$

which converges to 0 as $N \to \infty$ because $\{f(k)\}_{k \in \mathbb{Z}} \in \ell^2(\mathbb{Z})$ (we just saw that they are Fourier coefficients). Finally, that $\{\operatorname{sinc}(\cdot - k)\}_{k \in \mathbb{Z}}$ forms an orthonormal basis for PW follows from the fact that all equivalence classes in PW contain a continuous function. $\square$

Note that, via an appropriate scaling, the result in Theorem 2.8.1 can be extended to functions whose Fourier transform has support in an arbitrary fixed interval (Exercise 3.15).

We note that the Paley–Wiener space just is one particular Hilbert space, in which it makes sense to speak about sampling. In fact, assuming that $\phi \in L^2(\mathbb{R})$ is a continuous function for which

$$
\sum_{k \in \mathbb{Z}} \|\phi \, \chi_{[k,k+1[}\|_\infty < \infty,
$$

the Hilbert space

$$
\mathcal{H} := \left\{ \sum_{k \in \mathbb{Z}} c_k T_k \phi \; : \; \{c_k\}_{k \in \mathbb{Z}} \in \ell^2(\mathbb{Z}) \right\}
$$

consists of continuous functions. Thus, point-evaluations of functions in $\mathcal{H}$ make sense. A Hilbert space of this form is called a *shift-invariant space*. Shannon's sampling theorem can be considered as a special case of sampling in shift-invariant spaces; see the survey paper [1] for more information.

The principle in Shannon's sampling theorem is the basis for all modern communication. Most signals appearing in practice depend on a continuous variable (very often, the time). Processing of such a signal is facilitated greatly if it can be stored and handled in terms of a sequence of samples. As a concrete case, consider a piece of music. In principle, all frequencies might appear, but the human ear can only hear frequencies belonging to a certain range (at most up to 20.000 Hz). Thus, we can remove the high frequencies and consider the resulting signal as band-limited. Via an appropriate scaling, Theorem 2.8.1 shows that this signal can be recovered from its samples at sufficiently dense equidistant time intervals. This principle is used in CD players and other places where a conversion of an analog signal to a digital signal is needed.

3.9   Exercises

3.1 Assume that $\{f_k\}_{k=1}^{\infty}$ is a Bessel sequence with bound B. Prove that

(i) $||f_k||^2 \leq B$ for all $k \in \mathbb{N}$;

(ii) if $||f_k||^2 = B$ for some $k \in \mathbb{N}$, then $f_k \perp f_j$ for all $j \in \mathbb{N} \setminus \{k\}$.

3.2 Assume that $\{f_k\}_{k=1}^{\infty}$ is a Bessel sequence, and let $\{c_k\}_{k=1}^{\infty} \in \ell^2(\mathbb{N})$. The purpose of this exercise is to give a direct proof of the fact that $\sum_{k=1}^{\infty} c_k f_k$ is independent of the indexing of the sequences.

(i) Show that for any $f \in \mathcal{H}$, the series $\sum_{k=1}^{\infty} c_k \langle f_k, f \rangle$ is absolutely convergent.

(ii) Show that for any permutation σ of the natural numbers,

$$\langle \sum_{k=1}^{\infty} c_k f_k, f \rangle = \langle \sum_{k=1}^{\infty} c_{\sigma(k)} f_{\sigma(k)}, f \rangle.$$

(Hint: use that absolute convergence in $\mathbb{C}$ implies unconditional convergence).

(iii) Conclude that for any permutation σ of the natural numbers,

$$\sum_{k=1}^{\infty} c_k f_k = \sum_{k=1}^{\infty} c_{\sigma(k)} f_{\sigma(k)}.$$

3.3 Prove Lemma 3.1.6.

3.4 Prove that if $\{f_k\}_{k=1}^{\infty}$ is a sequence in a Hilbert space $\mathcal{H}$ and

$$\sum_{k=1}^{\infty} |\langle f, f_k \rangle|^2 < \infty, \ \forall f \in \mathcal{H},$$

then $\{f_k\}_{k=1}^{\infty}$ is a Bessel sequence.

3.5 Prove Proposition 3.2.8.

3.6 Prove Corollary 3.3.4

3.7 Prove that the upper and lower conditions in (3.16) are unrelated: there exists a sequence $\{f_k\}_{k=1}^\infty$ satisfying the upper condition for all finite sequences $\{c_k\}_{k=1}^\infty$, but not the lower condition; and vice versa.

3.8 Let $\{f_k\}_{k=1}^\infty$ be a sequence in a Hilbert space $\mathcal{H}$. Prove that

(i) If there exists $B > 0$ such that

$$\left\|\sum c_k f_k\right\|^2 \le B \sum |c_k|^2$$

for all finite sequences $\{c_k\}$, then $\sum_{k=1}^\infty c_k f_k$ converges for all $\{c_k\}_{k=1}^\infty \in \ell^2(\mathbb{N})$ and $\{f_k\}_{k=1}^\infty$ is a Bessel sequence with bound B.

(ii) If (3.16) holds for all finite scalar sequences $\{c_k\}$, then it holds for all $\{c_k\}_{k=1}^\infty \in \ell^2(\mathbb{N})$.

(iii) If $\{f_k\}_{k=1}^\infty$ is a Riesz basis, then

$$\sum_{k=1}^\infty c_k f_k \text{ is convergent} \Leftrightarrow \{c_k\}_{k=1}^\infty \in \ell^2(\mathbb{N}).$$

3.9 Consider the proof of Lemma 3.4.2. Where is the assumption

$$M_{j,k} = \overline{M_{k,j}}$$

used?

3.10 Consider the positive trigonometric polynomial

$$f(x) = 1 + \cos(2\pi x).$$

Find by direct calculation all trigonometric polynomials

$$g(x) = d_0 + d_1 e^{2\pi i x}, \quad d_0, d_1 \in \mathbb{R},$$

for which $|g(x)|^2 = f(x)$.

3.11 Let $\{e_k\}_{k=1}^\infty$ be an orthonormal basis for a Hilbert space $\mathcal{H}$, and define $\{f_k\}_{k=1}^\infty$ by $f_k = \frac{1}{k} e_k, k \in \mathbb{N}$.

(i) Prove that $\{f_k\}_{k=1}^\infty$ is a basis for $\mathcal{H}$, and find the biorthogonal system $\{g_k\}_{k=1}^\infty$.

(ii) Prove that the coefficient functionals associated with $\{f_k\}_{k=1}^\infty$ are not uniformly bounded.

(iii) Show that there exists $\{c_k\}_{k=1}^\infty \in \ell^2(\mathbb{N})$ for which $\sum_{k=1}^\infty c_k g_k$ is divergent.

3.12 Consider the proof of Theorem 3.7.2. Show that (3.48) defines a norm and that X is a Banach space with respect to that norm.

3.13 Prove Corollary 3.7.4.

3.14 Let $\{e_k\}_{k=1}^\infty$ be a basis for $\mathcal{H}$ and $\{g_k\}_{k=1}^\infty$ the associated biorthogonal system. Show that if $\{e_k\}_{k=1}^\infty$ is a Bessel sequence with bound B, then the following holds:

(i) $\frac{1}{B}\,||f||^2 \le \sum_{k=1}^\infty |\langle f, g_k\rangle|^2,\ \forall f \in \mathcal{H}$.

(ii) $\frac{1}{B}\sum_{k=1}^\infty |c_k|^2 \le ||\sum_{k=1}^\infty c_k g_k||^2$ for all finite sequences $\{c_k\}_{k=1}^\infty$.

3.15 Let $f \in L^2(\mathbb{R})$ be a continuous function for which

$$\text{supp}\,\hat{f} \subseteq [-\alpha/2, \alpha/2]$$

for some $\alpha > 0$. Show that f can be recovered from its samples $\{f(k/\alpha)\}_{k\in\mathbb{Z}}$ via

$$f(x) = \sum_{k\in\mathbb{Z}} f(\frac{k}{\alpha})\,\text{sinc}(\alpha x - k),\ x \in \mathbb{R}.$$

4

Bases and their Limitations

The next chapters will deal with frames, a generalization of the basis concept. It is natural to ask why they are needed. Bases exist in all separable Hilbert spaces, so why do we have to search for generalizations?

In this chapter, we will give some answers to this question. As we will see, the main point is the missing *flexibility*: the conditions for being a basis are so strong that

- it is often impossible to construct bases with special properties;

- even a slight modification of a basis might destroy the basis property.

In Section 4.1, we consider limitations in general Hilbert spaces and in the context of Fourier series in $L^2(0,1)$. Section 4.2 relates the issue to Gabor systems, and Section 4.3 deals with limitations within wavelet theory.

4.1 Bases in $L^2(0,1)$ and in general Hilbert spaces

The starting point for a more detailed discussion of the limitations of the basis concept must be to clarify why we are at all interested in bases! One reason is that a basis $\{e_k\}$ for a normed vector space X allows us to represent every $f \in X$ as a (maybe infinite) linear combination of the basis elements,

$$f = \sum c_k e_k, \tag{4.1}$$

O. Christensen, *Frames and Bases*. DOI: 10.1007/978-0-8176-4678-3_4,
© Springer Science+Business Media, LLC 2008

with coefficients $\{c_k\}$ that depend linearly on f. We will refer to this by saying that $\{e_k\}_{k=1}^{\infty}$ has the *expansion property*. This property makes it possible to reduce many questions about elements in X to the elements $\{e_k\}$ in the basis. For example, the action of a bounded operator U on f can be found if we know the representation (4.1) and the action of U on the basis $\{e_k\}$:

$$Uf = U\left(\sum c_k e_k\right) = \sum c_k U e_k.$$

Bases are characterized by the expansion property (4.1) with *unique* coefficients $\{c_k\}$ associated with each $f \in X$. One might ask whether uniqueness is really needed? Our answer is no: it is usually enough to know the *existence* of some usable coefficients, together with a recipe for finding them.

In this chapter, we discuss some cases where (4.1) holds without $\{e_k\}$ being a basis. We begin with the simple observation that if $\{e_k\}$ is a basis for X and ϕ is an arbitrary element in X, then $\{e_k\} \cup \{\phi\}$ is not a basis, despite the fact that each $f \in X$ has representations of the form

$$f = \sum c_k e_k + d\phi. \tag{4.2}$$

The reason is that $\{e_k\} \cup \{\phi\}$ is no longer independent, i.e., several choices for the coefficients $\{c_k\}$ and d are possible: one choice is to take $d = 0$ and let $\{c_k\}$ be the coefficients representing f in the basis $\{e_k\}$; another choice is to take $\{c_k\}$ such that $f - \phi = \sum c_k e_k$, and $d = 1$.

By this argument, the basis property is destroyed when an arbitrary non-empty collection of vectors is added to $\{e_k\}$, but the expansion property is preserved.

At first glance, the above construction might appear artificial: why would one like to add elements to a basis? One reason will be given in Section 5.9: having more elements than needed for a basis turns out to have a certain *noise suppressing* effect. Another reason is that we gain some freedom: the coefficients in (4.1) are unique, but in (4.2) we can *choose* between several options. We can actually say even more. In fact, we have seen in Theorem 3.3.2 and Theorem 3.7.5 that every basis has a unique dual basis; the frames introduced in the next chapter lead to expansions of a similar type, but usually with several possible choices for a dual. In Section 9.4, we will use this freedom to construct particulary convenient duals.

Non-bases with the expansion property also appear naturally in function spaces:

Example 4.1.1 Let us return to the orthonormal basis $\{e_k\}_{k\in\mathbb{Z}}$ for $L^2(0,1)$ considered in Section 3.5, i.e., the functions $e_k(x) = e^{2\pi i k x}$. Given an open subinterval $I \subset]0,1[$ with $|I| < 1$, we can identify $L^2(I)$ with the subspace of $L^2(0,1)$ consisting of the functions that are zero on $]0,1[\backslash I$. Hereby a function $f \in L^2(I)$ is identified with a function (which we still denote f) in $L^2(0,1)$, which has the expansion

$$f = \sum_{k \in \mathbb{Z}} \langle f, e_k \rangle e_k \quad \text{in} \ \ L^2(0,1). \tag{4.3}$$

Because

$$\left\| f - \sum_{|k| \le n} \langle f, e_k \rangle e_k \right\|_{L^2(I)}^2 = \int_I \left| f(x) - \sum_{k=-n}^{n} \langle f, e_k \rangle e^{2\pi i k x} \right|^2 dx$$

$$\le \int_0^1 \left| f(x) - \sum_{k=-n}^{n} \langle f, e_k \rangle e^{2\pi i k x} \right|^2 dx$$

$$\to 0 \ \text{as} \ n \to \infty,$$

we also have

$$f = \sum_{k \in \mathbb{Z}} \langle f, e_k \rangle e_k \quad \text{in} \ \ L^2(I). \tag{4.4}$$

That is, the (restrictions to I of the) functions $\{e_k\}_{k \in \mathbb{Z}}$ also have the expansion property in $L^2(I)$. However, they are not a basis for $L^2(I)$! To see this, define the function

$$\tilde{f}(x) = \begin{cases} f(x) & \text{if } x \in I, \\ 1 & \text{if } x \notin I. \end{cases}$$

Then $\tilde{f} \in L^2(0,1)$ and we have the representation

$$\tilde{f} = \sum_{k \in \mathbb{Z}} \langle \tilde{f}, e_k \rangle e_k \quad \text{in} \ \ L^2(0,1). \tag{4.5}$$

By restricting to I, the expansion (4.5) is also valid in $L^2(I)$; since $f = \tilde{f}$ on I, this shows that

$$f = \sum_{k \in \mathbb{Z}} \langle \tilde{f}, e_k \rangle e_k \quad \text{in} \ \ L^2(I). \tag{4.6}$$

Thus, (4.4) and (4.6) are both expansions of f in $L^2(I)$, and they are non-identical; the argument is that since $f \ne \tilde{f}$ in $L^2(0,1)$, the expansions (4.3) and (4.5) show that

$$\{\langle f, e_k \rangle\}_{k \in \mathbb{Z}} \ne \{\langle \tilde{f}, e_k \rangle\}_{k \in \mathbb{Z}}.$$

The conclusion is that the restriction of the functions $\{e_k\}_{k \in \mathbb{Z}}$ to I is not a basis for $L^2(I)$, but the expansion property is preserved. In Example 5.5.5, we prove that $\{e_k\}_{k \in \mathbb{Z}}$ is a frame for $L^2(I)$. □

In a finite-dimensional vector space X, we know that every family of vectors that spans X contains a basis (Exercise 1.3). In an infinite-dimensional

Hilbert space, the situation is dramatically different: one can prove the existence of families of vectors $\{f_k\}_{k=1}^{\infty}$ such that

- each $f \in \mathcal{H}$ has an unconditionally convergent expansion

$$f = \sum_{k=1}^{\infty} c_k f_k \quad \text{with } \{c_k\}_{k=1}^{\infty} \in \ell^2(\mathbb{N});$$

- no subsequence of $\{f_k\}_{k=1}^{\infty}$ is a basis for $\mathcal{H}$.

See [9] or [10]. Intuitively, this kind of example is difficult to understand: it shows that we might have the expansion property for families that have no relationship to a basis. This is also an argument for looking for generalizations of bases.

4.2   Gabor bases and the Balian–Low Theorem

In concrete Hilbert spaces like $L^2(\mathbb{R})$, we are able to give explicit constructions of bases, like the Gabor orthonormal basis

$$\{e^{2\pi i m x}\chi_{[0,1]}(x - n)\}_{m,n\in\mathbb{Z}} = \{E_m T_n \chi_{[0,1]}(x)\}_{m,n\in\mathbb{Z}}$$

in Example 3.5.3. However, this example touches one of the limitations on possible constructions of bases, as we will see now. Observe that the Fourier transform of $\chi_{[0,1]}$ is given by

$$\widehat{\chi}_{[0,1]}(\gamma) = \int_0^1 e^{-2\pi i x \gamma} \, dx = e^{-\pi i \gamma} \frac{\sin \pi \gamma}{\pi \gamma}.$$

The fact that $\chi_{[0,1]}$ is discontinuous, and the oscillations and slow decay of $\widehat{\chi}_{[0,1]}$, makes the characteristic function unattractive from the point of view of, e.g., time–frequency analysis. It is natural to ask whether more suitable Gabor bases could be obtained by replacing the function $\chi_{[0,1]}$ by a smoother function g? Unfortunately, one can prove that no continuous compactly supported function g can generate a Gabor Riesz basis (see Corollary 9.7.4).

Even if we give up the requirement that the function g should have compact support, the *Balian–Low Theorem* shows that there are limitations on the properties g can have if we want $\{E_m T_n g\}_{m,n\in\mathbb{Z}}$ to be a Riesz basis:

Theorem 4.2.1 Let $g \in L^2(\mathbb{R})$. If $\{E_m T_n g\}_{m,n\in\mathbb{Z}}$ is a Riesz basis for $L^2(\mathbb{R})$, then

$$\left(\int_{-\infty}^{\infty} |xg(x)|^2 \, dx\right)\left(\int_{-\infty}^{\infty} |\gamma\hat{g}(\gamma)|^2 \, d\gamma\right) = \infty. \tag{4.7}$$

In words, the Balian–Low theorem means that a function g generating a Gabor Riesz basis cannot be well localized in both time and frequency. For example, it is not possible that g and $\hat{g}$ satisfy decay conditions like

$$|g(x)| \le \frac{C}{1+x^2}, \ x \in \mathbb{R}, \ |\hat{g}(\gamma)| \le \frac{C}{1+\gamma^2}, \ \gamma \in \mathbb{R},$$

simultaneously. The reader can find proofs of the Balian–Low theorem in [25] and [41].

If fast decay of g and $\hat{g}$ is needed, we have to ask whether we need all the properties characterizing a Riesz basis or whether we can relax some of them. The property we want to keep is that every $f \in L^2(\mathbb{R})$ has an unconditionally convergent expansion in terms of modulated and translated versions of the function g; together with Lemma 3.7.6, this shows that we do not gain anything by asking for $\{E_m T_n g\}_{m,n \in \mathbb{Z}}$ being merely a basis instead of a Riesz basis. However, it turns out that the (unconditionally convergent) expansion property actually can be combined with g and $\hat{g}$ having very fast decay: the part of the definition of a basis that has to be given up is the *uniqueness* of such an expansion. This will bring us from bases to frames. The exact definition will be given in the next chapter, and frames having the Gabor structure will be the subject of Chapters 9–10.

4.3 Bases and wavelets

Wavelet orthonormal bases $\{\psi_{j,k}\}_{j,k \in \mathbb{Z}}$ form another important class of bases for $L^2(\mathbb{R})$. As discussed in Section 3.6, most of the important wavelet bases are constructed via multiresolution analysis. Some of the properties that are relevant for a basis $\{\psi_{j,k}\}_{j,k \in \mathbb{Z}}$ are

- that ψ has a computationally convenient form, for example that ψ is a piecewise polynomial (a spline);

- regularity of ψ;

- symmetry (or anti-symmetry) of ψ, i.e., that

$$\psi(x) = \psi(-x) \text{ or } \psi(x) = -\psi(-x), \ x \in \mathbb{R};$$

- compact support of ψ, or at least fast decay;

- that ψ has *vanishing moments*, i.e., that for a certain $m \in \mathbb{N}$,

$$\int_{-\infty}^{\infty} x^\ell \psi(x) \, dx = 0 \text{ for } \ell = 0, 1, \dots, m. \tag{4.8}$$

We will give a short description of the role played by these properties and how they motivated the development of wavelet theory. First, a large number of vanishing moments is important if we want to obtain smooth wavelets. In fact, if ψ is an m times differentiable function with bounded derivatives with reasonable decay and $\{\psi_{j,k}\}_{j,k \in \mathbb{Z}}$ is an orthonormal system, then (4.8) holds, see [26].

Vanishing moments are also essential in the context of *compression*. Assuming that $\{\psi_{j,k}\}_{j,k\in\mathbb{Z}}$ is an orthonormal basis for $L^2(\mathbb{R})$, every $f \in L^2(\mathbb{R})$ has the representation

$$f = \sum_{j,k\in\mathbb{Z}} \langle f, \psi_{j,k}\rangle \psi_{j,k}. \tag{4.9}$$

All information about f is stored in the coefficients $\{\langle f, \psi_{j,k}\rangle\}_{j,k\in\mathbb{Z}}$, and (4.9) tells us how to reconstruct f based on knowledge of the coefficients. However, in practice one cannot store an infinite sequence of non-zero numbers, so one has to select a finite number of the coefficients to keep. This is usually done by *thresholding*: one chooses a certain $\epsilon > 0$ and keeps only the coefficients $\langle f, \psi_{j,k}\rangle$ for which $|\langle f, \psi_{j,k}\rangle| \geq \epsilon$. Here the vanishing moments come in again. In fact, one can prove that if ψ has a large number of vanishing moments, then the coefficients $\langle f, \psi_{j,k}\rangle$ decay fast for $j \to \infty$:

Theorem 4.3.1 *Assume that the function $\psi \in L^2(\mathbb{R})$ is compactly supported and has $N-1$ vanishing moments, i.e., (4.8) holds with $m = N-1$. Then, for any N times differentiable function $f \in L^2(\mathbb{R})$ for which $f^{(N)}$ is bounded, there exists a constant $C > 0$ such that*

$$|\langle f, \psi_{j,k}\rangle| \leq C\, 2^{-jN} 2^{-j/2}, \ \forall j, k \in \mathbb{Z}. \tag{4.10}$$

For a proof, we refer to [64]. Assuming that (4.9) is available, we obtain an efficient compression of the signal f if we only keep the large coefficients $\langle f, \psi_{j,k}\rangle$ and throw the rest away.

It turns out that Theorem 4.3.1 also plays an important role for the wavelet constructions in Chapter 11. In fact, it turns out that the expansion coefficients associated with a so-called tight wavelet frame have exactly the same form as for an orthonormal basis. This implies that the estimate in (4.10) again can be used to determine which coefficients to keep.

Compact support (or at least fast decay) of ψ is essential for the use of computer-based methods, where a function with unbounded support always has to be truncated. For the same reason, we often want the support to be small. The condition of ψ being symmetric is less important (or even irrelevant) in many contexts, but there are cases where it is a helpful property; an example is image processing, where a non-symmetric wavelet will generate non-symmetric errors, which are more disturbing to the human eye than symmetric errors. The next result, which is also proved in [26], shows that it is difficult to combine the classical multiresolution analysis with the desire of having a symmetric wavelet ψ:

Proposition 4.3.2 *Assume that $\phi \in L^2(\mathbb{R})$ is real-valued and compactly supported, and let*

$$V_j = \overline{span}\{D^j T_k \phi\}_{k \in \mathbb{Z}}, \ j \in \mathbb{Z}.$$

Assume that $(\phi, \{V_j\})$ constitute a multiresolution analysis. Then, if the associated wavelet ψ is real-valued and compactly supported and has either a symmetry axis or an anti-symmetry axis, then ψ is necessarily the Haar wavelet.

Thus, under the above assumptions, we are back at the function we want to avoid! In Chapter 11, we will see how symmetry can be obtained at the price of looking at frames rather than bases.

The limitations on the possible constructions of bases give theoretical reasons to consider frames. In Section 5.9, we will describe cases where bases actually exist, but where frames simply perform better.

5

Frames in Hilbert Spaces

The main feature of a basis $\{f_k\}_{k=1}^{\infty}$ in a Hilbert space $\mathcal{H}$ is that every $f \in \mathcal{H}$ can be represented as an (infinite) linear combination of the elements f_k in the basis:

$$f = \sum_{k=1}^{\infty} c_k(f) f_k. \tag{5.1}$$

The coefficients $c_k(f)$ are unique. We now introduce the concept of *frames*. A frame is also a sequence of elements $\{f_k\}_{k=1}^{\infty}$ in $\mathcal{H}$, which allows every $f \in \mathcal{H}$ to be written as in (5.1). However, the corresponding coefficients are not necessarily unique. Thus a frame might not be a basis; arguments for generalizing the basis concept were given in Chapter 4.

Frames were introduced in 1952 by Duffin and Schaeffer in their fundamental paper [30]; they used frames as a tool in the study of *nonharmonic Fourier series*, i.e., sequences of the type $\{e^{i\lambda_n x}\}_{n\in\mathbb{Z}}$, where $\{\lambda_n\}_{n\in\mathbb{Z}}$ is a family of real or complex numbers. In 1985, as the wavelet era began, Daubechies, Grossmann, and Meyer [27] observed that frames can be used to find series expansions of functions in $L^2(\mathbb{R})$ that are very similar to the expansions using orthonormal bases. This was the time when many mathematicians started to see the potential of the topic. In this chapter, we present the general theory; the next chapters will go into detail with specific constructions.

Section 5.1 and Section 5.2 are instrumental for a good understanding of frames; here, their basic properties are presented, and the relationship between frames and Riesz basis is discussed. A reader who is mainly interested

O. Christensen, *Frames and Bases*. DOI: 10.1007/978-0-8176-4678-3_5, in Gabor frames or wavelet frames might go direct to the relevant later chapters in the book after reading these sections; in fact, the theory for these frames is to a large extent independent of the results for general frames. Section 5.3 deals with preservation of the frame property under the action of various operators. Section 5.4 characterizes the frame property, e.g., in terms of operators. Section 5.5 discusses various independency conditions and presents further relations between frames and Riesz bases. In Section 5.6, it is shown that the frame concept is stable, in the sense that sufficiently small "displacements" of a frame again is a frame. These results extend classical results known for bases. In Section 5.7, we characterize all dual frames associated with a given frame. Section 5.8 gives a short introduction to continuous frames. Finally, Section 5.9 relates frame theory to signal processing and signal transmission.

5.1   Frames and their properties

We are now ready to give the central definition. Let $\mathcal{H} \neq \{0\}$ denote a separable Hilbert space.

Definition 5.1.1 *A sequence $\{f_k\}_{k=1}^{\infty}$ of elements in $\mathcal{H}$ is a frame for $\mathcal{H}$ if there exist constants $A, B > 0$ such that*

$$A \, ||f||^2 \le \sum_{k=1}^{\infty} |\langle f, f_k \rangle|^2 \le B \, ||f||^2, \quad \forall f \in \mathcal{H}. \tag{5.2}$$

The numbers A and B are called *frame bounds*. They are not unique. The *optimal upper frame bound* is the infimum over all upper frame bounds, and the *optimal lower frame bound* is the supremum over all lower frame bounds. Note that the optimal bounds actually are frame bounds.

The following lemma shows that it is enough to check the frame condition on a dense set. The proof is left to the reader (Exercise 5.3).

Lemma 5.1.2 *Suppose that $\{f_k\}_{k=1}^{\infty}$ is a sequence of elements in $\mathcal{H}$ and that there exist constants $A, B > 0$ such that*

$$A \, ||f||^2 \le \sum_{k=1}^{\infty} |\langle f, f_k \rangle|^2 \le B \, ||f||^2 \tag{5.3}$$

for all f in a dense subset V of $\mathcal{H}$. Then $\{f_k\}_{k=1}^{\infty}$ is a frame for $\mathcal{H}$ with bounds A, B.

A special role is played by frames for which the optimal frame bounds coincide:

Definition 5.1.3 *A sequence $\{f_k\}_{k=1}^{\infty}$ in $\mathcal{H}$ is a tight frame if there exists a number $A > 0$ such that*

$$\sum_{k=1}^{\infty} |\langle f, f_k \rangle|^2 = A \, ||f||^2, \ \forall f \in \mathcal{H}.$$

The (exact) number A is called the frame bound.

It follows from the definition that if $\{f_k\}_{k=1}^{\infty}$ is a frame for $\mathcal{H}$, then

$$\overline{\text{span}}\{f_k\}_{k=1}^{\infty} = \mathcal{H};$$

in fact, if $f \in \mathcal{H}$ is perpendicular to all f_k, $k \in \mathbb{N}$, then (5.2) shows that $f = 0$. We often need to consider sequences that are not complete in $\mathcal{H}$; they cannot form frames for $\mathcal{H}$, but they can very well form frames for the closed linear span of their elements. For the purpose of considering such sequences, we need the following definition.

Definition 5.1.4 *Let $\{f_k\}_{k=1}^{\infty}$ be a sequence in $\mathcal{H}$. We say that $\{f_k\}_{k=1}^{\infty}$ is a frame sequence if it is a frame for $\overline{\text{span}}\{f_k\}_{k=1}^{\infty}$.*

Before we develop the theory for frames, we mention a few examples of frames. They might appear quite "constructed," but they are useful for the theoretical understanding of frames. In Chapters 7–11, we consider frames that are more interesting by themselves, for example frames in $L^2(\mathbb{R})$ having Gabor structure or wavelet structure.

Example 5.1.5 Let $\{e_k\}_{k=1}^{\infty}$ be an orthonormal basis for $\mathcal{H}$.
(i) By repeating each element in $\{e_k\}_{k=1}^{\infty}$ twice, we obtain

$$\{f_k\}_{k=1}^{\infty} = \{e_1, e_1, e_2, e_2, ..\},$$

which is a tight frame with frame bound $A = 2$. If only e_1 is repeated, we obtain

$$\{f_k\}_{k=1}^{\infty} = \{e_1, e_1, e_2, e_3, ..\},$$

which is a frame with bounds $A = 1, B = 2$.
(ii) Let

$$\{f_k\}_{k=1}^{\infty} := \{e_1, \frac{1}{\sqrt{2}}e_2, \frac{1}{\sqrt{2}}e_2, \frac{1}{\sqrt{3}}e_3, \frac{1}{\sqrt{3}}e_3, \frac{1}{\sqrt{3}}e_3, \cdots \};$$

that is, $\{f_k\}_{k=1}^{\infty}$ is the sequence where each vector $\frac{1}{\sqrt{\ell}}e_\ell, \ell \in \mathbb{N}$, is repeated ℓ times. Then, for each $f \in \mathcal{H}$,

$$\sum_{k=1}^{\infty} |\langle f, f_k \rangle|^2 = \sum_{\ell=1}^{\infty} \ell \, |\langle f, \frac{1}{\sqrt{\ell}}e_\ell \rangle|^2 = ||f||^2.$$

So $\{f_k\}_{k=1}^{\infty}$ is a tight frame for $\mathcal{H}$ with frame bound $A = 1$.

(iii) If $I \subset \mathbb{N}$ is a proper subset, then $\{e_k\}_{k \in I}$ is not complete in $\mathcal{H}$ and cannot be a frame for $\mathcal{H}$. However, $\{e_k\}_{k \in I}$ is a frame for $\overline{\text{span}}\{e_k\}_{k \in I}$, i.e., it is a frame sequence. $\qquad\square$

Since a frame $\{f_k\}_{k=1}^{\infty}$ is a Bessel sequence, the operator

$$T : \ell^2(\mathbb{N}) \to \mathcal{H}, \quad T\{c_k\}_{k=1}^{\infty} = \sum_{k=1}^{\infty} c_k f_k \tag{5.4}$$

is bounded by Theorem 3.1.3; T is called the *pre-frame operator* or the *synthesis operator*. By Lemma 3.1.1, the adjoint operator is given by

$$T^* : \mathcal{H} \to \ell^2(\mathbb{N}), \quad T^* f = \{\langle f, f_k \rangle\}_{k=1}^{\infty}. \tag{5.5}$$

T^* is called the *analysis operator*. Composing T and T^*, we obtain the *frame operator*

$$S : \mathcal{H} \to \mathcal{H}, \quad Sf = TT^* f = \sum_{k=1}^{\infty} \langle f, f_k \rangle f_k. \tag{5.6}$$

Note that because $\{f_k\}_{k=1}^{\infty}$ is a Bessel sequence, the series defining S converges unconditionally for all $f \in \mathcal{H}$ by Corollary 3.1.5. We state some of the important properties of S:

Lemma 5.1.6 *Let $\{f_k\}_{k=1}^{\infty}$ be a frame with frame operator S and frame bounds A, B. Then the following holds:*

(i) *S is bounded, invertible, self-adjoint, and positive.*

(ii) *$\{S^{-1} f_k\}_{k=1}^{\infty}$ is a frame with frame operator S^{-1} and frame bounds B^{-1}, A^{-1}.*

(iii) *If A, B are the optimal frame bounds for $\{f_k\}_{k=1}^{\infty}$, then the bounds B^{-1}, A^{-1} are optimal for $\{S^{-1} f_k\}_{k=1}^{\infty}$.*

Proof. (i) S is bounded, being a composition of two bounded operators. By Theorem 3.1.3,

$$\|S\| = \|TT^*\| = \|T\| \, \|T^*\| = \|T\|^2 \le B.$$

Because $S^* = (TT^*)^* = TT^* = S$, the operator S is self-adjoint. The inequality (5.2) means that

$$A\|f\|^2 \le \langle Sf, f \rangle \le B\|f\|^2, \quad \forall f \in \mathcal{H},$$

or, in the notation from Section 2.2,

$$AI \le S \le BI; \tag{5.7}$$

thus S is a positive operator. Furthermore,

$$0 \le I - B^{-1} S \le \frac{B - A}{B} I,$$

and consequently

$$\left\|I - B^{-1}S\right\| = \sup_{\|f\|=1} \left|\langle(I - B^{-1}S)f, f\rangle\right| \le \frac{B - A}{B} < 1.$$

By Theorem 2.2.3, this shows that S is invertible.

(ii) Because the operator S is self-adjoint, also S^{-1} is self-adjoint. Note that for $f \in \mathcal{H}$,

$$\sum_{k=1}^{\infty} |\langle f, S^{-1}f_k\rangle|^2 = \sum_{k=1}^{\infty} |\langle S^{-1}f, f_k\rangle|^2 \quad \le \quad B \, \|S^{-1}f\|^2$$

$$\le \quad B \, \|S^{-1}\|^2 \, \|f\|^2.$$

That is, $\{S^{-1}f_k\}_{k=1}^{\infty}$ is a Bessel sequence. It follows that the frame operator for $\{S^{-1}f_k\}_{k=1}^{\infty}$ is well defined. By definition, it acts on $f \in \mathcal{H}$ by

$$\sum_{k=1}^{\infty} \langle f, S^{-1}f_k\rangle S^{-1}f_k = S^{-1} \sum_{k=1}^{\infty} \langle S^{-1}f, f_k\rangle f_k \quad = \quad S^{-1}SS^{-1}f$$

$$= \quad S^{-1}f; \qquad (5.8)$$

this shows that the frame operator for $\{S^{-1}f_k\}_{k=1}^{\infty}$ equals S^{-1}. The operator S^{-1} commutes with both S and I, so using Theorem 2.4.2 we can "multiply" the inequality (5.7) with S^{-1}; this gives

$$B^{-1}I \le S^{-1} \le A^{-1}I,$$

i.e.,

$$B^{-1} \, \|f\|^2 \le \langle S^{-1}f, f\rangle \le A^{-1} \, \|f\|^2, \ \forall f \in \mathcal{H}.$$

Via (5.8), this means that

$$B^{-1} \, \|f\|^2 \le \sum_{k=1}^{\infty} |\langle f, S^{-1}f_k\rangle|^2 \le A^{-1} \, \|f\|^2, \ \forall f \in \mathcal{H};$$

thus $\{S^{-1}f_k\}_{k=1}^{\infty}$ is a frame with frame bounds B^{-1}, A^{-1}.

(iii) Let A be the optimal lower bound for $\{f_k\}_{k=1}^{\infty}$, and assume that the optimal upper bound for $\{S^{-1}f_k\}_{k=1}^{\infty}$ is $C < A^{-1}$. By applying (ii) to the frame $\{S^{-1}f_k\}_{k=1}^{\infty}$, we obtain that $\{f_k\}_{k=1}^{\infty} = \{(S^{-1})^{-1}S^{-1}f_k\}_{k=1}^{\infty}$ has the lower bound $C^{-1} > A$, but this is a contradiction. Thus $\{S^{-1}f_k\}_{k=1}^{\infty}$ has the optimal upper bound A^{-1}. The argument for the optimal lower bound of $\{S^{-1}f_k\}_{k=1}^{\infty}$ is similar. $\qquad \square$

The frame $\{S^{-1}f_k\}_{k=1}^{\infty}$ is called the *canonical dual frame* of $\{f_k\}_{k=1}^{\infty}$. The reason for the name will become clear soon; in fact, Theorem 5.1.7 will show that $\{S^{-1}f_k\}_{k=1}^{\infty}$ plays the same role in frame theory as the dual in the theory of bases, see Theorem 3.3.2 and Theorem 3.7.5.

The *frame decomposition*, stated in (5.9) below, is the most important frame result. It shows that if $\{f_k\}_{k=1}^{\infty}$ is a frame for $\mathcal{H}$, then every element in $\mathcal{H}$ has a representation as an infinite linear combination of the frame elements. Thus it is natural to view a frame as some kind of "generalized basis."

Theorem 5.1.7 *Let* $\{f_k\}_{k=1}^{\infty}$ *be a frame with frame operator* S. *Then*

$$f = \sum_{k=1}^{\infty} \langle f, S^{-1}f_k \rangle f_k, \quad \forall f \in \mathcal{H}, \tag{5.9}$$

and

$$f = \sum_{k=1}^{\infty} \langle f, f_k \rangle S^{-1}f_k, \quad \forall f \in \mathcal{H}. \tag{5.10}$$

Both series converge unconditionally for all $f \in \mathcal{H}$.

Proof. Let $f \in \mathcal{H}$. Using the properties of the frame operator in Lemma 5.1.6,

$$f = SS^{-1}f = \sum_{k=1}^{\infty} \langle S^{-1}f, f_k \rangle f_k = \sum_{k=1}^{\infty} \langle f, S^{-1}f_k \rangle f_k.$$

Because $\{f_k\}_{k=1}^{\infty}$ is a Bessel sequence and $\{\langle f, S^{-1}f_k \rangle\}_{k=1}^{\infty} \in \ell^2(\mathbb{N})$, the fact that the series converges unconditionally follows from Corollary 3.1.5. The expansion (5.10) is proved similarly, using that $f = S^{-1}Sf$. $\square$

Theorem 5.1.7 shows that all information about a given vector $f \in \mathcal{H}$ is contained in the sequence $\{\langle f, S^{-1}f_k \rangle\}_{k=1}^{\infty}$. The numbers $\langle f, S^{-1}f_k \rangle$ are called *frame coefficients*.

Theorem 5.1.7 also immediately reveals one of the main difficulties in frame theory. In fact, in order for the expansions (5.9) and (5.10) to be applicable in practice, we need to be able to find the operator S^{-1}, or at least to calculate its action on all f_k, $k \in \mathbb{N}$. In general, this is a major problem. One way of circumventing the problem is to consider only tight frames:

Corollary 5.1.8 *If* $\{f_k\}_{k=1}^{\infty}$ *is a tight frame with frame bound* A, *then the canonical dual frame is* $\{A^{-1}f_k\}_{k=1}^{\infty}$, *and*

$$f = \frac{1}{A} \sum_{k=1}^{\infty} \langle f, f_k \rangle f_k, \quad \forall f \in \mathcal{H}. \tag{5.11}$$

Proof. If $\{f_k\}_{k=1}^{\infty}$ is a tight frame with frame bound A and frame operator S, the definition shows that

$$\langle Sf, f \rangle = \sum_{k=1}^{\infty} |\langle f, f_k \rangle|^2 = A \, ||f||^2 = \langle Af, f \rangle, \quad \forall f \in \mathcal{H}.$$

By Lemma 2.4.3, this implies that $S = AI$; thus, S^{-1} acts by multiplication by A^{-1}, and the result follows from (5.9).   □

By a suitable scaling of the vectors $\{f_k\}_{k=1}^{\infty}$ in a tight frame, we can always obtain that $A = 1$; in that case, (5.11) has exactly the same form as the representation via an orthonormal basis, see (3.7). Thus, such frames can be used without any additional computational effort compared with the use of orthonormal bases.

Tight frames have other advantages. For the design of frames with pre-scribed properties, it is essential to control the behavior of the canonical dual frame, but the complicated structure of the frame operator and its in-verse makes this difficult. If, for example, we consider a frame $\{f_k\}_{k=1}^{\infty}$ for $L^2(\mathbb{R})$ consisting of functions with exponential decay, nothing guarantees that the functions in the canonical dual frame $\{S^{-1}f_k\}_{k=1}^{\infty}$ have exponen-tial decay. However, for tight frames, questions of this type trivially have satisfying answers. Also, for a tight frame, the canonical dual frame au-tomatically has the same structure as the frame itself: if the frame has wavelet structure or Gabor structure as described in Section 3.5 and Sec-tion 3.6, the same is the case for the canonical dual frame. In contrast, the canonical dual frame of a non-tight wavelet frame might not have the wavelet structure; a concrete example is given later, see Example 11.1.2.

Later we will discuss another way to avoid the problem of inverting the frame operator S. In fact, for frames $\{f_k\}_{k=1}^{\infty}$ that are *not* bases, we prove in Theorem 5.2.3 that one can find other frames $\{g_k\}_{k=1}^{\infty}$ than $\{S^{-1}f_k\}_{k=1}^{\infty}$, for which

$$f = \sum_{k=1}^{\infty} \langle f, g_k \rangle f_k, \quad \forall f \in \mathcal{H}. \tag{5.12}$$

Such a frame $\{g_k\}_{k=1}^{\infty}$ is called a *dual frame* of $\{f_k\}_{k=1}^{\infty}$. Now, there is a chance that even if the canonical dual frame is difficult to find, there exist other duals that are easy to find! In Section 7.4 and Section 9.4, we discuss such cases. For general frames in Hilbert spaces, all duals are characterized in Section 5.7.

A note on terminology is in order. In Lemma 5.7.1, we prove that if $\{g_k\}_{k=1}^{\infty}$ is a dual frame of $\{f_k\}_{k=1}^{\infty}$, then $\{f_k\}_{k=1}^{\infty}$ is also a dual of $\{g_k\}_{k=1}^{\infty}$. For this reason, we will usually call $\{f_k\}_{k=1}^{\infty}$ and $\{g_k\}_{k=1}^{\infty}$ a *pair of dual frames* or a *dual frame pair* when (5.12) holds.

Example 5.1.9 Let $\{e_k\}_{k=1}^{\infty}$ be an orthonormal basis for $\mathcal{H}$ and consider the frame

$$\{f_k\}_{k=1}^{\infty} = \{e_1, e_1, e_2, e_3, ..\},$$

see Example 5.1.5(ii). The canonical dual frame is given by

$$\{S^{-1}f_k\}_{k=1}^{\infty} = \{\frac{1}{2}e_1, \frac{1}{2}e_1, e_2, e_3, ..\}.$$

As examples of non-canonical dual frames we mention

$$\{g_k\}_{k=1}^{\infty} = \{0, e_1, e_2, e_3, ..\}$$

and

$$\{g_k\}_{k=1}^{\infty} = \{\frac{1}{3}e_1, \frac{2}{3}e_1, e_2, e_3, ..\}.$$

We leave the verifications to the reader (Exercise 5.8).   □

We end this section with a warning. Misled by the situation in the finite-dimensional setting, one could expect that if $\{f_k\}_{k=1}^{\infty}$ is a sequence in $\mathcal{H}$ for which $\overline{\text{span}}\{f_k\}_{k=1}^{\infty} = \mathcal{H}$, then every $f \in \mathcal{H}$ would have an expansion

$$f = \sum_{k=1}^{\infty} c_k f_k$$

for certain scalar coefficients $\{c_k\}_{k=1}^{\infty}$. However, in an infinite-dimensional Hilbert space this does not necessarily hold. We give an example below. The example will be used several times in the sequel, and we extend it later in Example 5.5.6.

Example 5.1.10 Let $\{e_k\}_{k=1}^{\infty}$ be an orthonormal basis for $\mathcal{H}$ and define

$$f_k := e_k + e_{k+1}, \ k \in \mathbb{N}.$$

We will show that

(i) $\overline{\text{span}}\{f_k\}_{k=1}^{\infty} = \mathcal{H}$;

(ii) $\{f_k\}_{k=1}^{\infty}$ is a Bessel sequence, but not a frame;

(iii) There exists $f \in \mathcal{H}$ that cannot be written on the form $\sum_{k=1}^{\infty} c_k f_k$ for any choice of the coefficients c_k.

To prove that $\{f_k\}_{k=1}^{\infty}$ is complete, assume that $f \in \mathcal{H}$ and that

$$\langle f, f_k \rangle = 0 \text{ for all } k \in \mathbb{N}.$$

Then $\langle f, e_k \rangle = -\langle f, e_{k+1} \rangle$ for all $k \in \mathbb{N}$, implying that $|\langle f, e_k \rangle|$ is a constant. Since

$$\sum_{k=1}^{\infty} |\langle f, e_k \rangle|^2 = ||f||^2 < \infty,$$

we conclude that $\langle f, e_k \rangle = 0$, $\forall k$, so $f = 0$. Thus $\{f_k\}_{k=1}^{\infty}$ is complete.

We now prove that $\{f_k\}_{k=1}^{\infty}$ is a Bessel sequence. For that purpose, we will use that for any $a, b \in \mathbb{R}$, the inequality $(a + b)^2 \le 2(a^2 + b^2)$ holds;

the inequality is a consequence of $2ab \leq a^2 + b^2$, which again follows from $a^2 + b^2 - 2ab = (a - b)^2 \geq 0$. Now, for any $f \in \mathcal{H}$,

$$
\begin{aligned}
\sum_{k=1}^{\infty} |\langle f, e_k + e_{k+1} \rangle|^2 &= \sum_{k=1}^{\infty} |\langle f, e_k \rangle + \langle f, e_{k+1} \rangle|^2 \\
&\leq \sum_{k=1}^{\infty} (|\langle f, e_k \rangle| + |\langle f, e_{k+1} \rangle|)^2 \\
&\leq 2 \sum_{k=1}^{\infty} |\langle f, e_k \rangle|^2 + 2 \sum_{k=1}^{\infty} |\langle f, e_{k+1} \rangle|^2 \\
&\leq 4 \, ||f||^2.
\end{aligned}
$$

This proves that $\{f_k\}_{k=1}^{\infty}$ is a Bessel sequence. However, $\{f_k\}_{k=1}^{\infty}$ does not satisfy the lower frame condition. To see this, consider the vectors

$$
g_j := \sum_{n=1}^{j} (-1)^{n+1} e_n, \ j \in \mathbb{N}.
$$

We note that $||g_j||^2 = j$ for all $j \in \mathbb{N}$. Let us now calculate the inner products $\langle g_j, f_k \rangle$. Considering a fixed $j \in \mathbb{N}$, we see that

$$
\langle g_j, f_k \rangle = \langle e_1 - e_2 + \cdots + (-1)^{j+1} e_j, e_k + e_{k+1} \rangle = \begin{cases} 0 & \text{if } k > j; \\ (-1)^{j+1} & \text{if } k = j; \\ 0 & \text{if } k < j. \end{cases}
$$

Therefore

$$
\sum_{k=1}^{\infty} |\langle g_j, f_k \rangle|^2 = 1 = \frac{1}{j} \, ||g_j||^2.
$$

Since this holds for all $j \in \mathbb{N}$, we see that $\{f_k\}_{k=1}^{\infty}$ does not satisfy the lower frame condition. We finally notice that, despite the fact that $\{f_k\}_{k=1}^{\infty}$ is complete, there exists $f \in \mathcal{H}$ that cannot be written as $f = \sum_{k=1}^{\infty} c_k f_k$ for *any* choice of the coefficients $\{c_k\}_{k=1}^{\infty}$. As a concrete example, take $f = e_1$. $\square$

5.2   Frames and Riesz bases

As we have seen, a frame $\{f_k\}_{k=1}^{\infty}$ in a Hilbert space $\mathcal{H}$ has one of the main properties of a basis: given $f \in \mathcal{H}$, there exist coefficients $\{c_k\}_{k=1}^{\infty} \in \ell^2(\mathbb{N})$ such that $f = \sum_{k=1}^{\infty} c_k f_k$. This makes it natural to study the relationship between frames and bases. In this section, we notice that all Riesz bases are frames and characterize the frames that are actually Riesz bases.

Theorem 5.2.1 *A Riesz basis $\{f_k\}_{k=1}^{\infty}$ for $\mathcal{H}$ is a frame for $\mathcal{H}$, and the Riesz basis bounds coincide with the frame bounds. The dual Riesz basis equals the canonical dual frame $\{S^{-1}f_k\}_{k=1}^{\infty}$.*

Proof. By Proposition 3.3.5, a Riesz basis $\{f_k\}_{k=1}^{\infty}$ for $\mathcal{H}$ is also a frame for $\mathcal{H}$; if we also involve Proposition 3.4.5, we obtain the statement about the bounds. The rest follows from the frame decomposition combined with the uniqueness part of Theorem 3.3.2. $\square$

A frame that is *not* a Riesz basis is said to be *overcomplete;* in the literature, the term *redundant frame* is also used. Theorem 5.2.2 will explain why the word "overcomplete" is used: in fact, if $\{f_k\}_{k=1}^{\infty}$ is a frame that is not a Riesz basis, there exist coefficients $\{c_k\}_{k=1}^{\infty} \in \ell^2(\mathbb{N}) \setminus \{0\}$ for which

$$\sum_{k=1}^{\infty} c_k f_k = 0. \tag{5.13}$$

That is, for such frames there is some dependency between the frame elements.

Theorem 5.2.2 *Let $\{f_k\}_{k=1}^{\infty}$ be a frame for $\mathcal{H}$. Then the following are equivalent:*

(i) $\{f_k\}_{k=1}^{\infty}$ *is a Riesz basis for $\mathcal{H}$.*

(ii) *If $\sum_{k=1}^{\infty} c_k f_k = 0$ for some $\{c_k\}_{k=1}^{\infty} \in \ell^2(\mathbb{N})$, then $c_k = 0$, $\forall k \in \mathbb{N}$.*

Proof.
(i)$\Rightarrow$(ii). Assume that $\{f_k\}_{k=1}^{\infty}$ is a Riesz basis and that $\sum_{k=1}^{\infty} c_k f_k = 0$ for a sequence $\{c_k\}_{k=1}^{\infty} \in \ell^2(\mathbb{N})$. Writing $\{f_k\}_{k=1}^{\infty} = \{Ue_k\}_{k=1}^{\infty}$ for a certain orthonormal basis for $\mathcal{H}$ and an appropriate bounded bijective operator $\mathcal{H}$, it follows that $U \sum_{k=1}^{\infty} c_k e_k = 0$. Because U is injective, this implies that $\sum_{k=1}^{\infty} c_k e_k = 0$, and therefore $c_k = 0$ for all k.

(ii)$\Rightarrow$(i). Let $\{\delta_k\}_{k=1}^{\infty}$ be the canonical orthonormal basis for $\ell^2(\mathbb{N})$. The assumption (ii) assures that the pre-frame operator T associated with $\{f_k\}_{k=1}^{\infty}$ is injective, and T is also surjective because $\{f_k\}_{k=1}^{\infty}$ is a frame. Since $T\delta_k = f_k$, $\forall k$, the result follows from the definition of a Riesz basis. $\square$

Much more can be said about the relationship between frames and Riesz bases – see Theorem 5.5.4. For now, we just notice that if $\{f_k\}_{k=1}^{\infty}$ is an overcomplete frame, then a given element $f \in \mathcal{H}$ has *many* representations in terms of the frame elements f_k, $k \in \mathbb{N}$. In fact, for any sequence $\{c_k\}_{k=1}^{\infty} \in \ell^2(\mathbb{N}) \setminus \{0\}$ for which (5.13) holds, the frame decomposition shows that

$$f = \sum_{k=1}^{\infty} \langle f, S^{-1} f_k \rangle f_k = \sum_{k=1}^{\infty} \left(\langle f, S^{-1} f_k \rangle + c_k \right) f_k.$$

We can actually go one important step further: we will now prove that every overcomplete frame has other dual frames than the canonical. This result should be compared with the uniqueness statement for Riesz bases in Theorem 3.3.2.

Theorem 5.2.3 *Assume that $\{f_k\}_{k=1}^{\infty}$ is an overcomplete frame. Then there exist frames $\{g_k\}_{k=1}^{\infty} \neq \{S^{-1} f_k\}_{k=1}^{\infty}$ for which*

$$f = \sum_{k=1}^{\infty} \langle f, g_k \rangle f_k, \ \forall f \in \mathcal{H}. \tag{5.14}$$

Proof. We split the proof in two cases and assume first that $f_\ell = 0$ for some $\ell \in \mathbb{N}$; in this case $S^{-1} f_\ell = 0$. Letting $g_k := S^{-1} f_k$ for $k \neq \ell$ and choosing g_ℓ to be an arbitrary non-zero vector, the frame decomposition shows that (5.14) holds, and $\{g_k\}_{k=1}^{\infty} \neq \{S^{-1} f_k\}_{k=1}^{\infty}$.

Now we consider the case where $f_k \neq 0$ for all $k \in \mathbb{N}$. By Theorem 5.2.2, there exists a sequence $\{c_k\}_{k=1}^{\infty} \in \ell^2(\mathbb{N}) \setminus \{0\}$ such that

$$0 = \sum_{k=1}^{\infty} c_k f_k.$$

For a certain $\ell \in \mathbb{N}$ we have $c_\ell \neq 0$, and we can write

$$f_\ell = \frac{-1}{c_\ell} \sum_{k \neq \ell} c_k f_k.$$

We now show that $\{f_k\}_{k \neq \ell}$ is a frame for $\mathcal{H}$; we only have to prove that $\{f_k\}_{k \neq \ell}$ satisfies the lower frame condition. In order to do so, observe that for any $f \in \mathcal{H}$, Cauchy–Schwarz' inequality shows that

$$\begin{aligned}
|\langle f, f_\ell \rangle|^2 &= \left| \frac{-1}{c_\ell} \sum_{k \neq \ell} c_k \langle f, f_k \rangle \right|^2 \\
&\leq \frac{1}{|c_\ell|^2} \sum_{k \neq \ell} |c_k|^2 \sum_{k \neq \ell} |\langle f, f_k \rangle|^2 \\
&= C \sum_{k \neq \ell} |\langle f, f_k \rangle|^2,
\end{aligned}$$

where $C := \frac{1}{|c_\ell|^2} \sum_{k \neq \ell} |c_k|^2$. Letting A denote a lower frame bound for the frame $\{f_k\}_{k=1}^\infty$, this implies that

$$
\begin{aligned}
A \, \|f\|^2 &\leq \sum_{k=1}^\infty |\langle f, f_k \rangle|^2 \\
&= \sum_{k \neq \ell} |\langle f, f_k \rangle|^2 + |\langle f, f_\ell \rangle|^2 \\
&\leq (1 + C) \sum_{k \neq \ell} |\langle f, f_k \rangle|^2.
\end{aligned}
$$

This shows that $\{f_k\}_{k \neq \ell}$ indeed satisfies the lower frame condition.

Denoting the canonical dual frame of $\{f_k\}_{k \neq \ell}$ by $\{g_k\}_{k \neq \ell}$ and defining $g_\ell = 0$, we have found a frame $\{g_k\}_{k=1}^\infty$ for which (5.14) holds; it is different from the canonical dual of $\{f_k\}_{k=1}^\infty$ because $S^{-1} f_\ell \neq 0$. $\qquad\square$

All dual frames associated with a given frame $\{f_k\}_{k=1}^\infty$ are characterized in Section 5.7. At the moment, it is not clear that there are cases where it is an advantage to consider a dual frame different from the canonical one; such cases will appear in Section 7.4 and Section 9.4.

5.3 Frames and operators

Lemma 5.1.6 shows that if $\{f_k\}_{k=1}^\infty$ is a frame, then $\{S^{-1} f_k\}_{k=1}^\infty$ is also a frame. This is a special case of a much more general result: $\{U f_k\}_{k=1}^\infty$ is actually a frame for a large class of operators U. For later reference, we state some general versions of this result, where we assume that U is a bounded operator with closed range $\mathcal{R}_U$. We denote the *pseudo-inverse* (see Lemma 2.5.1) of such an operator U by $U^\dagger$.

Proposition 5.3.1 *Let $\{f_k\}_{k=1}^\infty$ be a frame for $\mathcal{H}$ with bounds A, B, and let $U : \mathcal{H} \to \mathcal{H}$ be a bounded operator with closed range. Then $\{U f_k\}_{k=1}^\infty$ is a frame sequence with frame bounds $A \, \|U^\dagger\|^{-2}, \; B \, \|U\|^2$.*

Proof. If $f \in \mathcal{H}$, then

$$
\sum_{k=1}^\infty |\langle f, U f_k \rangle|^2 \leq B \, \|U^* f\|^2 \leq B \, \|U\|^2 \, \|f\|^2,
$$

which proves that $\{U f_k\}_{k=1}^\infty$ is a Bessel sequence. For the lower frame condition, let $g \in \text{span}\{U f_k\}_{k=1}^\infty$; we can write $g = U f$ for some $f \in \text{span}\{f_k\}_{k=1}^\infty$. By Lemma 2.5.2, the operator $U U^\dagger$ is the orthogonal projection onto $\mathcal{R}_U$ and therefore self-adjoint. This implies that

$$
g = U f = (U U^\dagger)^* U f = (U^\dagger)^* U^* U f.
$$

Thus

$$\begin{aligned}
||g||^2 &\leq ||(U^\dagger)^*||^2 \, ||U^*Uf||^2 \\
&\leq \frac{||(U^\dagger)^*||^2}{A} \sum_{k=1}^{\infty} |\langle U^*Uf, f_k\rangle|^2 \\
&= \frac{||U^\dagger||^2}{A} \sum_{k=1}^{\infty} |\langle g, Uf_k\rangle|^2.
\end{aligned}$$

Consequently, the lower frame condition is satisfied for all $g \in \mathrm{span}\{Uf_k\}_{k=1}^{\infty}$. Via Lemma 5.1.2, we conclude that the condition holds on $\overline{\mathrm{span}}\{Uf_k\}_{k=1}^{\infty}$, i.e., that $\{Uf_k\}_{k=1}^{\infty}$ is a frame sequence. $\qquad\square$

Exercise 5.13 shows that the conclusion in Proposition 5.3.1 might fail if U does not have closed range. And even if U has closed range, it is not enough to assume that $\{f_k\}_{k=1}^{\infty}$ is a frame sequence (Exercise 5.14).

Corollary 5.3.2 *Assume that $\{f_k\}_{k=1}^{\infty}$ is a frame for $\mathcal{H}$ with bounds A, B and that $U : \mathcal{H} \to \mathcal{H}$ is a bounded surjective operator. Then $\{Uf_k\}_{k=1}^{\infty}$ is a frame for $\mathcal{H}$ with frame bounds $A \, ||U^\dagger||^{-2}, \ B \, ||U||^2$.*

In the next result, it is enough to assume that $\{f_k\}_{k=1}^{\infty}$ is a frame sequence. We leave the proof to the reader (Exercise 5.15).

Lemma 5.3.3 *If $\{f_k\}_{k=1}^{\infty}$ is a frame sequence with frame bounds A, B and $U : \mathcal{H} \to \mathcal{H}$ is a unitary operator, then $\{Uf_k\}_{k=1}^{\infty}$ is a frame sequence with frame bounds A, B. If $\{f_k\}_{k=1}^{\infty}$ is a frame for $\mathcal{H}$, then $\{Uf_k\}_{k=1}^{\infty}$ is also a frame for $\mathcal{H}$.*

The kind of stability discussed here can often be used to construct frames with special properties. For example, it is important to notice that we to every frame can associate a *canonical tight frame* with frame bound $A = 1$:

Theorem 5.3.4 *Let $\{f_k\}_{k=1}^{\infty}$ be a frame for $\mathcal{H}$ with frame operator S. Denote the positive square root of S^{-1} by $S^{-1/2}$. Then $\{S^{-1/2}f_k\}_{k=1}^{\infty}$ is a tight frame with frame bound equal to 1, and*

$$f = \sum_{k=1}^{\infty} \langle f, S^{-1/2}f_k\rangle S^{-1/2}f_k, \ \forall f \in \mathcal{H}.$$

Proof. The existence of a unique positive square root of S^{-1} follows from Lemma 2.4.4. Since $S^{-1/2}$ is a limit of a sequence of polynomials in S^{-1}, it commutes with S^{-1} and therefore with S. Therefore every $f \in \mathcal{H}$ can be written

$$f = S^{-1/2}SS^{-1/2}f$$

$$= \sum_{k=1}^{\infty} \langle S^{-1/2}f, f_k \rangle S^{-1/2} f_k$$

$$= \sum_{k=1}^{\infty} \langle f, S^{-1/2} f_k \rangle S^{-1/2} f_k.$$

This implies that

$$\|f\|^2 = \langle f, f \rangle = \sum_{k=1}^{\infty} |\langle S^{-1/2}f, f_k \rangle|^2,$$

i.e., that $\{S^{-1/2}f_k\}_{k=1}^{\infty}$ is a tight frame with frame bound equal to 1.   □

Orthogonal projections play a special role in many contexts. We state a few relationships between frames and orthogonal projections.

Proposition 5.3.5 *Let $\{f_k\}_{k=1}^{\infty}$ be a sequence in a Hilbert space $\mathcal{H}$, and let P denote the orthogonal projection of $\mathcal{H}$ onto a closed subspace V. Then the following holds:*

(i) *If $\{f_k\}_{k=1}^{\infty}$ is a frame for $\mathcal{H}$ with frame bounds A, B, then $\{Pf_k\}_{k=1}^{\infty}$ is a frame for V with frame bounds A, B.*

(ii) *If $\{f_k\}_{k=1}^{\infty}$ is a frame for V with frame operator S, then the orthogonal projection of $\mathcal{H}$ onto V is given by*

$$Pf = \sum_{k=1}^{\infty} \langle f, S^{-1}f_k \rangle f_k, \ f \in \mathcal{H}. \tag{5.15}$$

The proof of (i) is left to the reader (Exercise 5.16), and the proof of (ii) is identical to the proof of Theorem 1.1.11.

In Section 5.2, we saw that if $\{f_k\}_{k=1}^{\infty}$ is an overcomplete frame, there exist several sets of coefficients $\{c_k\}_{k=1}^{\infty} \in \ell^2(\mathbb{N})$ for which $f = \sum_{k=1}^{\infty} c_k f_k$. The frame coefficients $\{\langle f, S^{-1}f_k \rangle\}_{k=1}^{\infty}$ have minimal ℓ^2-norm among all sequences representing f:

Lemma 5.3.6 *Let $\{f_k\}_{k=1}^{\infty}$ be a frame for $\mathcal{H}$ and let $f \in \mathcal{H}$. If f has a representation $f = \sum_{k=1}^{\infty} c_k f_k$ for some coefficients $\{c_k\}_{k=1}^{\infty}$, then*

$$\sum_{k=1}^{\infty} |c_k|^2 = \sum_{k=1}^{\infty} |\langle f, S^{-1}f_k \rangle|^2 + \sum_{k=1}^{\infty} |c_k - \langle f, S^{-1}f_k \rangle|^2.$$

The proof is identical to the proof of Theorem 1.1.5(iii). As a consequence of Lemma 5.3.6, we obtain an explicit expression for the pseudo-inverse $T^{\dagger}$ of the pre-frame operator. Recall that $T^{\dagger}$ is defined in Section 2.5.

Theorem 5.3.7 *Let* $\{f_k\}_{k=1}^{\infty}$ *be a frame with pre-frame operator T and frame operator S. Then the pseudo-inverse of T is given by*

$$T^{\dagger} : \mathcal{H} \to \ell^2(\mathbb{N}), \quad T^{\dagger}f = \{\langle f, S^{-1}f_k\rangle\}_{k=1}^{\infty}.$$

Proof. In terms of the pre-frame operator T, the equation $f = \sum_{k=1}^{\infty} c_k f_k$ means that $T\{c_k\}_{k=1}^{\infty} = f$. Now the result follows from Lemma 5.3.6 combined with Theorem 2.5.3. $\qquad\square$

The optimal frame bounds can be expressed in terms of the operators T and S and their inverses/pseudo-inverses:

Proposition 5.3.8 *The optimal frame bounds A, B for a frame $\{f_k\}_{k=1}^{\infty}$ are given by*

$$A = ||S^{-1}||^{-1} = ||T^{\dagger}||^{-2}, \quad B = ||S|| = ||T||^2.$$

Proof. By definition, the optimal upper frame bound is given by

$$B = \sup_{||f||=1} \sum_{k=1}^{\infty} |\langle f, f_k\rangle|^2 = \sup_{||f||=1} \langle Sf, f\rangle = ||S||.$$

The last equality follows from S being self-adjoint. Also, $S = TT^*$ implies that $||S|| = ||TT^*|| = ||T||^2$.

In order to prove the results about the lower frame bound, we use that the dual frame $\{S^{-1}f_k\}_{k=1}^{\infty}$ has frame operator S^{-1} and the optimal upper bound A^{-1} by Lemma 5.1.6. Thus, according to what we just proved, we know that $A^{-1} = ||S^{-1}||$. Finally, via Theorem 5.3.7,

$$||S^{-1}|| = \sup_{||f||=1} \sum_{k=1}^{\infty} |\langle f, S^{-1}f_k\rangle|^2 = \sup_{||f||=1} ||T^{\dagger}f||^2 = ||T^{\dagger}||^2. \qquad\square$$

Because a frame $\{f_k\}_{k=1}^{\infty}$ might be overcomplete, it is possible that removal of an element f_j leaves us with a sequence $\{f_k\}_{k\neq j}$ that is still a frame. It turns out that whether this happens or not can be determined based on the value of the frame coefficient $\langle f_j, S^{-1}f_j\rangle$:

Theorem 5.3.9 *The removal of a vector f_j from a frame $\{f_k\}_{k=1}^{\infty}$ for $\mathcal{H}$ leaves either a frame or an incomplete set. More precisely, the following holds:*

(i) If $\langle f_j, S^{-1}f_j\rangle \neq 1$, then $\{f_k\}_{k\neq j}$ is a frame for $\mathcal{H}$;

(ii) If $\langle f_j, S^{-1}f_j\rangle = 1$, then $\{f_k\}_{k\neq j}$ is incomplete.

Proof. Choose $j \in \mathbb{N}$ arbitrarily. By the frame decomposition,

$$f_j = \sum_{k=1}^{\infty} \langle f_j, S^{-1}f_k\rangle f_k.$$

Define, for notational convenience, $a_k = \langle f_j, S^{-1} f_k \rangle$, so $f_j = \sum_{k=1}^{\infty} a_k f_k$. Clearly, we also have $f_j = \sum_{k=1}^{\infty} \delta_{j,k} f_k$, so Lemma 5.3.6 yields the following relation between $\delta_{j,k}$ and a_k:

$$
\begin{aligned}
1 = \sum_{k=1}^{\infty} |\delta_{j,k}|^2 &= \sum_{k=1}^{\infty} |a_k|^2 + \sum_{k=1}^{\infty} |a_k - \delta_{j,k}|^2 \\
&= |a_j|^2 + \sum_{k \neq j} |a_k|^2 + |a_j - 1|^2 + \sum_{k \neq j} |a_k|^2. \quad (5.16)
\end{aligned}
$$

We consider the cases $a_j = 1$ and $a_j \neq 1$ separately. First, suppose that $a_j \neq 1$; then $f_j = \frac{1}{1-a_j} \sum_{k \neq j} a_k f_k$. Exactly like in the proof of Theorem 5.2.3, this implies that $\{f_k\}_{k \neq j}$ is a frame for $\mathcal{H}$; this proves (i).

Suppose now that $a_j = 1$. The calculation in (5.16) implies that $\sum_{k \neq j} |a_k|^2 = 0$, so that

$$
a_k = \langle S^{-1} f_j, f_k \rangle = 0 \quad \text{for all} \ \ k \neq j. \quad (5.17)
$$

Since $a_j = \langle S^{-1} f_j, f_j \rangle = 1$ we know that $S^{-1} f_j \neq 0$. Thus we have found a non-zero element $S^{-1} f_j$ that is orthogonal to $\{f_k\}_{k \neq j}$, so $\{f_k\}_{k \neq j}$ is incomplete. This proves (ii). $\qquad\square$

5.4   Characterization of frames

Let us for a moment go back to the definition of a frame. In order to check that a sequence $\{f_k\}_{k=1}^{\infty}$ is a frame, we have to verify the existence of a positive lower frame bound A and a finite upper frame bound B. Intuitively, the lower frame condition is the most critical one to verify: bad upper estimates on $\sum_{k=1}^{\infty} |\langle f, f_k \rangle|^2$ will sometimes force us to take a larger value for B than necessary, but bad lower estimates can easily make it impossible to find a value for $A > 0$ that can be used for all $f \in \mathcal{H}$. We now give a characterization of frames in terms of the pre-frame operator. It does not involve any knowledge of the frame bounds.

Theorem 5.4.1 *A sequence $\{f_k\}_{k=1}^{\infty}$ in $\mathcal{H}$ is a frame for $\mathcal{H}$ if and only if*

$$
T : \{c_k\}_{k=1}^{\infty} \mapsto \sum_{k=1}^{\infty} c_k f_k
$$

is a well-defined mapping of $\ell^2(\mathbb{N})$ onto $\mathcal{H}$.

Proof. First, suppose that $\{f_k\}_{k=1}^{\infty}$ is a frame. By Theorem 3.1.3, T is a well-defined bounded operator from $\ell^2(\mathbb{N})$ into $\mathcal{H}$, and by Lemma 5.1.6(i), the frame operator $S = TT^*$ is surjective. Thus T is surjective. For the opposite implication, suppose that T is a well-defined operator from $\ell^2(\mathbb{N})$ onto $\mathcal{H}$. Then Lemma 3.1.1 shows that T is bounded and that $\{f_k\}_{k=1}^\infty$ is a Bessel sequence. Let $T^\dagger : \mathcal{H} \to \ell^2(\mathbb{N})$ denote the pseudo-inverse of T. For $f \in \mathcal{H}$, we have that

$$f = TT^\dagger f = \sum_{k=1}^\infty (T^\dagger f)_k f_k,$$

where $(T^\dagger f)_k$ denotes the k-th coordinate of $T^\dagger f$. Thus

$$
\begin{aligned}
||f||^4 &= |\langle f, f\rangle|^2 \\
&= |\langle \sum_{k=1}^\infty (T^\dagger f)_k f_k, f\rangle|^2 \\
&\leq \sum_{k=1}^\infty |(T^\dagger f)_k|^2 \sum_{k=1}^\infty |\langle f, f_k\rangle|^2 \\
&\leq ||T^\dagger||^2 \, ||f||^2 \sum_{k=1}^\infty |\langle f, f_k\rangle|^2.
\end{aligned}
$$

We conclude that

$$\sum_{k=1}^\infty |\langle f, f_k\rangle|^2 \geq \frac{1}{||T^\dagger||^2} \, ||f||^2, \ \forall f \in \mathcal{H},$$

i.e., that $\{f_k\}_{k=1}^\infty$ is a frame. $\square$

For an arbitrary sequence $\{f_k\}_{k=1}^\infty$ in a Hilbert space, $\overline{\text{span}}\{f_k\}_{k=1}^\infty$ is itself a Hilbert space, and Theorem 5.4.1 leads to a statement about frame sequences:

Corollary 5.4.2 *A sequence $\{f_k\}_{k=1}^\infty$ in $\mathcal{H}$ is a frame sequence if and only if the pre-frame operator is well-defined on $\ell^2(\mathbb{N})$ and has closed range.*

In terms of the adjoint of the pre-frame operator we have:

Corollary 5.4.3 *For a sequence $\{f_k\}_{k=1}^\infty$ in $\mathcal{H}$ the following holds:*

(i) *$\{f_k\}_{k=1}^\infty$ is a frame sequence if and only if*

$$f \mapsto \{\langle f, f_k\rangle\}_{k=1}^\infty \tag{5.18}$$

defines a map from $\mathcal{H}$ onto a closed subspace of $\ell^2(\mathbb{N})$.

(ii) *If $\{f_k\}_{k=1}^\infty$ is a frame sequence, it is a frame for $\mathcal{H}$ if and only if the map (5.18) is injective.*

Proof. The proof of (i) uses that a bounded operator has closed range if and only if its adjoint operator has closed range, see Lemma 2.4.1. First, assume that $\{f_k\}_{k=1}^{\infty}$ is a frame sequence. Then the pre-frame operator T is well-defined and bounded, and the range $\mathcal{R}_T$ is closed. Therefore T^*, i.e., the operator in (5.18), is well-defined, and has closed range. For the opposite implication, if the operator in (5.18) maps $\mathcal{H}$ into $\ell^2(\mathbb{N})$, then $\{f_k\}_{k=1}^{\infty}$ is a Bessel sequence (Exercise 3.4). Thus, the pre-frame operator T is well defined and bounded; furthermore, if the range of the map in (5.18) is closed, the same is true for T. This implies by Corollary 5.4.2 that $\{f_k\}_{k=1}^{\infty}$ is a frame sequence.

For the proof of (ii), we note that $\overline{\mathcal{R}_T} = (\mathcal{N}_{T^*})^{\perp}$. Thus, if $\{f_k\}_{k=1}^{\infty}$ is a frame for $\mathcal{H}$, then T^* is injective. On the other hand, if (5.18) defines an injective mapping, then $\{f_k\}_{k=1}^{\infty}$ is complete in $\mathcal{H}$; thus, if $\{f_k\}_{k=1}^{\infty}$ is a frame sequence, it is a frame for $\mathcal{H}$. □

Recall that Riesz bases for $\mathcal{H}$ are characterized as the families $\{Ue_k\}_{k=1}^{\infty}$, where $\{e_k\}_{k=1}^{\infty}$ is an orthonormal basis for $\mathcal{H}$ and $U : \mathcal{H} \to \mathcal{H}$ is bounded and invertible. We can now give a similar characterization of frames:

Theorem 5.4.4 Let $\{e_k\}_{k=1}^{\infty}$ be an arbitrary orthonormal basis for $\mathcal{H}$. The frames for $\mathcal{H}$ are precisely the families $\{Ue_k\}_{k=1}^{\infty}$, where $U : \mathcal{H} \to \mathcal{H}$ is a bounded and surjective operator.

Proof. Let $\{\delta_k\}_{k=1}^{\infty}$ be the canonical basis for $\ell^2(\mathbb{N})$ and $\{e_k\}_{k=1}^{\infty}$ an orthonormal basis for $\mathcal{H}$. Let $\Phi : \mathcal{H} \to \ell^2(\mathbb{N})$ be the isometric isomorphism defined by $\Phi e_k = \delta_k$. If $\{f_k\}_{k=1}^{\infty}$ is a frame, then the pre-frame operator T is bounded and surjective, and $T\delta_k = f_k$. With $U := T\Phi$, we have $\{f_k\}_{k=1}^{\infty} = \{Ue_k\}_{k=1}^{\infty}$, and U is bounded and surjective. That every family $\{Ue_k\}_{k=1}^{\infty}$ of the described type is a frame follows from Theorem 5.4.1 (see Exercise 5.19). Alternatively, we can observe that

$$\sum_{k=1}^{\infty} |\langle f, Ue_k \rangle|^2 = ||U^*f||^2,$$

and refer to Lemma 2.4.1. □

Via Theorem 5.4.1, the question of existence of an upper and a lower frame bound is replaced by an investigation of infinite series: we have to check that $\sum_{k=1}^{\infty} c_k f_k$ converges for all $\{c_k\}_{k=1}^{\infty} \in \ell^2(\mathbb{N})$ and that each $f \in \mathcal{H}$ can be represented via such an infinite series. The above consequences of Theorem 5.4.1 do not involve the frame bounds either. We now state a characterization of frames that keeps the information about the frame bounds. The obtained result is probably most useful for theoretical considerations; see the proof of Theorem 7.1.7 for an application.

Lemma 5.4.5 *A sequence $\{f_k\}_{k=1}^{\infty}$ in $\mathcal{H}$ is a frame for $\mathcal{H}$ with bounds A, B if and only if the following two conditions are satisfied:*

(i) $\{f_k\}_{k=1}^{\infty}$ is complete in $\mathcal{H}$.

(ii) The pre-frame operator T is well defined on $\ell^2(\mathbb{N})$ and

$$A \sum_{k=1}^{\infty} |c_k|^2 \leq ||T\{c_k\}_{k=1}^{\infty}||^2 \leq B \sum_{k=1}^{\infty} |c_k|^2, \ \forall \{c_k\}_{k=1}^{\infty} \in \mathcal{N}_T^{\perp}. \quad (5.19)$$

In particular, $\{f_k\}_{k=1}^{\infty}$ is a frame sequence if and only if (ii) holds.

Proof. Theorem 3.1.3 gives the first part: the upper frame condition with bound B is equivalent to the right-hand inequality in (5.19) (it is clear that it is enough to check the condition for $\{c_k\}_{k=1}^{\infty} \in \mathcal{N}_T^{\perp}$). We therefore assume that $\{f_k\}_{k=1}^{\infty}$ is a Bessel sequence and prove the equivalence of the lower frame condition with the two conditions formed by (i) and the left-hand inequality in (5.19).

First, assume that $\{f_k\}_{k=1}^{\infty}$ satisfies the lower frame condition with bound A. Then (i) is satisfied. Note that $\mathcal{R}_{T^*}$ is closed because $\mathcal{R}_T$ is closed (the latter is equal to $\mathcal{H}$ because $\{f_k\}_{k=1}^{\infty}$ is a frame). Therefore

$$\mathcal{N}_T^{\perp} = \overline{\mathcal{R}_{T^*}} = \mathcal{R}_{T^*},$$

i.e., $\mathcal{N}_T^{\perp}$ consists of all sequences of the form $\{\langle f, f_k \rangle\}_{k=1}^{\infty}$, $f \in \mathcal{H}$. Now, given $f \in \mathcal{H}$,

$$\left(\sum_{k=1}^{\infty} |\langle f, f_k \rangle|^2 \right)^2 = |\langle Sf, f \rangle|^2 \ \leq \ ||Sf||^2 \, ||f||^2$$

$$\leq \ ||Sf||^2 \, \frac{1}{A} \sum_{k=1}^{\infty} |\langle f, f_k \rangle|^2.$$

This implies that

$$A \sum_{k=1}^{\infty} |\langle f, f_k \rangle|^2 \leq ||Sf||^2 = ||T\{\langle f, f_k \rangle\}_{k=1}^{\infty}||^2,$$

as desired. For the other implication, assume that $\{f_k\}_{k=1}^{\infty}$ is complete and that the left-hand inequality in (5.19) is satisfied. We first prove that $\mathcal{R}_T = \mathcal{H}$. Since $\text{span}\{f_k\}_{k=1}^{\infty} \subset \mathcal{R}_T$, it is enough to prove that $\mathcal{R}_T$ is closed. Let $\{y_n\}$ be a sequence in $\mathcal{R}_T$ that converges to some $y \in \mathcal{H}$. Take a sequence $\{x_n\}$ in $\mathcal{N}_T^{\perp}$ such that $y_n = Tx_n$; then (5.19) implies that $\{x_n\}$ is a Cauchy sequence. Therefore $\{x_n\}$ converges to some x, which by continuity of T satisfies that $Tx = y$. Thus $\mathcal{R}_T$ is closed and hence $\mathcal{R}_T = \mathcal{H}$. Let $T^{\dagger}$ denote the pseudo-inverse of T. By Lemma 2.5.2 and (2.10), we know that the operator $T^{\dagger}T$ is the orthogonal projection onto $\mathcal{N}_T^{\perp}$, and that $TT^{\dagger}$ is the orthogonal projection onto $\mathcal{R}_T = \mathcal{H}$. Thus, for any $\{c_k\}_{k=1}^{\infty} \in \ell^2(\mathbb{N})$, the inequality (5.19) has the consequence that

$$A \, ||T^{\dagger}T\{c_k\}_{k=1}^{\infty}||^2 \leq ||TT^{\dagger}T\{c_k\}_{k=1}^{\infty}||^2 = ||T\{c_k\}_{k=1}^{\infty}||^2. \qquad (5.20)$$

Again by (2.10), we have $\mathcal{N}_{T^{\dagger}} = \mathcal{R}_T^{\perp}$; therefore (5.20) implies that

$$||T^{\dagger}||^2 \leq A^{-1}.$$

Using Lemma 2.5.2, we also have $||(T^*)^{\dagger}||^2 \leq A^{-1}$. But $(T^*)^{\dagger}T^*$ is the orthogonal projection onto

$$\mathcal{R}_{(T^*)^{\dagger}} = \mathcal{R}_{(T^{\dagger})^*} = \mathcal{N}_{T^{\dagger}}^{\perp} = \mathcal{R}_T = \mathcal{H},$$

so for all $f \in \mathcal{H}$,

$$\begin{aligned} ||f||^2 = ||(T^*)^{\dagger}T^*f||^2 &\leq \frac{1}{A} \, ||T^*f||^2 \\ &= \frac{1}{A} \sum_{k=1}^{\infty} |\langle f, f_k \rangle|^2. \end{aligned}$$

This shows that $\{f_k\}_{k=1}^{\infty}$ satisfies the lower frame condition as desired. $\square$

5.5   Various independency conditions

From linear algebra in finite-dimensional vector spaces, we know that a basis is a linearly independent set. Linear independence is also necessary for an infinite set to be a basis in an infinite-dimensional Hilbert space, but it is not sufficient. We now discuss some of the relevant independency conditions in infinite-dimensional spaces and their relationships.

Definition 5.5.1 *Let $\{f_k\}_{k=1}^{\infty}$ be a sequence in $\mathcal{H}$. We say that*

(i) *$\{f_k\}_{k=1}^{\infty}$ is linearly independent if every finite subset of $\{f_k\}_{k=1}^{\infty}$ is linearly independent;*

(ii) *$\{f_k\}_{k=1}^{\infty}$ is ω-independent if whenever the series $\sum_{k=1}^{\infty} c_k f_k$ is convergent and equal to zero for some scalar coefficients $\{c_k\}_{k=1}^{\infty}$, then necessarily $c_k = 0$ for all $k \in \mathbb{N}$.*

(iii) *$\{f_k\}_{k=1}^{\infty}$ is minimal if $f_j \notin \overline{span}\{f_k\}_{k \neq j}, \, \forall j \in \mathbb{N}$.*

The relationship between the definitions is as follows:

Lemma 5.5.2 *Let $\{f_k\}_{k=1}^{\infty}$ be a sequence in $\mathcal{H}$. Then the following holds:*

(i) *If $\{f_k\}_{k=1}^{\infty}$ is minimal, then $\{f_k\}_{k=1}^{\infty}$ is ω-independent.*

(ii) *If $\{f_k\}_{k=1}^{\infty}$ is ω-independent, then $\{f_k\}_{k=1}^{\infty}$ is linearly independent.*

The opposite implications in (i) and (ii) are not valid.

Proof. For the proof of (i), assume that $\{f_k\}_{k=1}^{\infty}$ is not ω-independent. Choose scalar coefficients $\{c_k\}_{k=1}^{\infty}$ with $c_j \neq 0$ for some j, such that $\sum_{k=1}^{\infty} c_k f_k = 0$; then $f_j = \sum_{k \neq j} \frac{-c_k}{c_j} f_k$, implying that $f_j \in \overline{\text{span}}\{f_k\}_{k \neq j}$. That is, $\{f_k\}_{k=1}^{\infty}$ is not minimal. The statement (ii) is obvious, and the fact that the opposite implications are not valid is demonstrated by examples in Exercise 5.22. $\qquad\square$

Recall from Theorem 3.4.4 that the existence of a complete biorthogonal sequence is necessary for a sequence $\{f_k\}_{k=1}^{\infty}$ to be a Riesz basis. This implies that $\{f_k\}_{k=1}^{\infty}$ must be minimal:

Lemma 5.5.3 *Let $\{f_k\}_{k=1}^{\infty}$ be a sequence in $\mathcal{H}$. Then the following holds:*

(i) $\{f_k\}_{k=1}^{\infty}$ has a biorthogonal sequence $\{g_k\}_{k=1}^{\infty}$ if and only if $\{f_k\}_{k=1}^{\infty}$ is minimal.

(ii) If a biorthogonal sequence for $\{f_k\}_{k=1}^{\infty}$ exists, it is uniquely determined if and only if $\{f_k\}_{k=1}^{\infty}$ is complete in $\mathcal{H}$.

Proof. For the proof of (i), suppose first that $\{f_k\}_{k=1}^{\infty}$ has a biorthogonal system $\{g_k\}_{k=1}^{\infty}$. Then, for any given $j \in \mathbb{N}$,

$$\langle f_j, g_j \rangle = 1 \text{ and } \langle f_k, g_j \rangle = 0 \text{ for } k \neq j.$$

Therefore $f_j \notin \overline{\text{span}}\{f_k\}_{k \neq j}$, i.e., $\{f_k\}_{k=1}^{\infty}$ is minimal. For the other implication in (i), assume that $\{f_k\}_{k=1}^{\infty}$ is minimal. Given $j \in \mathbb{N}$, let P_j denote the orthogonal projection of $\mathcal{H}$ onto $\overline{\text{span}}\{f_k\}_{k \neq j}$. Then it follows that $(I - P_j)f_j \neq 0$, and

$$\langle f_j, (I - P_j)f_j \rangle = \langle P_j f_j + (I - P_j)f_j, (I - P_j)f_j \rangle = ||(I - P_j)f_j||^2 \neq 0.$$

For $k \neq j$, clearly $\langle f_k, (I - P_j)f_j \rangle = 0$. Defining

$$g_j := \frac{(I - P_j)f_j}{||(I - P_j)f_j||^2}, \quad j \in \mathbb{N},$$

we obtain that $\{g_k\}_{k=1}^{\infty}$ is a biorthogonal system for $\{f_k\}_{k=1}^{\infty}$.

For the proof of (ii), assume that $\{f_k\}_{k=1}^{\infty}$ has a biorthogonal system $\{g_k\}_{k=1}^{\infty}$. If $\{f_k\}_{k=1}^{\infty}$ is not complete, then it has several biorthogonal systems. In fact, letting

$$\mathcal{H}_0 := \overline{\text{span}}\{f_k\}_{k=1}^{\infty},$$

we can replace $\{g_k\}_{k=1}^{\infty}$ by $\{g_k + h_k\}_{k=1}^{\infty}$ for some $h_k \in \mathcal{H}_0^{\perp} \setminus \{0\}$ and hereby obtain a new biorthogonal system for $\{f_k\}_{k=1}^{\infty}$. We leave it to the reader to verify that if $\{f_k\}_{k=1}^{\infty}$ is complete, then the biorthogonality condition can at most be satisfied for one family $\{g_k\}_{k=1}^{\infty}$. $\qquad\square$

Depending on the frame at hand, it might happen that removal of a particular element destroys the frame property, or that the frame property is preserved. A frame that ceases to be a frame when an *arbitrary element* is removed is called an *exact frame*.

Lemma 5.5.2 shows that ω-independence and minimality are two different concepts. We now give some equivalent conditions for a frame to be a Riesz basis; in particular, we prove that for a frame, ω-independence is equivalent to minimality.

Theorem 5.5.4 *Let* $\{f_k\}_{k=1}^{\infty}$ *be a frame for* $\mathcal{H}$*. Then the following are equivalent.*

(i) $\{f_k\}_{k=1}^{\infty}$ *is a Riesz basis for* $\mathcal{H}$*.*

(ii) *If* $\sum_{k=1}^{\infty} c_k f_k = 0$ *for some* $\{c_k\}_{k=1}^{\infty} \in \ell^2(\mathbb{N})$*, then* $c_k = 0$*,* $\forall k \in \mathbb{N}$*.*

(iii) $\{f_k\}_{k=1}^{\infty}$ *is an exact frame.*

(iv) $\{f_k\}_{k=1}^{\infty}$ *and* $\{S^{-1}f_k\}_{k=1}^{\infty}$ *are biorthogonal.*

(v) $\{f_k\}_{k=1}^{\infty}$ *has a biorthogonal sequence.*

(vi) $\{f_k\}_{k=1}^{\infty}$ *is minimal.*

(vii) $\{f_k\}_{k=1}^{\infty}$ *is a basis.*

(viii) $\{f_k\}_{k=1}^{\infty}$ *is* ω-*independent.*

Proof.

Note that (i)$\Leftrightarrow$(ii) is proved in Theorem 5.2.2. To prove the rest of the equivalences, we proceed with the following steps:

(a) (i)$\Rightarrow$(iii);

(b) (iii)$\Leftrightarrow$(iv)$\Leftrightarrow$(v)$\Leftrightarrow$(vi);

(c) (iv)$\Rightarrow$(vii) and (vii)$\Rightarrow$(ii);

(d) (i)$\Rightarrow$(viii) and (viii)$\Rightarrow$(ii).

Step (a):

(i)$\Rightarrow$(iii). Let $\{f_k\}_{k=1}^{\infty}$ be a Riesz basis. If an arbitrary element is removed, the remaining family is not complete, and therefore not a frame; thus $\{f_k\}_{k=1}^{\infty}$ is an exact frame.

Step (b):

(iii)$\Rightarrow$(iv). Assume that $\{f_k\}_{k=1}^{\infty}$ is an exact frame and fix $j \in \mathbb{N}$. Then $\{f_k\}_{k \neq j}$ is not a frame, implying by Theorem 5.3.9 that

$$a_j := \langle f_j, S^{-1}f_j \rangle = 1.$$

In the proof of Theorem 5.3.9, the condition $a_j = 1$ was sufficient to derive (5.17), which shows that $\langle S^{-1}f_j, f_k \rangle = \delta_{j,k}$. We conclude that $\{f_k\}_{k=1}^{\infty}$ and $\{S^{-1}f_k\}_{k=1}^{\infty}$ are biorthogonal.

(iv) $\Rightarrow$ (v). Clear.

(v)$\Rightarrow$(vi). This is proved in Lemma 5.5.3 (i).

(vi)⇒(iii). Assume that $\{f_k\}_{k=1}^{\infty}$ is minimal. Then, for an arbitrary $j \in \mathbb{N}$, the family $\{f_k\}_{k \neq j}$ is incomplete in $\mathcal{H}$, and therefore not a frame for $\mathcal{H}$. **Step(c):** only the first of these implications needs an argument.

(iv)⇒(vii). Assume that $\{f_k\}_{k=1}^{\infty}$ and $\{S^{-1}f_k\}_{k=1}^{\infty}$ are biorthogonal. By the frame decomposition, we have that every $f \in \mathcal{H}$ can be expressed as $f = \sum_{k=1}^{\infty}\langle f, S^{-1}f_k\rangle f_k$. In order to show that $\{f_k\}_{k=1}^{\infty}$ is a basis, it is enough to show that this representation is unique. But if $f = \sum_{k=1}^{\infty} b_k f_k$ for some coefficients b_k, then

$$\langle f, S^{-1}f_k\rangle = \left\langle \sum_{j=1}^{\infty} b_j f_j, S^{-1}f_k \right\rangle = \sum_{j=1}^{\infty} b_j \langle f_j, S^{-1}f_k\rangle = b_k.$$

Step (d): only the implication (i)⇒(viii) needs an argument. But if $\{f_k\}_{k=1}^{\infty}$ is a Riesz basis, Exercise 3.8 shows that $\sum_{k=1}^{\infty} c_k f_k$ is convergent if and only if $\{c_k\}_{k=1}^{\infty} \in \ell^2(\mathbb{N})$; now the result follows from the equivalence between (i) and (ii). □

The next example is very illustrative: it exhibits a concrete frame, which is overcomplete despite the fact that the elements are linearly independent.

Example 5.5.5 Let us return to Example 4.1.1, where we considered the orthonormal basis $\{e_k\}_{k \in \mathbb{Z}} = \{e^{2\pi i k x}\}_{k \in \mathbb{Z}}$ for $L^2(0, 1)$. Let $I \subset [0, 1]$ be a proper subinterval, $|I| < 1$. We know that

$$\sum_{k \in \mathbb{Z}} |\langle f, e_k\rangle|^2 = ||f||^2, \ \forall f \in L^2(0, 1);$$

identifying $L^2(I)$ with a subspace of $L^2(0, 1)$, it follows (Exercise 5.25) that the restriction of the functions $\{e_k\}_{k \in \mathbb{Z}}$ to I form a tight frame for $L^2(I)$. We have already in Example 4.1.1 proved that it is overcomplete. However, recall from Exercise 1.21 that $\{e_k\}_{k \in \mathbb{Z}}$ is linearly independent. □

Example 5.5.5 points to a central property of frames: they can be overcomplete and linearly independent at the same time. The reason for this is the difference between linear independence (i.e., independence of all finite subsets) and ω-independence. For frames, the most suitable notion of independence is that of ω-independence.

Intuitively, we think about a frame as some kind of "overcomplete basis," so a natural question is the following: given a frame $\{f_k\}_{k=1}^{\infty}$, is it possible to extract a basis $\{f_k\}_{k \in J}$, $J \subseteq \mathbb{N}$, from $\{f_k\}_{k=1}^{\infty}$, i.e., does $\{f_k\}_{k=1}^{\infty}$ contain a basis as a subset? A part of the answer is given already in Example 5.1.5(ii), which exhibits a concrete frame for which no subset forms a Riesz basis (Exercise 5.4). The answer to the general question turns out to be surprising: there even exist frames for which no subset is a Schauder basis. The proof is much more involved that the above example – see [9] or [10].

We now return to Example 5.1.10; we show that the sequence considered there is minimal and we calculate its biorthogonal system.

Example 5.5.6 Let $\{e_k\}_{k=1}^\infty$ be an orthonormal basis for $\mathcal{H}$ and define

$$f_k := e_k + e_{k+1}, \ k \in \mathbb{N}.$$

We will show that $\{f_k\}_{k=1}^\infty$ is minimal and that its unique biorthogonal sequence $\{g_k\}_{k=1}^\infty$ is given by

$$g_k = (-1)^k \sum_{j=1}^{k} (-1)^j e_j, \ k \in \mathbb{N}.$$

To prove that $\{f_k\}_{k=1}^\infty$ is minimal, assume the opposite, i.e., that for some $j \in \mathbb{N}$,

$$\begin{aligned} f_j \ &\in \ \overline{\operatorname{span}}\{f_k\}_{k \neq j} \\ &= \ \overline{\operatorname{span}}\{e_1 + e_2, e_2 + e_3, \dots, e_{j-1} + e_j, e_{j+1} + e_{j+2}, \dots\}. \end{aligned} \quad (5.21)$$

We now consider the component of $f_j = e_j + e_{j+1}$ separately. From Example 5.1.10, we know that

$$\begin{aligned} e_{j+1} \ &\in \ \overline{\operatorname{span}}\{e_{j+k}\}_{k=1}^\infty \\ &\subseteq \ \overline{\operatorname{span}}\{e_1 + e_2, e_2 + e_3, \dots, e_{j-1} + e_j, e_{j+1} + e_{j+2}, \dots\}; \end{aligned}$$

thus, (5.21) implies that

$$e_j = f_j - e_{j+1} \in \overline{\operatorname{span}}\{e_1 + e_2, e_2 + e_3, \dots, e_{j-1} + e_j, e_{j+1} + e_{j+2}, \dots\}.$$

That would imply that we for any given $\epsilon > 0$ could find coefficients $\{c_k\}_{k=1}^{j-1}$ and $\{d_k\}_{k=1}^N$ for some $N \in \mathbb{N}$, such that

$$\left\| e_j - \left(\sum_{k=1}^{j-1} c_k(e_k + e_{k+1}) + \sum_{k=1}^{N} d_k(e_{j+k} + e_{j+k+1}) \right) \right\| \le \epsilon.$$

But

$$\left\| e_j - \left(\sum_{k=1}^{j-1} c_k(e_k + e_{k+1}) + \sum_{k=1}^{N} d_k(e_{j+k} + e_{j+k+1}) \right) \right\|$$

$$\ge \ \left\| e_j - \sum_{k=1}^{j-1} c_k(e_k + e_{k+1}) \right\|,$$

so then

$$e_j \in \operatorname{span}\{e_1 + e_2, e_2 + e_3, \dots, e_{j-1} + e_j\},$$

a conclusion that certainly does not hold. Thus $\{f_k\}_{k=1}^\infty$ is minimal. By Lemma 5.5.3, $\{f_k\}_{k=1}^\infty$ has a unique biorthogonal sequence $\{g_k\}_{k=1}^\infty$, which is determined by the conditions

$$\langle g_k, e_k + e_{k+1} \rangle = 1, \quad \langle g_k, e_j + e_{j+1} \rangle = 0 \text{ for } j \neq k. \quad (5.22)$$

In order to find $\{g_k\}_{k=1}^\infty$, fix $k \in \mathbb{N}$, and let $C := \langle g_k, e_k \rangle$. Then the first condition in (5.22) implies that $\langle g_k, e_{k+1} \rangle = 1 - C$; now the second condition implies that, in general for $j > k$, $|\langle g_k, e_j \rangle| = |1 - C|$. Because $\{e_j\}_{j=1}^\infty$ is a Bessel sequence, we know that

$$\sum_{j=k+1}^\infty |\langle g_k, e_j \rangle|^2 \leq \sum_{j=1}^\infty |\langle g_k, e_j \rangle|^2 < \infty,$$

so it follows that $C = 1$, i.e.,

$$\langle g_k, e_k \rangle = 1 \text{ and } \langle g_k, e_j \rangle = 0 \text{ for } j > k.$$

Now apply the second condition in (5.22) for $j = k-1, k-2, \ldots, 1$; this shows that

$$\langle g_k, e_j \rangle = (-1)^{k-j}, \ j = 1, \ldots, k.$$

We now put all the information together, and conclude that

$$g_k = \sum_{j=1}^\infty \langle g_k, e_j \rangle e_j = \sum_{j=1}^k (-1)^{k-j} e_j = (-1)^k \sum_{j=1}^k (-1)^j e_j. \qquad \square$$

5.6   Perturbation of frames

In applications where bases appear, the question of *stability* plays an important role. That is, if $\{f_k\}_{k=1}^\infty$ is a basis and $\{g_k\}_{k=1}^\infty$ in some sense is "close" to $\{f_k\}_{k=1}^\infty$, does it follow that $\{g_k\}_{k=1}^\infty$ also is a basis? A classical result states that if $\{f_k\}_{k=1}^\infty$ is a basis for a Banach space X, then a sequence $\{g_k\}_{k=1}^\infty$ in X is a basis if there exists a constant $\lambda \in]0, 1[$ such that

$$\left\| \sum c_k(f_k - g_k) \right\| \leq \lambda \left\| \sum c_k f_k \right\| \tag{5.23}$$

for all finite sequences of scalars $\{c_k\}$. The result is usually attributed to Paley and Wiener, but it can actually be traced back to Neumann: in fact, it is an almost immediate consequence of Theorem 2.2.3 with $U f_k := g_k$.

We will now discuss a natural extension of this result to the frame setting. That is, assuming that $\{f_k\}_{k=1}^\infty$ is a frame for a Hilbert space $\mathcal{H}$, we want to find conditions on a perturbed family $\{g_k\}_{k=1}^\infty$ that imply that it is a frame. As a convention, we denote the pre-frame operators for $\{f_k\}_{k=1}^\infty$ and $\{g_k\}_{k=1}^\infty$ by T and U respectively, i.e.,

$$T, U : \ell^2(\mathbb{N}) \to \mathcal{H}, \ T\{c_k\}_{k=1}^\infty = \sum_{k=1}^\infty c_k f_k, \ U\{c_k\}_{k=1}^\infty = \sum_{k=1}^\infty c_k g_k.$$

Note that T is well defined by assumption; the pre-frame operator U is at least well defined on finite sequences, but we have to *prove* that $\{g_k\}_{k=1}^{\infty}$ is a Bessel sequence before we know that U is well defined on $\ell^2(\mathbb{N})$. See Theorem 3.1.3.

We first note that the condition (5.23) with $\lambda < 1$ is too restrictive if $\{f_k\}_{k=1}^{\infty}$ is an overcomplete frame. In fact, if (5.23) holds for all finite sequences $\{c_k\}$ and some $\lambda \in]0,1[$, then for all such sequences it holds that

$$\sum c_k f_k = 0 \Leftrightarrow \sum c_k g_k = 0;$$

thus, the condition can only handle perturbations $\{g_k\}_{k=1}^{\infty}$ that have the "same linear dependence" as $\{f_k\}_{k=1}^{\infty}$. A much more flexible result can be obtained by adding an extra term in the perturbation condition as in the following Theorem 5.6.1. The original reference is [12].

Theorem 5.6.1 *Let $\{f_k\}_{k=1}^{\infty}$ be a frame for $\mathcal{H}$ with bounds A, B. Let $\{g_k\}_{k=1}^{\infty}$ be a sequence in $\mathcal{H}$ and assume that there exist constants $\lambda, \mu \geq 0$ such that $\lambda + \frac{\mu}{\sqrt{A}} < 1$ and*

$$\left\|\sum c_k(f_k - g_k)\right\| \leq \lambda \left\|\sum c_k f_k\right\| + \mu \left(\sum |c_k|^2\right)^{1/2} \qquad (5.24)$$

for all finite scalar sequences $\{c_k\}$. Then $\{g_k\}_{k=1}^{\infty}$ is a frame for $\mathcal{H}$ with bounds

$$A\left(1 - \left(\lambda + \frac{\mu}{\sqrt{A}}\right)\right)^2, \quad B\left(1 + \lambda + \frac{\mu}{\sqrt{B}}\right)^2.$$

Moreover, if $\{f_k\}_{k=1}^{\infty}$ is a Riesz basis, then $\{g_k\}_{k=1}^{\infty}$ is a Riesz basis.

Proof. $\{f_k\}_{k=1}^{\infty}$ is assumed to be a frame, so by Theorem 3.1.3, the pre-frame operator T is bounded and $\|T\| \leq \sqrt{B}$. The condition (5.24) implies that for all finite sequences $\{c_k\}$,

$$\begin{aligned}
\left\|\sum c_k g_k\right\| &= \left\|-\sum c_k(f_k - g_k) + \sum c_k f_k\right\| \\
&\leq \left\|-\sum c_k(f_k - g_k)\right\| + \left\|\sum c_k f_k\right\| \\
&\leq (1 + \lambda)\left\|\sum c_k f_k\right\| + \mu\left(\sum |c_k|^2\right)^{1/2}.
\end{aligned}$$

This calculation actually holds for all $\{c_k\}_{k=1}^{\infty} \in \ell^2(\mathbb{N})$. To see this, we first have to prove that $\sum_{k=1}^{\infty} c_k g_k$ is convergent for any given $\{c_k\}_{k=1}^{\infty} \in \ell^2(\mathbb{N})$. Given $n, m \in \mathbb{N}$ with $n > m$,

$$\begin{aligned}
\left\|\sum_{k=1}^{n} c_k g_k - \sum_{k=1}^{m} c_k g_k\right\| &= \left\|\sum_{k=m+1}^{n} c_k g_k\right\| \\
&\leq (1 + \lambda)\left\|\sum_{k=m+1}^{n} c_k f_k\right\| + \mu\left(\sum_{k=m+1}^{n} |c_k|^2\right)^{1/2};
\end{aligned}$$

since $\{c_k\}_{k=1}^{\infty} \in \ell^2(\mathbb{N})$ and $\sum_{k=1}^{\infty} c_k f_k$ is convergent, this implies that $\{\sum_{k=1}^{n} c_k g_k\}_{n=1}^{\infty}$ is a Cauchy sequence in $\mathcal{H}$ and therefore convergent. Thus the pre-frame operator U is well defined on $\ell^2(\mathbb{N})$; it follows that for all $\{c_k\}_{k=1}^{\infty} \in \ell^2(\mathbb{N})$,

$$\left\| \sum_{k=1}^{\infty} c_k g_k \right\| \leq (1+\lambda) \left\| \sum_{k=1}^{\infty} c_k f_k \right\| + \mu \left(\sum_{k=1}^{\infty} |c_k|^2 \right)^{1/2}. \quad (5.25)$$

In terms of the operators T, U, (5.25) states that

$$\begin{aligned}
\|U\{c_k\}_{k=1}^{\infty}\| &\leq (1+\lambda) \|T\{c_k\}_{k=1}^{\infty}\| + \mu \left(\sum_{k=1}^{\infty} |c_k|^2 \right)^{1/2} \\
&\leq \left((1+\lambda)\sqrt{B} + \mu \right) \left(\sum_{k=1}^{\infty} |c_k|^2 \right)^{1/2}, \quad \forall \{c_k\}_{k=1}^{\infty} \in \ell^2(\mathbb{N}).
\end{aligned}$$

Via Theorem 3.1.3, this estimate shows that $\{g_k\}_{k=1}^{\infty}$ is a Bessel sequence with bound

$$\left((1+\lambda)\sqrt{B} + \mu \right)^2 = B \left(1 + \lambda + \frac{\mu}{\sqrt{B}} \right)^2.$$

Now we prove that $\{g_k\}_{k=1}^{\infty}$ has a lower frame bound. Because $\{f_k\}_{k=1}^{\infty}$ is a frame, the frame operator $S = TT^*$ is invertible by Lemma 5.1.6, and we can define an operator (the pseudo-inverse of T, see Section 2.5) by

$$T^{\dagger} : \mathcal{H} \to \ell^2(\mathbb{N}), \quad T^{\dagger}f := T^*(TT^*)^{-1}f = \{\langle f, (TT^*)^{-1}f_k \rangle\}_{k=1}^{\infty}. \quad (5.26)$$

Note that $\{(TT^*)^{-1}f_k\}_{k=1}^{\infty}$ is the canonical dual frame of $\{f_k\}_{k=1}^{\infty}$, so by Lemma 5.1.6,

$$\begin{aligned}
\|T^{\dagger}f\|^2 &= \sum_{k=1}^{\infty} |\langle f, (TT^*)^{-1}f_k \rangle|^2 \\
&\leq \frac{1}{A} \|f\|^2, \quad \forall f \in \mathcal{H}.
\end{aligned}$$

Since $\sum_{k=1}^{\infty} c_k f_k$ and $\sum_{k=1}^{\infty} c_k g_k$ are convergent for all $\{c_k\}_{k=1}^{\infty} \in \ell^2(\mathbb{N})$, the inequality (5.24) holds for all $\{c_k\}_{k=1}^{\infty} \in \ell^2(\mathbb{N})$. In terms of the operators T and U,

$$\|T\{c_k\}_{k=1}^{\infty} - U\{c_k\}_{k=1}^{\infty}\| \leq \lambda \|T\{c_k\}_{k=1}^{\infty}\| + \mu \left(\sum_{k=1}^{\infty} |c_k|^2 \right)^{1/2}, \quad (5.27)$$

for all $\{c_k\}_{k=1}^{\infty} \in \ell^2(\mathbb{N})$. Note that for $f \in \mathcal{H}$,

$$TT^{\dagger}f = TT^*(TT^*)^{-1}f = f,$$

$$UT^{\dagger}f = \sum_{k=1}^{\infty}(T^{\dagger}f)_k g_k = \sum_{k=1}^{\infty}\langle f, (TT^*)^{-1}f_k\rangle g_k.$$

Using (5.27) on the sequence $\{c_k\}_{k=1}^{\infty} = T^{\dagger}f$ yields

$$\|f - UT^{\dagger}f\| \leq \lambda \|f\| + \mu \|T^{\dagger}f\|$$

$$\leq \left(\lambda + \frac{\mu}{\sqrt{A}}\right)\|f\|, \ \forall f \in \mathcal{H}.$$

Since we have assumed that $\lambda + \frac{\mu}{\sqrt{A}} < 1$, this implies that the operator $UT^{\dagger}$ is invertible, and that we have the estimate (Exercise 5.27)

$$\|UT^{\dagger}\| \leq 1 + \lambda + \frac{\mu}{\sqrt{A}}, \quad \|(UT^{\dagger})^{-1}\| \leq \frac{1}{1 - \left(\lambda + \frac{\mu}{\sqrt{A}}\right)}. \tag{5.28}$$

Now, any $f \in \mathcal{H}$ can be written as

$$f = UT^{\dagger}(UT^{\dagger})^{-1}f$$

$$= \sum_{k=1}^{\infty}\langle (UT^{\dagger})^{-1}f, (TT^*)^{-1}f_k\rangle g_k;$$

inserting this in the first entry of $\langle f, f\rangle$ leads to

$$\|f\|^4 = |\langle f, f\rangle|^2$$

$$= \left|\sum_{k=1}^{\infty}\langle (UT^{\dagger})^{-1}f, (TT^*)^{-1}f_k\rangle\langle g_k, f\rangle\right|^2$$

$$\leq \sum_{k=1}^{\infty}|\langle (UT^{\dagger})^{-1}f, (TT^*)^{-1}f_k\rangle|^2 \sum_{k=1}^{\infty}|\langle g_k, f\rangle|^2$$

$$\leq \frac{1}{A}\|(UT^{\dagger})^{-1}f\|^2 \sum_{k=1}^{\infty}|\langle g_k, f\rangle|^2$$

$$\leq \frac{1}{A}\left(\frac{1}{1 - \left(\lambda + \frac{\mu}{\sqrt{A}}\right)}\right)^2 \|f\|^2 \sum_{k=1}^{\infty}|\langle g_k, f\rangle|^2.$$

So

$$\sum_{k=1}^{\infty}|\langle g_k, f\rangle|^2 \geq A\left(1 - \left(\lambda + \frac{\mu}{\sqrt{A}}\right)\right)^2\|f\|^2, \ \forall f \in \mathcal{H},$$

i.e., $\{g_k\}_{k=1}^{\infty}$ is a frame for $\mathcal{H}$.

For the rest of the proof, we now assume that $\{f_k\}_{k=1}^{\infty}$ is a Riesz basis. To prove that $\{g_k\}_{k=1}^{\infty}$ is a Riesz basis, we use Theorem 5.2.2 and assume that $\sum_{k=1}^{\infty} c_k g_k = 0$ for some coefficients $\{c_k\}_{k=1}^{\infty} \in \ell^2(\mathbb{N})$. By Theorem 5.2.1 the lower frame bound for $\{f_k\}_{k=1}^{\infty}$ is also a lower Riesz basis bound, so (5.27) implies that

$$\left\| \sum_{k=1}^{\infty} c_k f_k \right\| \leq \lambda \left\| \sum_{k=1}^{\infty} c_k f_k \right\| + \mu \left(\sum_{k=1}^{\infty} |c_k|^2 \right)^{1/2}$$

$$\leq \left(\lambda + \frac{\mu}{\sqrt{A}} \right) \left\| \sum_{k=1}^{\infty} c_k f_k \right\|.$$

Because $\lambda + \frac{\mu}{\sqrt{A}} < 1$, it follows that $\sum_{k=1}^{\infty} c_k f_k = 0$. Using Theorem 5.2.2 on the Riesz basis $\{f_k\}_{k=1}^{\infty}$, we conclude that $c_k = 0$ for all $k \in \mathbb{N}$; therefore $\{g_k\}_{k=1}^{\infty}$ is a Riesz basis.   □

We already argued for the role of the μ-term in the condition (5.24). Most applications of Theorem 5.6.1 actually take place with $\lambda = 0$, so a natural question is whether the appearance of the λ-term improves the result. In fact, it does: in Exercise 5.29, we consider an example where the λ-term guarantees the frame property for a larger class of sequences than the corresponding result without the λ-term.

We now illustrate Theorem 5.6.1 by an example in a general Hilbert space. In particular, the example shows that the conclusion in Theorem 5.6.1 might fail if the condition $\lambda + \frac{\mu}{\sqrt{A}} < 1$ is replaced by $\lambda + \frac{\mu}{\sqrt{A}} = 1$. In that sense, Theorem 5.6.1 is the best possible perturbation result. The reader is asked to provide the details of the argument in Exercise 5.30.

Example 5.6.2 Let $\{e_k\}_{k=1}^{\infty}$ be an orthonormal basis for $\mathcal{H}$. Given a sequence $\{a_k\}_{k=1}^{\infty}$ of complex numbers, we consider the family of vectors $\{g_k\}_{k=1}^{\infty}$ defined by

$$g_k = e_k + a_k e_{k+1}, \ k \in \mathbb{N}.$$

If $a := \sup_k |a_k| < 1$, Theorem 5.6.1 shows that $\{g_k\}_{k=1}^{\infty}$ is a frame (in fact, a Riesz basis) with bounds $(1-a)^2, (1+a)^2$.

On the other hand, by taking $a_k = 1$ for all $k \in \mathbb{N}$, we obtain the family

$$g_k = e_k + e_{k+1}, \ k \in \mathbb{N},$$

which was considered in Example 5.1.10. In particular, we know that $\{g_k\}_{k=1}^{\infty}$ is not a frame. Letting $f_k = e_k$, one can check that the condition (5.24) is satisfied with $(\lambda, \mu) = (1, 0)$, or $(\lambda, \mu) = (0, 1)$; in either case, it shows that the condition $\lambda + \frac{\mu}{\sqrt{A}} = 1$ together with (5.24) in Theorem 5.6.1 does not imply that $\{g_k\}_{k=1}^{\infty}$ is a frame.   □

An important special case of Theorem 5.6.1 is given by

Corollary 5.6.3 *Let* $\{f_k\}_{k=1}^{\infty}$ *be a frame for* $\mathcal{H}$ *with bounds* $A, B,$ *and let* $\{g_k\}_{k=1}^{\infty}$ *be a sequence in* $\mathcal{H}$. *If there exists a constant* $R < A$ *such that*

$$\sum_{k=1}^{\infty} |\langle f, f_k - g_k \rangle|^2 \le R \, ||f||^2, \ \forall f \in \mathcal{H}, \tag{5.29}$$

then $\{g_k\}_{k=1}^{\infty}$ *is a frame for* $\mathcal{H}$ *with bounds*

$$A \left(1 - \sqrt{\frac{R}{A}} \right)^2, \ B \left(1 + \sqrt{\frac{R}{B}} \right)^2.$$

If $\{f_k\}_{k=1}^{\infty}$ *is a Riesz basis, then* $\{g_k\}_{k=1}^{\infty}$ *is a Riesz basis.*

Proof. The condition (5.29) corresponds to the condition in Theorem 5.6.1 with $\lambda = 0$, $\mu = \sqrt{R}$, just formulated in terms of the adjoint of the pre-frame operator instead of the pre-frame operator itself. However, an easier way to prove the frame part is to apply the triangle inequality in $\ell^2(\mathbb{N})$ to the sequence

$$\{\langle f, g_k \rangle\}_{k=1}^{\infty} = \{\langle f, f_k \rangle\}_{k=1}^{\infty} - \{\langle f, f_k - g_k \rangle\}_{k=1}^{\infty}. \qquad \square$$

5.7 The dual frames

We now aim at a characterization of all dual frames $\{g_k\}_{k=1}^{\infty}$ associated with a given frame $\{f_k\}_{k=1}^{\infty}$. The result is originally due to Li. Since $\{f_k\}_{k=1}^{\infty}$ and $\{g_k\}_{k=1}^{\infty}$ are assumed to be Bessel sequences, we can consider the associated pre-frame operators; we will (as usual) denote the pre-frame operator for $\{f_k\}_{k=1}^{\infty}$ by T, and the pre-frame operator for $\{g_k\}_{k=1}^{\infty}$ by U.

As we have seen in Section 5.2, two Bessel sequences $\{f_k\}_{k=1}^{\infty}$ and $\{g_k\}_{k=1}^{\infty}$ are dual frames if (5.14) holds, i.e., if

$$f = \sum_{k=1}^{\infty} \langle f, g_k \rangle f_k, \ \forall f \in \mathcal{H}. \tag{5.30}$$

In terms of the operators T and U, (5.30) means that

$$TU^* = I.$$

We begin with a lemma, which shows that the roles of $\{f_k\}_{k=1}^{\infty}$ and $\{g_k\}_{k=1}^{\infty}$ can be interchanged and that the lower frame condition automatically is satisfied for Bessel sequences $\{f_k\}_{k=1}^{\infty}, \{g_k\}_{k=1}^{\infty}$ if (5.30) holds.

Lemma 5.7.1 *Assume that* $\{f_k\}_{k=1}^{\infty}$ *and* $\{g_k\}_{k=1}^{\infty}$ *are Bessel sequences in* $\mathcal{H}$. *Then the following are equivalent:*

(i) $f = \sum_{k=1}^{\infty}\langle f, g_k\rangle f_k, \ \forall f \in \mathcal{H}$.

(ii) $f = \sum_{k=1}^{\infty}\langle f, f_k\rangle g_k, \ \forall f \in \mathcal{H}$.

(iii) $\langle f, g\rangle = \sum_{k=1}^{\infty}\langle f, f_k\rangle\langle g_k, g\rangle, \ \forall f, g \in \mathcal{H}$.

In case the equivalent conditions are satisfied, then $\{f_k\}_{k=1}^{\infty}$ *and* $\{g_k\}_{k=1}^{\infty}$ *are dual frames for* $\mathcal{H}$; *and if* B *denotes an upper frame bound for* $\{f_k\}_{k=1}^{\infty}$, *then* B^{-1} *is a lower frame bound for* $\{g_k\}_{k=1}^{\infty}$.

Proof. In terms of the pre-frame operators, (i) means that $TU^* = I$; this is equivalent to

$$UT^* = I, \tag{5.31}$$

which is identical to the statement in (ii). It is also clear that (ii) implies (iii). To prove that (iii) implies (ii), we fix $f \in \mathcal{H}$ and note that $\sum_{k=1}^{\infty}\langle f, f_k\rangle g_k$ is well defined as an element in $\mathcal{H}$ because $\{f_k\}_{k=1}^{\infty}$ and $\{g_k\}_{k=1}^{\infty}$ are Bessel sequences. Now the assumption in (iii) shows that

$$\left\langle f - \sum_{k=1}^{\infty}\langle f, f_k\rangle g_k, g \right\rangle = 0, \ \forall g \in \mathcal{H},$$

and (ii) follows.

In case the equivalent conditions are satisfied, we can write

$$\|f\|^2 = \langle f, f\rangle = \sum_{k=1}^{\infty}\langle f, g_k\rangle\langle f_k, f\rangle, \ \forall f \in \mathcal{H}.$$

Using Cauchy–Schwarz' inequality and that *one* of the families $\{f_k\}_{k=1}^{\infty}$, $\{g_k\}_{k=1}^{\infty}$ is a Bessel sequence, we obtain that the *other* family satisfies the lower frame condition, with the relationship between the frame bounds as stated in the lemma. $\qquad\square$

When (5.31) is satisfied, we say that U is a *left-inverse* of T^*.

Lemma 5.7.2 *Let* $\{f_k\}_{k=1}^{\infty}$ *be a frame for* $\mathcal{H}$ *and* $\{\delta_k\}_{k=1}^{\infty}$ *be the canonical orthonormal basis for* $\ell^2(\mathbb{N})$. *The dual frames for* $\{f_k\}_{k=1}^{\infty}$ *are precisely the families* $\{g_k\}_{k=1}^{\infty} = \{V\delta_k\}_{k=1}^{\infty}$, *where* $V : \ell^2(\mathbb{N}) \to \mathcal{H}$ *is a bounded left-inverse of* T^*.

Proof. If V is a bounded left-inverse of T^*, then V is surjective; by Theorem 5.4.1 it follows that $\{g_k\}_{k=1}^{\infty} := \{V\delta_k\}_{k=1}^{\infty}$ is a frame. Note that in terms of $\{\delta_k\}_{k=1}^{\infty}$,

$$T^*f = \{\langle f, f_k\rangle\}_{k=1}^{\infty} = \sum_{k=1}^{\infty}\langle f, f_k\rangle\delta_k;$$

thus, for all $f \in \mathcal{H}$,

$$f = VT^*f = \sum_{k=1}^{\infty} \langle f, f_k \rangle g_k,$$

i.e., $\{g_k\}_{k=1}^{\infty}$ is a dual frame of $\{f_k\}_{k=1}^{\infty}$. For the other implication, assume that $\{g_k\}_{k=1}^{\infty}$ is a dual frame of $\{f_k\}_{k=1}^{\infty}$. Then the pre-frame operator U for $\{g_k\}_{k=1}^{\infty}$ satisfies the conditions: in fact, $\{g_k\}_{k=1}^{\infty} = \{U\delta_k\}_{k=1}^{\infty}$, and by Lemma 5.7.1, $UT^* = I$. $\qquad\square$

Lemma 5.7.3 *Let* $\{f_k\}_{k=1}^{\infty}$ *be a frame with pre-frame operator* T. *The bounded left-inverses of* T^* *are precisely the operators having the form* $S^{-1}T + W(I - T^*S^{-1}T)$, *where* $W : \ell^2(\mathbb{N}) \to \mathcal{H}$ *is a bounded operator, and* I *denotes the identity operator on* $\ell^2(\mathbb{N})$.

Proof. Straightforward calculation gives that an operator of the given form is a left-inverse of T^*. On the other hand, if U is a given left-inverse of T^*, then by taking $W = U$,

$$S^{-1}T + W(I - T^*S^{-1}T) = S^{-1}T + U - UT^*S^{-1}T = U. \qquad\square$$

We are now ready for the announced characterization of all dual frames associated with a given frame.

Theorem 5.7.4 *Let* $\{f_k\}_{k=1}^{\infty}$ *be a frame for* $\mathcal{H}$. *The dual frames of* $\{f_k\}_{k=1}^{\infty}$ *are precisely the families*

$$\{g_k\}_{k=1}^{\infty} = \left\{ S^{-1}f_k + h_k - \sum_{j=1}^{\infty} \langle S^{-1}f_k, f_j \rangle h_j \right\}_{k=1}^{\infty}, \qquad (5.32)$$

where $\{h_k\}_{k=1}^{\infty}$ *is a Bessel sequence in* $\mathcal{H}$.

Proof. By Lemma 5.7.2 and Lemma 5.7.3, we can characterize the dual frames as all families of the form

$$\{g_k\}_{k=1}^{\infty} = \{S^{-1}T\delta_k + W(I - T^*S^{-1}T)\delta_k\}_{k=1}^{\infty}, \qquad (5.33)$$

where $W : \ell^2(\mathbb{N}) \to \mathcal{H}$ is a bounded operator, or, equivalently, an operator of the form $W\{c_j\}_{j=1}^{\infty} = \sum_{j=1}^{\infty} c_j h_j$ where $\{h_k\}_{k=1}^{\infty}$ is a Bessel sequence. Inserting this expression for W in (5.33), we get

$$\begin{aligned} \{g_k\}_{k=1}^{\infty} &= \{S^{-1}f_k + W\delta_k - WT^*S^{-1}T\delta_k\}_{k=1}^{\infty} \\ &= \left\{ S^{-1}f_k + h_k - \sum_{j=1}^{\infty} \langle S^{-1}f_k, f_j \rangle h_j \right\}_{k=1}^{\infty}. \end{aligned}$$

$\qquad\square$

Note that if $\{f_k\}_{k=1}^{\infty}$ is a Riesz basis, then $\{f_k\}_{k=1}^{\infty}$ and $\{S^{-1}f_k\}_{k=1}^{\infty}$ are biorthogonal by Theorem 5.2.1. Thus, independently of the choice of $\{h_k\}_{k=1}^{\infty}$, the element g_k in (5.32) is given by

$$g_k = S^{-1}f_k + h_k - \sum_{j=1}^{\infty} \langle S^{-1}f_k, f_j \rangle h_j = S^{-1}f_k + h_k - h_k = S^{-1}f_k.$$

This shows that a Riesz basis $\{f_k\}_{k=1}^{\infty}$ has a unique dual frame, in accordance with Theorem 3.3.2.

5.8   Continuous frames

The frames discussed so far all lead to expansions of elements in Hilbert spaces in terms of infinite sums. One can consider these frames as manifestations of a broader theory, which in general leads to integral representations in Hilbert spaces. The following generalization of frames was proposed by Kaiser and independently by Ali, Antoine, and Gazeau. In the current book, it will only play a role in Section 9.9 and Section 11.8.

Definition 5.8.1 *Let $\mathcal{H}$ be a complex Hilbert space and M a measure space with a positive measure μ. A continuous frame is a family of vectors $\{f_k\}_{k \in M}$ for which the following hold:*

(i) For all $f \in \mathcal{H}$, the mapping $k \mapsto \langle f, f_k \rangle$ is a measurable function on M;

(ii) There exist constants $A, B > 0$ such that

$$A \, ||f||^2 \leq \int_M |\langle f, f_k \rangle|^2 d\mu(k) \leq B \, ||f||^2, \ \forall f \in \mathcal{H}.$$

Note that Kaiser used the terminology *generalized frames*. Also, because $\{f_k\}_{k \in M}$ being a continuous frame or not depends on the measure space, it would be more exact to speak about a continuous frame for $\mathcal{H}$ *with respect to the measure space* (M, μ).

In order to distinguish them from the continuous frames, the frames $\{f_k\}_{k=1}^{\infty}$ considered so far are frequently called *discrete frames*. The discrete frames $\{f_k\}_{k=1}^{\infty}$ are actually a special case of the continuous frames, corresponding to the case where $M = \mathbb{N}$, equipped with the counting measure. An important feature of continuous frames is that the theory for discrete frames and some results on the continuous Gabor transformation and the wavelet transform can be considered as different manifestations of a single theory. We come back to this in Section 9.9 and Section 11.8.

Let us derive the basic results for a continuous frame $\{f_k\}_{k \in M}$. First, Cauchy–Schwarz' inequality shows that the integral $\int_M \langle f, f_k \rangle \langle f_k, g \rangle \, d\mu(k)$

is well defined for all $f, g \in \mathcal{H}$. For a fixed $f \in \mathcal{H}$, the mapping

$$g \mapsto \int_M \langle f, f_k \rangle \langle f_k, g \rangle \, d\mu(k)$$

is conjugated linear, and bounded because

$$\left| \int_M \langle f, f_k \rangle \langle f_k, g \rangle \, d\mu(k) \right|^2 \leq \int_M |\langle f, f_k \rangle|^2 d\mu(k) \int_M |\langle f_k, g \rangle|^2 d\mu(k)$$
$$\leq B^2 \, \|f\|^2 \, \|g\|^2. \tag{5.34}$$

By Theorem 2.3.2, there exists a unique element in $\mathcal{H}$ – we call it $\int_M \langle f, f_k \rangle f_k \, d\mu(k)$ – such that

$$\left\langle \int_M \langle f, f_k \rangle f_k \, d\mu(k), g \right\rangle = \int_M \langle f, f_k \rangle \langle f_k, g \rangle \, d\mu(k)$$

for all $g \in \mathcal{H}$. By this procedure, we have defined a mapping

$$S : \mathcal{H} \to \mathcal{H}, \quad Sf = \int_M \langle f, f_k \rangle f_k \, d\mu(k).$$

It is easy to check that S is a linear operator. Using that

$$\|Sf\| = \sup_{\|g\|=1} |\langle Sf, g \rangle|,$$

it follows from (5.34) that S is bounded and that $\|S\| \leq B$. By definition, S is positive, and

$$A \, \|f\|^2 \leq \langle Sf, f \rangle \leq B \, \|f\|^2, \ \forall f \in \mathcal{H}.$$

Exactly as in the proof of Lemma 5.1.6, one can now prove that S is invertible. Thus, every $f \in \mathcal{H}$ has the representations

$$f = S^{-1} S f = \int_M \langle f, f_k \rangle S^{-1} f_k \, d\mu(k),$$
$$f = S S^{-1} f = \int_M \langle f, S^{-1} f_k \rangle f_k \, d\mu(k).$$

Remember that these representations have to be interpreted in the weak sense. Sometimes stronger results can be obtained in concrete cases.

5.9   Frames and signal processing

The frame theory described so far takes place in an ideal world, which can hardly be realized in, e.g., signal processing. In this section, we describe some of the steps that have to be taken in order to apply the abstract results in practice. Much more can of course be said about this important subject, and we refer to the books [53] by Mallat and [63] by Vetterli and Kovačević for more detailed information.

Some of the problems appear before one even thinks about frames. In fact, even the most basic ingredient in mathematics, the real numbers, are disturbed when we move away from the abstract level: every number has to be replaced by a number with finitely many digits before any processing can take place. In practice, this means that we represent all numbers in an interval (for example, $[1, 1 + 10^{-18}[$) by the same number (in this case the number 1). The consequence is an inaccuracy, which is called the *quantization error.*

The basic limitation in applications of the frame results is that any type of signal processing has to be performed on finite sequences of numbers. For example, this implies that the frame representation (5.1.7) has to be *truncated:* we can only aim at calculating a finite number of frame coefficients, say, $\{\langle f, S^{-1} f_k \rangle\}_{k=1}^N$, and the exact representation in (5.1.7) has to be replaced by

$$f \sim \sum_{k=1}^{N} \langle f, S^{-1} f_k \rangle f_k.$$

Even calculation of the frame coefficients $\langle f, S^{-1} f_k \rangle$ can in general only be done with finite precision. That is, the outcome of a calculation will be

$$\langle f, S^{-1} f_k \rangle + w_k \qquad (5.35)$$

for some (hopefully small) error term w_k. All types of transmission or further processing will introduce extra inaccuracies. One says that the frame coefficients $\langle f, S^{-1} f_k \rangle$ have been *contaminated by the noise* w_k.

Already on page 28, we gave a rather intuitive argument that overcompleteness of frames might reduce the influence of noise, compared with use of an orthonormal basis. To support this further, we now discuss a result that is borrowed from [53].

Let us again use the example of signal transmission, as on page 28. That is, we assume that one wants to transmit the signal f from $\mathcal{A}$ to $\mathcal{R}$ by sending the frame coefficients $\{\langle f, S^{-1} f_k \rangle\}_{k=1}^\infty$. Because of quantization, the coefficients will be contaminated by some noise $\{w_k\}_{k=1}^\infty$, and $\mathcal{R}$ will receive the coefficients $\{\langle f, S^{-1} f_k \rangle + w_k\}_{k=1}^\infty$; we assume that $\{w_k\}_{k=1}^\infty \in \ell^2(\mathbb{N})$. The receiver $\mathcal{R}$ will believe that the transmitted function was

$$\sum_{k=1}^{\infty} \left(\langle f, S^{-1} f_k \rangle + w_k \right) f_k = f + \sum_{k=1}^{\infty} w_k f_k$$

rather than f.

Note that $\mathcal{R}$ actually knows that the transmitted sequence was supposed to be a sequence of frame coefficients, i.e., a sequence of the form $\{\langle g, f_k \rangle\}_{k=1}^\infty$ for some $g \in \mathcal{H}$ (namely, $g = S^{-1} f$); that is, the sequence belongs to the range of the operator T^*. This might not be the case for the perturbed coefficients $\{\langle f, S^{-1} f_k \rangle + w_k\}_{k=1}^\infty$, so it is natural to compensate for this by projecting that sequence onto the range of the operator T^*. Denoting the projection operator by Q, Exercise 5.17 shows that the outcome is

$$Q\{\langle f, S^{-1}f_k\rangle + w_k\}_{k=1}^{\infty} = \{\langle f, S^{-1}f_k\rangle\}_{k=1}^{\infty} + Q\{w_k\}_{k=1}^{\infty}. \qquad (5.36)$$

Let $w = \{w_k\}_{k=1}^{\infty}$. Based on (5.36), $\mathcal{R}$ will reconstruct the transmitted signal as

$$\sum_{k=1}^{\infty} \left(\langle f, S^{-1}f_k\rangle + (Qw)_k\right) f_k = f + \sum_{k=1}^{\infty} (Qw)_k f_k.$$

We will assume that the quantization error is *white noise*. This means that the components w_k are random variables with zero mean, variance σ^2 independent of k, and that

$$E[w_j w_\ell] = \sigma^2 \delta_{j,\ell}. \qquad (5.37)$$

We now prove that increased redundancy of the frame, measured by a larger lower frame bound, will decrease the energy of the coefficients in the "projected noise" Qw, i.e., the mean of the random variables $|(Qw)_k|^2$. We return to this result in a concrete setting in Section 7.5.

Proposition 5.9.1 *Suppose that the frame* $\{f_k\}_{k=1}^{\infty}$ *has lower frame bound* A *and consists of normalized vectors. If* w *is white noise, then for each* $k \in \mathbb{N}$,

$$E|(Qw)_k|^2 \le \frac{\sigma^2}{A},$$

with equality if $\{f_k\}_{k=1}^{\infty}$ *is a tight frame.*

Proof. According to Exercise 5.17, the k-th component of Qw is given by

$$(Qw)_k = \sum_{j=1}^{\infty} w_j \langle S^{-1}f_j, f_k\rangle.$$

Via (5.37), this implies that

$$
\begin{aligned}
E|(Qw)_k|^2 &= E\left[(Qw)_k \overline{(Qw)_k}\right] \\
&= E\left[\sum_{j=1}^{\infty} w_j\langle S^{-1}f_j, f_k\rangle \overline{\sum_{\ell=1}^{\infty} w_\ell\langle S^{-1}f_\ell, f_k\rangle}\right] \\
&= \sum_{j=1}^{\infty}\sum_{\ell=1}^{\infty} E[w_j\overline{w_\ell}]\langle S^{-1}f_j, f_k\rangle\overline{\langle S^{-1}f_\ell, f_k\rangle} \\
&= \sum_{\ell=1}^{\infty} E|w_\ell|^2|\langle S^{-1}f_\ell, f_k\rangle|^2 \\
&= \sigma^2 \sum_{\ell=1}^{\infty} |\langle S^{-1}f_\ell, f_k\rangle|^2.
\end{aligned}
$$

Using that $\{S^{-1}f_\ell\}_{\ell=1}^\infty$ is a frame with upper frame bound A^{-1}, we finally arrive at

$$E|(Qw)_k|^2 \le \frac{\sigma^2}{A}||f_k||^2 = \frac{\sigma^2}{A};$$

the inequality is an equality if $\{S^{-1}f_k\}_{k=1}^\infty$ is a tight frame. By Lemma 5.1.6, the frame operator for $\{S^{-1}f_k\}_{k=1}^\infty$ is S^{-1}; thus, $\{S^{-1}f_k\}_{k=1}^\infty$ being a tight frame is equivalent with S^{-1} being a multiple of the identity. But this is equivalent with S being a multiple of the identity, i.e., with $\{f_k\}_{k=1}^\infty$ being a tight frame. □

Quantization errors and noise during transmission are just some of the obstacles for frames in real life. Depending on the underlying Hilbert space $\mathcal{H}$, there might be additional complications. In many cases $\mathcal{H}$ will be a function space like $L^2(\mathbb{R})$, and even finite-dimensional subspaces hereof cannot be processed directly: a discretization step is needed in order to transfer the setting to a finite-dimensional sequence space like $\mathbb{C}^n$. This is exactly the point where the importance of the Gram matrix becomes clear: whereas the frame operator $S = TT^*$ maps $\mathcal{H}$ onto itself, the Gram matrix T^*T is an operator on the sequence space $\ell^2(\mathbb{N})$, i.e., the only step that is needed is a truncation. For this reason, it is an advantage to formulate algorithms involving frames in terms of the Gram matrix rather than the frame operator, if possible.

5.10 Exercises

5.1 Find an example of a sequence in a Hilbert space that is a basis but not a frame.

5.2 Prove that the upper and lower frame conditions are unrelated: in an arbitrary Hilbert space $\mathcal{H}$, there exists a sequence $\{f_k\}_{k=1}^\infty$ satisfying the upper condition for all $f \in \mathcal{H}$, but not the lower condition; and vice versa.

5.3 Prove Lemma 5.1.2 (use Lemma 3.1.6).

5.4 Find a frame that contains a Schauder basis as a subset, but not a Riesz basis. (Hint: use Example 5.1.5.)

5.5 Let $\{e_k\}_{k=1}^\infty$ be an orthonormal basis and consider the family $\{f_k\}_{k=1}^\infty := \{e_1 + \frac{1}{k}e_k, e_k\}_{k=2}^\infty$.

(i) Prove that $\{f_k\}_{k=1}^\infty$ is not a Bessel sequence.

(ii) Find all possible representations of e_1 as infinite linear combinations of $\{f_k\}_{k=1}^\infty$.

(iii) Prove that there exists no set of coefficients having minimal ℓ^1-norm among all sequences representing e_1.

5.6 Give an example of a frame $\{f_k\}_{k=1}^\infty$, for which $\sum_{k=1}^\infty c_k f_k$ converges for some $\{c_k\}_{k=1}^\infty \notin \ell^2(\mathbb{N})$ (compare with Exercise 3.8!).

5.7 Let $\{f_k\}_{k=1}^\infty$ be a frame. Show that the following are equivalent:

(i) $\{f_k\}_{k=1}^\infty$ is tight.

(ii) $\{f_k\}_{k=1}^\infty$ has a dual of the form $g_k = Cf_k$ for some constant $C > 0$.

5.8 Verify the statements in Example 5.1.9.

5.9 Let $\{f_k\}_{k=1}^\infty$ be a frame sequence in $\mathcal{H}$, with pre-frame operator $T : \ell^2(\mathbb{N}) \to \mathcal{H}$. Prove that $\{f_k\}_{k=1}^\infty$ is a frame for $\mathcal{H}$ if and only if T^* is injective.

5.10 Show that the family $\{e_k + e_{k+1}\}_{k=1}^\infty$ in Example 5.1.10 and Example 5.5.6 cannot be extended to a basis for $\mathcal{H}$.

5.11 Let $\tilde{\mathcal{H}}$ be the complexification of a real Hilbert space $\mathcal{H}$. Prove that a frame for $\mathcal{H}$ also is a frame for $\tilde{\mathcal{H}}$.

5.12 Let $\{f_k\}_{k=1}^\infty$ be a Riesz basis with bounds A, B. Prove that

$$A \le ||f_k||^2 \le B \text{ for all } k \in \mathbb{N},$$

and that the elements in the dual Riesz basis $\{g_k\}_{k=1}^\infty$ satisfy

$$\frac{1}{B} \le ||g_k||^2 \le \frac{1}{A} \text{ for all } k \in \mathbb{N}.$$

5.13 Prove that the conclusion in Proposition 5.3.1 might fail if U is not assumed to have closed range. (Hint: let $\{e_k\}_{k=1}^\infty$ be an orthonormal basis and define U by $Ue_k = e_k + e_{k+1}$.)

5.14 Let $\{e_k\}_{k=1}^\infty$ be an orthonormal basis for $\mathcal{H}$, and define an operator U on $\mathcal{H}$ by

$$Ue_{2k} = e_{2k}, \quad Ue_{2k-1} = \frac{1}{k}e_{2k}, \ k \in \mathbb{N}.$$

Prove that

(i) U is a well-defined bounded operator on $\mathcal{H}$, and $\mathcal{R}_U$ is closed.

(ii) $\{e_{2k-1}\}_{k=1}^{\infty}$ is a frame sequence, but $\{Ue_{2k-1}\}_{k=1}^{\infty}$ is not.

Thus, Proposition 5.3.1 does not extend to frame sequences.

5.15 Prove Lemma 5.3.3.

5.16 Prove Proposition 5.3.5.

5.17 Let $\{f_k\}_{k=1}^{\infty}$ be a frame sequence in $\mathcal{H}$. Show that the orthogonal projection Q of a sequence $\{c_k\}_{k=1}^{\infty} \in \ell^2(\mathbb{N})$ onto the range of T^* is given by

$$Q\{c_k\}_{k=1}^{\infty} = \left\{ \left\langle \sum_{j=1}^{\infty} c_j S^{-1} f_j, f_k \right\rangle \right\}_{k=1}^{\infty}. \tag{5.38}$$

5.18 Let $\{f_k\}_{k \in \mathbb{Z}}$ be a Riesz basis. Our purpose is to show that

$$\{f_k + f_{k+1}\}_{k \in \mathbb{Z}}$$

cannot be a frame. Let $\{g_k\}_{k \in \mathbb{Z}}$ be the biorthogonal basis associated with $\{f_k\}_{k \in \mathbb{Z}}$, and let

$$h_j = \sum_{k=1}^{j} (-1)^k g_k, \ j \in \mathbb{Z}.$$

(i) Prove that $\sum_{k \in \mathbb{Z}} |\langle h_j, f_k + f_{k+1} \rangle|^2 = 2$.

(ii) Prove that $||h_j||^2 \geq j/B$, where B is an upper frame bound for $\{f_k\}_{k \in \mathbb{Z}}$.

(iii) Conclude that $\{f_k + f_{k+1}\}_{k \in \mathbb{Z}}$ is not a frame.

5.19 Prove via Theorem 5.4.1 that $\{Ue_k\}_{k=1}^{\infty}$ is a frame whenever $\{e_k\}_{k=1}^{\infty}$ is an orthonormal basis and U is a bounded surjective operator.

5.20 Let $\{f_k\}_{k=1}^{\infty} = \{Ue_k\}_{k=1}^{\infty}$ be a Riesz basis for $\mathcal{H}$ as in Definition 3.3.1. Prove that the frame operator for $\{f_k\}_{k=1}^{\infty}$ is given by

$$S = UU^*.$$

5.21 Let $\{f_k\}_{k=1}^{\infty}$ be a frame with frame bounds A, B. Show that the frame operator S satisfies the inequalities

$$A\,||f|| \leq ||Sf|| \leq B\,||f||, \ \forall f \in \mathcal{H}.$$

5.22 Let $\{e_k\}_{k=1}^{\infty}$ be an orthonormal basis for a Hilbert space $\mathcal{H}$.

    (i) Prove that $\{\sum_{\ell=1}^{\infty} \frac{1}{\ell} e_\ell\} \cup \{e_k\}_{k=1}^{\infty}$ is linearly independent, but not ω-independent.

    (ii) Prove that $\{e_1\} \cup \{e_k + e_{k+1}\}_{k=1}^{\infty}$ is ω-independent, but not minimal. (Hint: In Example 5.1.10, we proved that $\{e_k + e_{k+1}\}_{k=1}^{\infty}$ is complete.)

5.23 Prove that a basis in a Hilbert space is minimal.

5.24 Definition 5.5.1 applies word by word with the Hilbert space $\mathcal{H}$ replaced by a Banach space X. Extend Lemma 5.5.2 to that setting.

5.25 Prove that the sequence $\{e_k\}_{k\in\mathbb{Z}}$ in Example 5.5.5 is a tight frame for $L^2(I)$.

5.26 Let U be a bounded operator between Hilbert spaces. Prove that if at least one of the spaces $\mathcal{R}_U$ and $\mathcal{R}_{UU^*}$ is closed, then

$$\mathcal{R}_U = \mathcal{R}_{UU^*}.$$

5.27 Prove the estimates (5.28).

5.28 Let $\{f_k\}_{k=1}^{\infty}$ be a frame for $\mathcal{H}$ with frame bounds A, B, and let $\{g_k\}_{k=1}^{\infty}$ be a sequence in $\mathcal{H}$. Prove that if

$$\sum_{k=1}^{\infty} ||f_k - g_k||^2 < A,$$

then $\{g_k\}_{k=1}^{\infty}$ is a frame for $\mathcal{H}$.

5.29 Let $\{e_1, e_2\}$ be an orthonormal basis for $\mathbb{C}^2$, and consider the frame $\{f_1, f_2\}$ given by

$$f_1 = e_1, f_2 = 2e_2.$$

Given a number $c \in \mathbb{C}$, let

$$g_1 = e_1, g_2 = ce_2.$$

Based on the frame $\{f_1, f_2\}$ we want to find the range of parameter c for which $\{g_1, g_2\}$ is also a frame.

    (i) Apply Theorem 5.6.1 with $\lambda = 0$ – for which $c \in \mathbb{C}$ does the result guarantee that $\{g_1, g_2\}$ is a frame?

    (ii) Apply Theorem 5.6.1 with $\mu = 0$ – for which $c \in \mathbb{C}$ does the result guarantee that $\{g_1, g_2\}$ is a frame?

    (iii) What is the exact range of $c \in \mathbb{C}$ for which $\{g_1, g_2\}$ is a frame?

5.30 Provide the details in Example 5.6.2.

5.31 Extend Corollary 5.6.3 to Riesz sequences.

5.32 Let $\{f_k\}_{k=1}^\infty$ and $\{g_k\}_{k=1}^\infty$ be dual frames for a Hilbert space $\mathcal{H}$, and $U : \mathcal{H} \to \mathcal{H}$ a unitary operator. Show that $\{Uf_k\}_{k=1}^\infty$ and $\{Ug_k\}_{k=1}^\infty$ also form a pair of dual frames for $\mathcal{H}$.

5.33 Find a tight frame $\{f_k\}_{k=1}^\infty$ for which dual frames $\{g_k\}_{k=1}^\infty$ with arbitrary large optimal frame Bessel bound exist. (This shows that for non-canonical dual frames, no expression for the upper frame bound in terms of the frame bounds for $\{f_k\}_{k=1}^\infty$ exists.)

5.34 This exercise concerns the question of finding *generalized duals* that are not frames.

(i) Find an overcomplete frame $\{f_k\}_{k=1}^\infty$, for which a family $\{g_k\}_{k=1}^\infty$ satisfying (5.14) automatically is a frame.

(ii) Find an overcomplete frame $\{f_k\}_{k=1}^\infty$ and a non-Bessel sequence $\{g_k\}_{k=1}^\infty$ such that (5.14) is satisfied.

6
B-splines

In Chapters 1–5, we described frames in abstract Hilbert spaces. The purpose of the rest of the book is to provide concrete frame constructions in $L^2(\mathbb{R})$ and subspaces hereof. The intention is that the constructions should be convenient to apply in practice, so it is important that the frames are generated by functions that are easy to deal with. For this reason, several of the frames will be based on B-splines; and this is the reason for including an introductory chapter on these.

In short, *splines* are functions that are piecewise polynomials; in the one-dimensional case, this means that one can split the domain of a spline into intervals in such a way that the function is a polynomial on each interval. The points where the function changes from one polynomial to another polynomial are called *knots*. In principle, no assumptions on the knots need to be made, but it is convenient to restrict attention to splines which are continuous – or even differentiable – at the knots.

We will not discuss general splines, but only some special splines, called *B-splines*. In Section 6.1, we introduce B-splines supported on compact subintervals of the positive real axis. We derive explicit expressions for these functions and their Fourier transforms. Furthermore, we prove that the integer-translates of any B-spline satisfies a so-called partition of unity property, which will play a crucial role in the constructions of Gabor frames and their duals in Chapter 9. The symmetric counterparts of the B-splines are treated in Section 6.2.

O. Christensen, *Frames and Bases*. DOI: 10.1007/978-0-8176-4678-3_6,
© Springer Science+Business Media, LLC 2008

6.1   The B-splines

The B-splines are defined inductively: the first is simply

$$N_1(x) = \chi_{[0,1]}(x), \tag{6.1}$$

and, assuming that we have defined N_n for some $n \in \mathbb{N}$, the next is defined by a convolution:

$$N_{n+1}(x) = N_n * N_1(x) \;=\; \int_{-\infty}^{\infty} N_n(x-t)N_1(t)\, dt$$

$$=\; \int_0^1 N_n(x-t)\, dt. \tag{6.2}$$

The functions N_n defined by (6.1) and (6.2) are called *B-splines*, and n is the *order*. See Figures 6.1 and 6.2 for graphs of the first few *B*-splines. We collect some of their fundamental properties:

Theorem 6.1.1 *Given $n \in \mathbb{N}$, the B-spline N_n has the following properties:*

(i) *supp $N_n = [0,n]$ and $N_n > 0$ on $]0,n[$.*

(ii) *$\int_{-\infty}^{\infty} N_n(x)\, dx = 1$.*

(iii) *For $n \geq 2$,*

$$\sum_{k \in \mathbb{Z}} N_n(x-k) = 1 \text{ for all } x \in \mathbb{R}; \tag{6.3}$$

for $n = 1$, the formula (6.3) holds for a.e. $x \in \mathbb{R}$.

(iv) *For any continuous function $f : \mathbb{R} \to \mathbb{C}$,*

$$\int_{-\infty}^{\infty} N_n(x)f(x)\, dx = \int_{[0,1]^n} f(x_1 + \cdots + x_n)\, dx_1 \cdots dx_n. \tag{6.4}$$

Proof.   All the statements can be proved by induction based on (6.2); we prove (i) and leave the rest to the reader (Exercise 6.1). It is clear that (i) holds for $n = 1$. Assuming that (i) holds for some $n \in \mathbb{N}$, we now consider the B-spline N_{n+1}. Whenever $t \in [0,1]$, it is only possible for $N_n(x-t)$ to be non-zero if $x \in [0, n+1]$, so (6.2) shows that supp $N_{n+1} \subseteq [0, n+1]$. On the other hand, if $x \in]0, n+1[$, then there exists $t \in]0,1[$ such that $x - t \in]0, n[$; by the induction hypothesis this implies that $N_n(x-t) > 0$. Since N_n is non-negative, we conclude by (6.2) that $N_{n+1}(x) > 0$. This also shows that actually supp $N_{n+1} = [0, n+1]$. $\square$

The formula (6.3) shows that the integer-translates of any B-spline point-wise adds up to 1; we say that they form a *partition of unity*. This property will be of central importance in Chapter 9.

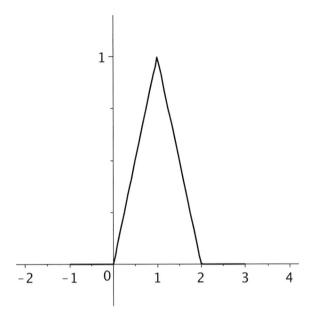

Figure 6.1. The B-spline N_2.

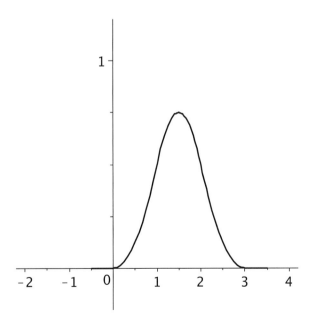

Figure 6.2. The B-spline N_3.

Via (6.4), we can find the Fourier transform of N_n:

Corollary 6.1.2 *For any $n \in \mathbb{N}$,*

$$\widehat{N_n}(\gamma) = \left(\frac{1 - e^{-2\pi i \gamma}}{2\pi i \gamma}\right)^n. \tag{6.5}$$

Proof.

$$\widehat{N_n}(\gamma) = \int_{-\infty}^{\infty} N_n(x) e^{-2\pi i x \gamma}\, dx = \int_{[0,1]^n} e^{-2\pi i (x_1 + \cdots + x_n)\gamma}\, dx_1 \cdots dx_n$$

$$= \left(\int_0^1 e^{-2\pi i x \gamma} dx\right)^n = \left(\frac{1 - e^{-2\pi i \gamma}}{2\pi i \gamma}\right)^n. \qquad \square$$

We will now derive an alternative expression for the B-splines N_n. For a real-valued function f, let

$$f(x)_+ := \max\{0, f(x)\}.$$

Also, for any non-negative integer n, let

$$f(x)_+^n := (f(x)_+)^n.$$

Finally, for $n \in \mathbb{N}$ and $j = 0, 1, \ldots, n$, let

$$\binom{n}{j} := \frac{n!}{j!(n-j)!}.$$

Theorem 6.1.3 *For each $n = 2, 3, \ldots,$ the B-spline N_n can be written*

$$N_n(x) = \frac{1}{(n-1)!} \sum_{j=0}^{n} (-1)^j \binom{n}{j} (x-j)_+^{n-1}, \quad x \in \mathbb{R}. \tag{6.6}$$

Proof. We prove (6.6) by induction. For $n = 2$, the result can be proved by a direct calculation (Exercise 6.2). Now, assume that (6.6) holds for the B-spline N_n for some $n \in \mathbb{N}$, and consider the B-spline N_{n+1}; we want to show that

$$N_{n+1}(x) = \frac{1}{n!} \sum_{j=0}^{n+1} (-1)^j \binom{n+1}{j} (x-j)_+^n, \quad x \in \mathbb{R}. \tag{6.7}$$

First we notice that for $x < 0$, we have $N_{n+1}(x) = 0$ and $(x-j)_+ = 0$ for all $j = 0, \ldots, n+1$; thus, the equation in (6.7) holds. Let us now consider

$x \in [0, n+1]$. Via the induction hypothesis, we derive that

$$
\begin{aligned}
N_{n+1}(x) &= \int_0^1 N_n(x-t)\, dt \\
&= \frac{1}{(n-1)!} \sum_{j=0}^n (-1)^j \binom{n}{j} \int_0^1 (x-t-j)_+^{n-1}\, dt. \quad (6.8)
\end{aligned}
$$

For technical reasons, we will now split the interval $[0, n+1]$ into subintervals and consider $x \in [J, J+1]$ for some arbitrary but fixed $J \in \{0, 1, \ldots, n\}$; if we can prove (6.7) for such x, the result holds for all $x \in [0, n+1]$. In order to calculate the integrals in (6.8), we split the index set $j = 0, 1, \ldots, n$ into 3 groups:

- For $j = J+1, J+2, \ldots, n$,

$$
\int_0^1 (x-t-j)_+^{n-1}\, dt = 0.
$$

- For $j = J$,

$$
\begin{aligned}
\int_0^1 (x-t-J)_+^{n-1}\, dt &= \int_0^{x-J} (x-t-J)^{n-1}\, dt \\
&= \frac{1}{n}(x-J)^n.
\end{aligned}
$$

- For $j = 0, 1, \ldots, J-1$,

$$
\begin{aligned}
\int_0^1 (x-t-j)_+^{n-1}\, dt &= \int_0^1 (x-t-j)^{n-1}\, dt \\
&= \frac{1}{n}\left((x-j)^n - (x-1-j)^n\right).
\end{aligned}
$$

We now have all the information needed to calculate the sum in (6.8). Let us first consider the partial sum corresponding to $j = 0, \ldots, J-1$:

$$
\begin{aligned}
& \sum_{j=0}^{J-1} (-1)^j \binom{n}{j} \int_0^1 (x-t-j)_+^{n-1}\, dt \\
&= \frac{1}{n} \sum_{j=0}^{J-1} (-1)^j \binom{n}{j} \left((x-j)^n - (x-1-j)^n\right) \\
&= \frac{1}{n} \sum_{j=0}^{J-1} (-1)^j \binom{n}{j} (x-j)^n - \frac{1}{n} \sum_{j=0}^{J-1} (-1)^j \binom{n}{j} (x-1-j)^n = (*).
\end{aligned}
$$

Splitting of the sum into two and reordering of the terms leads to

$$
\begin{aligned}
(*) &= \frac{1}{n}\sum_{j=0}^{J-1}(-1)^j\binom{n}{j}(x-j)^n + \frac{1}{n}\sum_{j=1}^{J}(-1)^j\binom{n}{j-1}(x-j)^n \\
&= \frac{1}{n}x^n + \frac{1}{n}\sum_{j=1}^{J-1}(-1)^j\left(\binom{n}{j}+\binom{n}{j-1}\right)(x-j)^n \\
&\quad + \frac{1}{n}(-1)^J\binom{n}{J-1}(x-J)^n.
\end{aligned}
$$

Using that

$$
\binom{n}{j}+\binom{n}{j-1}=\binom{n+1}{j}, \tag{6.9}
$$

this implies that

$$
\sum_{j=0}^{J-1}(-1)^j\binom{n}{j}\int_0^1(x-t-j)_+^{n-1}\,dt
$$

$$
= \frac{1}{n}x^n + \frac{1}{n}\sum_{j=1}^{J-1}(-1)^j\binom{n+1}{j}(x-j)^n + \frac{1}{n}(-1)^J\binom{n}{J-1}(x-J)^n.
$$

We can now find N_{n+1} using (6.8):

$$
\begin{aligned}
N_{n+1}&(x) \\
&= \frac{1}{(n-1)!}\sum_{j=0}^{n}(-1)^j\binom{n}{j}\int_0^1(x-t-j)_+^{n-1}\,dt \\
&= \frac{1}{(n-1)!}\sum_{j=0}^{J}(-1)^j\binom{n}{j}\int_0^1(x-t-j)_+^{n-1}\,dt \\
&= \frac{1}{(n-1)!}\left(\frac{1}{n}x^n + \frac{1}{n}\sum_{j=1}^{J-1}(-1)^j\binom{n+1}{j}(x-j)^n\right) \\
&\quad + \frac{1}{(n-1)!}\frac{1}{n}(-1)^J\binom{n}{J-1}(x-J)^n \\
&\quad + \frac{1}{(n-1)!}\frac{1}{n}(-1)^J\binom{n}{J}(x-J)^n \\
&= \frac{1}{n!}x^n + \frac{1}{n!}\sum_{j=1}^{J-1}(-1)^j\binom{n+1}{j}(x-j)^n \\
&\quad + \frac{1}{n!}(-1)^J\left(\binom{n}{J-1}+\binom{n}{J}\right)(x-J)^n.
\end{aligned}
$$

Using (6.9) again, this leads to

$$N_{n+1}(x) = \frac{1}{n!} \sum_{j=0}^{J} (-1)^j \binom{n+1}{j} (x-j)^n$$

$$= \frac{1}{n!} \sum_{j=0}^{n+1} (-1)^j \binom{n+1}{j} (x-j)_+^n.$$

This proves (6.7) for $x \in [0, n+1]$. The proof that (6.7) holds for $x > n+1$ is similar and is left to the reader (Exercise 6.3). □

Theorem 6.6 has some direct consequences:

Corollary 6.1.4 *For* $n = 2, 3, \ldots,$ *the B-spline* N_n *has the following properties:*

(i) $N_n \in C^{n-2}(\mathbb{R})$.

(ii) *The restriction of* N_n *to each interval* $[k, k+1]$, $k \in \mathbb{Z}$, *is a polynomial of degree at most* $n - 1$.

We now state a lemma concerning linear independence of translated versions of a B-spline. We will only need the lemma in Section 7.4.

Lemma 6.1.5 *Let* $n \in \mathbb{N}$. *Then the functions* $N_n(\cdot + k)$, $k = 0, \ldots, n-1$, *are linearly independent on* $[0, 1]$.

Proof. For $0 \le x \le 1$ and $k = 0, \ldots, n-1$, it follows from (6.6) that

$$N_n(x+k) = \frac{1}{(n-1)!} \sum_{j=0}^{n} (-1)^j \binom{n}{j} (x+k-j)_+^{n-1}$$

$$= \frac{1}{(n-1)!} \sum_{j=0}^{k} (-1)^j \binom{n}{j} (x+k-j)^{n-1}$$

$$= \frac{1}{(n-1)!} \sum_{\ell=0}^{k} (-1)^{k-\ell} \binom{n}{k-\ell} (x+\ell)^{n-1}$$

$$= \frac{(-1)^k}{(n-1)!} \sum_{\ell=0}^{k} (-1)^\ell \binom{n}{k-\ell} (x+\ell)^{n-1}.$$

This calculation shows that the linear operator that maps the functions

$$(\cdot + \ell)^{n-1}, \ell = 0, 1, \ldots, n-1,$$

onto the functions

$$N_n(\cdot + k), \ k = 0, 1, \ldots, n-1$$

is lower triangular, with non-zero diagonal entries; thus, the transformation is invertible. Consider the operator

$$\nabla f(x) := f(x+1) - f(x);$$

in the literature, this is often called the *forward difference operator*. For $k = 0, \ldots, n-1$, $\nabla^k(x^{n-1})$ is a polynomial of exact degree $n-1-k$, which is a linear combination of $x^{n-1}, (x+1)^{n-1}, \cdots, (x+k)^{n-1}$. It follows that

$$\text{span}\{(\cdot + k)^{n-1} : k = 0, \ldots, n-1\}$$

equals the space of all polynomials of degree less than n. Therefore, the polynomials $(\cdot + k)^{n-1}, k = 0, \ldots, n-1$ are linearly independent on $[0,1]$. As we have seen, $(\cdot + k)^{n-1}, k = 0, \ldots, n-1$ and $N_n(\cdot + k), k = 0, 1, \ldots, n-1$ are related by an invertible operator; as a consequence, we infer that the functions $N_n(\cdot + k), k = 0, 1, \ldots, n-1$ are linearly independent on $[0,1]$. $\square$

6.2   Symmetric B-splines

We will now consider a symmetric version of the B-splines discussed in Section 6.1. For $n \in \mathbb{N}$, let

$$B_n(x) := T_{-\frac{n}{2}} N_n(x) = N_n(x + \frac{n}{2}). \tag{6.10}$$

We will also call the functions B_n for B-splines. Alternatively, one can define these functions by

$$B_1 := \chi_{[-1/2, 1/2]}, \quad B_{n+1} := B_n * B_1, n \in \mathbb{N}, \tag{6.11}$$

see Exercise 6.6. Thus, for any $n \in \mathbb{N}$, we have that

$$B_{n+1}(x) = \int_{-\frac{1}{2}}^{\frac{1}{2}} B_n(x-t)\, dt.$$

Note that an explicit expression for B_n can be obtained via the definition (6.10) together with Theorem 6.6.

It is clear that B_n has support on the interval $[-n/2, n/2]$. We state the following direct consequences of Theorem 6.1.1 and Corollary 6.1.2:

Corollary 6.2.1 *For $n \in \mathbb{N}$, the B-spline B_n has the following properties:*

(i) For $n \geq 2$,

$$\sum_{k \in \mathbb{Z}} B_n(x-k) = 1 \quad \text{for all } x \in \mathbb{R}.$$

For $n = 1$, the formula holds for a.e. $x \in \mathbb{R}$.

(ii) $\widehat{B_n}(\gamma) = \left(\dfrac{e^{\pi i \gamma} - e^{-\pi i \gamma}}{2\pi i \gamma} \right)^n = \left(\dfrac{\sin(\pi \gamma)}{\pi \gamma} \right)^n.$

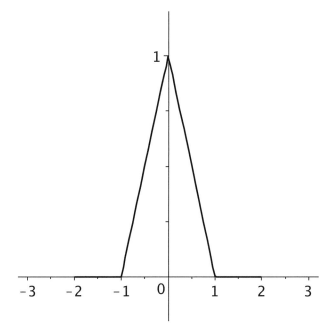

Figure 6.3. The B-spline B_2.

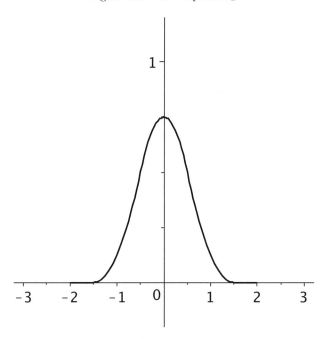

Figure 6.4. The B-spline B_3.

6.3 Exercises

6.1 Prove Theorem 6.1.1 (ii)–(iv).

6.2 (i) Show via the definition that the B-spline N_2 is given by

$$N_2(x) = \begin{cases} x & \text{if } x \in [0,1], \\ 2-x & \text{if } x \in [1,2], \\ 0 & \text{otherwise,} \end{cases}$$

(ii) Use this to show that (6.6) holds for $n = 2$.

(iii) Show that

$$N_3(x) = \begin{cases} \frac{1}{2}x^2 & \text{if } x \in [0,1], \\ -x^2 + 3x - \frac{3}{2} & \text{if } x \in [1,2], \\ \frac{1}{2}x^2 - 3x + \frac{9}{2} & \text{if } x \in [2,3], \\ 0 & \text{otherwise.} \end{cases}$$

6.3 Show that (6.6) holds for $x > n + 1$.

6.4 Consider the B-spline N_n, $n \in \mathbb{N}$.

(i) Show that

$$\widehat{N_n}(\gamma) = e^{-\pi i n \gamma} \left(\frac{\sin \pi \gamma}{\pi \gamma} \right)^n.$$

(ii) Show that

$$\widehat{N_n}(2\gamma) = e^{-\pi i n \gamma} \left(\cos \pi \gamma \right)^n \widehat{N_n}(\gamma).$$

(iii) Show that the function

$$H_0(\gamma) := e^{-\pi i n \gamma} \left(\cos \pi \gamma \right)^n$$

is 1-periodic.

6.5 Show that the B-splines B_2 and B_3 are given by

$$B_2(x) = \begin{cases} 1+x & \text{if } x \in [-1,0], \\ 1-x & \text{if } x \in [0,1], \\ 0 & \text{otherwise,} \end{cases}$$

$$B_3(x) = \begin{cases} \frac{1}{2}x^2 + \frac{3}{2}x + \frac{9}{8} & \text{if } x \in [-\frac{3}{2}, -\frac{1}{2}], \\ -x^2 + \frac{3}{4} & \text{if } x \in [-\frac{1}{2}, \frac{1}{2}], \\ \frac{1}{2}x^2 - \frac{3}{2}x + \frac{9}{8} & \text{if } x \in [\frac{1}{2}, \frac{3}{2}], \\ 0 & \text{otherwise.} \end{cases}$$

6.6 Show that the definitions of the symmetric B-splines B_n in (6.10) and (6.11) coincide.

6.7 Consider the B-spline B_n, $n \in \mathbb{N}$.

(i) Show that

$$\widehat{B_n}(2\gamma) = (\cos \pi\gamma)^n \, \widehat{B_n}(\gamma).$$

(ii) Show that the function

$$H_0(\gamma) := (\cos \pi\gamma)^n$$

is 1-periodic if and only if n is even.

7

Frames of Translates

The previous chapters have concentrated on general frame theory. We have only seen a few concrete frames, and most of them were constructed via manipulations on an orthonormal basis for an arbitrary separable Hilbert space. An advantage of this approach is that we obtain universal constructions, valid in all Hilbert spaces.

In order to make frame theory useful in engineering or signal processing, it is necessary to be more specific and construct frames in concrete Hilbert spaces consisting of functions or sequences. This will be the central theme in the following chapters, where we show how to do this in $L^2(\mathbb{R})$ and subspaces hereof. We will exclusively consider *coherent frames*, i.e., frames $\{f_k\}_{k=1}^{\infty}$ for which all the elements f_k appear by the action of some operators (belonging to a special class) on a single element f in the Hilbert space. This feature is essential for applications: it simplifies manipulations on the frame and makes it easier to store information about the frame.

In this chapter, we consider the case where the operators act by translation. That is, we consider families of functions of the form $\{\phi(\cdot - k)\}_{k\in\mathbb{Z}}$ for some $\phi \in L^2(\mathbb{R})$. We characterize the frame properties for such systems and find an expression for the canonical dual frame.

Frames of translates are natural examples of frame sequences. In fact, we show that $\{\phi(\cdot - k)\}_{k\in\mathbb{Z}}$ at most can be a frame for a proper subspace of $L^2(\mathbb{R})$. It turns out that this feature gives us some additional flexibility compared with the frame theory described in the previous chapters: we have the option to consider "dual frames" just belonging to $L^2(\mathbb{R})$, but *outside* the space where we have the frame expansion. We will show that

O. Christensen, *Frames and Bases*. DOI: 10.1007/978-0-8176-4678-3_7,

this freedom allows us to construct duals with better properties than the dual frames considered so far.

The next chapters will show that sequences of translates are of fundamental importance also for construction of frames for all of $L^2(\mathbb{R})$.

7.1   Frames of translates

For $b \in \mathbb{R}$, we define the translation operator $T_b : L^2(\mathbb{R}) \to L^2(\mathbb{R})$ by

$$(T_b f)(x) = f(x - b), \ x \in \mathbb{R}.$$

Some of the most important properties of the translation operator are described in Section 2.9. We will now consider systems of functions in $L^2(\mathbb{R})$ of the form $\{T_{kb}\phi\}_{k \in \mathbb{Z}}$, where ϕ is a fixed function and $b > 0$ is given. In the current section, our goal is to characterize Riesz sequences and frames of this form, but we need some preparation before we can do that. Our presentation is inspired by the paper [50].

The Fourier transform and the modulation operators E_b, defined in Section 2.9, will play central roles in this chapter. Recall that for $b \in \mathbb{R}$ and $f \in L^2(\mathbb{R})$,

$$E_b f(x) = e^{2\pi i b x} f(x), \ x \in \mathbb{R},$$

and that we also use the notation E_b for the function $E_b(x) = e^{2\pi i b x}$. The commutator relationship

$$\mathcal{F}T_b = E_{-b}\mathcal{F}$$

will be used repeatedly.

In order to simplify the notation, we will assume that $b = 1$. This is not a restriction; in fact, the following lemma shows that we can always obtain this by a scaling of the function ϕ. Remember that for $a > 0$, the scaling operator $D_a : L^2(\mathbb{R}) \to L^2(\mathbb{R})$ is defined by

$$(D_a f)(x) = \frac{1}{\sqrt{a}} f(\frac{x}{a}), \ x \in \mathbb{R}.$$

Lemma 7.1.1 *Let $\phi \in L^2(\mathbb{R})$ and $a, b > 0$ be given. Then $\{T_{kb}\phi\}_{k \in \mathbb{Z}}$ is a frame sequence (Riesz sequence) if and only if $\{T_{kba}D_a\phi\}_{k \in \mathbb{Z}}$ is a frame sequence (Riesz sequence). In the affirmative case, the two sequences have the same frame bounds.*

Proof. By the commutator relations stated in (2.20),

$$D_a T_{kb} = T_{kba}D_a.$$

Since D_a is a unitary operator, the frame-case in Lemma 7.1.1 follows from Lemma 5.3.3. The Riesz-basis case also follows from the stated commutator relation, together with the definition of a Riesz basis. $\square$

The following lemma will be used throughout the book. For this reason, we state it slightly more general than needed in the current section.

Lemma 7.1.2 *Let* $a > 0$ *be given, and* $f : \mathbb{R} \rightarrow \mathbb{C}$ *be a bounded, measurable, and* a-*periodic function. Then, for* $g \in L^1(\mathbb{R})$,

$$\int_{-\infty}^{\infty} f(x)g(x)\,dx = \int_0^a f(x) \sum_{k \in \mathbb{Z}} g(x - ka)\,dx. \tag{7.1}$$

Proof. We first show that

$$\int_0^a |f(x)| \sum_{k \in \mathbb{Z}} |g(x - ka)|\,dx < \infty. \tag{7.2}$$

For positive functions, sums and integrals can be interchanged, so

$$\int_0^a |f(x)| \sum_{k \in \mathbb{Z}} |g(x - ka)|\,dx = \sum_{k \in \mathbb{Z}} \int_0^a |f(x)|\,|g(x - ka)|\,dx = (*).$$

Using that f is assumed to be a-periodic,

$$(*) = \sum_{k \in \mathbb{Z}} \int_0^a |f(x - ka)|\,|g(x - ka)|\,dx = \int_{-\infty}^{\infty} |f(x)|\,|g(x)|\,dx,$$

which is finite because f is bounded and $g \in L^1(\mathbb{R})$. This proves (7.2). As a consequence, Lebesgue's dominated convergence theorem now allows us to exchange the order of the sum and the integral on the right-hand side in (7.1). Repeating the above calculations now leads to the result.   □

We will often need a variant of this result. The proof is similar and is left to the reader as Exercise 7.1.

Lemma 7.1.3 *Let* $a > 0$ *be given, and* $f, g \in L^2(\mathbb{R})$. *Then the series* $\sum_{k \in \mathbb{Z}} f(x - ka)g(x - ka)$ *is absolutely convergent for a.e.* $x \in \mathbb{R}$, *and*

$$\int_{-\infty}^{\infty} f(x)g(x)\,dx = \int_0^a \sum_{k \in \mathbb{Z}} f(x - ka)g(x - ka)\,dx.$$

We know from Chapter 5 that the pre-frame operator plays a central role in the analysis of the frame-properties for a collection of elements in a Hilbert space. For a system of translates $\{T_k\phi\}_{k \in \mathbb{Z}}$, the associated pre-frame operator is

$$T : \{c_k\}_{k \in \mathbb{Z}} \rightarrow \sum_{k \in \mathbb{Z}} c_k T_k \phi. \tag{7.3}$$

If we do not assume that $\{T_k\phi\}_{k \in \mathbb{Z}}$ is a Bessel sequence, the operator T might not be defined on all sequences in $\ell^2(\mathbb{Z})$. However, it is always well defined as a map from all *finite sequences* in $\ell^2(\mathbb{Z})$ to $L^2(\mathbb{R})$. We will need the following result.

Lemma 7.1.4 *Let $\phi \in L^2(\mathbb{R})$, and consider the operator T defined in (7.3) for finite sequences $\{c_k\}_{k\in\mathbb{Z}}$. Then the following holds:*

(i) For all finite sequences $\{c_k\}_{k\in\mathbb{Z}}$,

$$||T\{c_k\}_{k\in\mathbb{Z}}||^2 = \int_0^1 \left|\sum_{k\in\mathbb{Z}} c_k e^{-2\pi i k\gamma}\right|^2 \sum_{k\in\mathbb{Z}} |\hat{\phi}(\gamma+k)|^2 \, d\gamma. \quad (7.4)$$

(ii) $\{T_k\phi\}_{k\in\mathbb{Z}}$ is a Bessel sequence with bound B if and only if

$$\sum_{k\in\mathbb{Z}} |\hat{\phi}(\gamma+k)|^2 \leq B, \ a.e. \ \gamma \in \mathbb{R}. \quad (7.5)$$

Proof. Let $\{c_k\}_{k\in\mathbb{Z}}$ be a finite sequence. Using that the Fourier transform is unitary, we obtain that

$$\begin{aligned}
||T\{c_k\}_{k\in\mathbb{Z}}||^2 &= \left|\left|\sum_{k\in\mathbb{Z}} c_k T_k\phi\right|\right|^2 \\
&= \left|\left|\mathcal{F}\sum_{k\in\mathbb{Z}} c_k T_k\phi\right|\right|^2 \\
&= \int_{-\infty}^{\infty} \left|\sum_{k\in\mathbb{Z}} c_k e^{-2\pi i k\gamma}\hat{\phi}(\gamma)\right|^2 \, d\gamma \\
&= \int_{-\infty}^{\infty} \left|\sum_{k\in\mathbb{Z}} c_k e^{-2\pi i k\gamma}\right|^2 |\hat{\phi}(\gamma)|^2 \, d\gamma.
\end{aligned}$$

Via Lemma 7.1.2, we can continue with

$$||T\{c_k\}_{k\in\mathbb{Z}}||^2 = \int_0^1 \left|\sum_{k\in\mathbb{Z}} c_k e^{-2\pi i k\gamma}\right|^2 \sum_{k\in\mathbb{Z}} \left|\hat{\phi}(\gamma+k)\right|^2 \, d\gamma.$$

Thus we have obtained the result in (i). To prove (ii), we first assume that (7.5) holds. Then (i) implies that for all finite sequences $\{c_k\}_{k\in\mathbb{Z}}$,

$$||T\{c_k\}_{k\in\mathbb{Z}}||^2 \leq B \int_0^1 \left|\sum_{k\in\mathbb{Z}} c_k e^{-2\pi i k\gamma}\right|^2 \, d\gamma = B\sum_{k\in\mathbb{Z}} |c_k|^2.$$

Exercise 3.8 now shows that $\{T_k\phi\}_{k\in\mathbb{Z}}$ is a Bessel sequence with bound B.

In order to prove the other implication in (ii), assume that $\{T_k\phi\}_{k\in\mathbb{Z}}$ is a Bessel sequence with bound B. Using Plancherel's equation we see that, for any $f \in L^2(\mathbb{R})$,

$$
\begin{aligned}
\sum_{k \in \mathbb{Z}} |\langle f, T_k \phi \rangle|^2 &= \sum_{k \in \mathbb{Z}} \left| \int_{-\infty}^{\infty} f(x) \overline{T_k \phi(x)} \, d\gamma \right|^2 \\
&= \sum_{k \in \mathbb{Z}} \left| \int_{-\infty}^{\infty} \hat{f}(\gamma) \overline{\mathcal{F} T_k \phi} \, d\gamma \right|^2 \\
&= \sum_{k \in \mathbb{Z}} \left| \int_{-\infty}^{\infty} \hat{f}(\gamma) \overline{\hat{\phi}(\gamma)} e^{2\pi i k \gamma} \, d\gamma \right|^2.
\end{aligned}
$$

Using Lemma 7.1.3, this shows that

$$
\sum_{k \in \mathbb{Z}} |\langle f, T_k \phi \rangle|^2 = \sum_{k \in \mathbb{Z}} \left| \int_0^1 \sum_{n \in \mathbb{Z}} \hat{f}(\gamma + n) \overline{\hat{\phi}(\gamma + n)} e^{2\pi i k(\gamma + n)} \, d\gamma \right|^2. \tag{7.6}
$$

Note that

$$
\int_0^1 \sum_{n \in \mathbb{Z}} \hat{f}(\gamma + n) \overline{\hat{\phi}(\gamma + n)} e^{2\pi i k(\gamma + n)} \, d\gamma = \int_0^1 \sum_{n \in \mathbb{Z}} \hat{f}(\gamma + n) \overline{\hat{\phi}(\gamma + n)} e^{2\pi i k \gamma} \, d\gamma;
$$

this expression is the $-k$-th Fourier coefficients for the 1-periodic function

$$
\gamma \mapsto \sum_{n \in \mathbb{Z}} \hat{f}(\gamma + n) \overline{\hat{\phi}(\gamma + n)}.
$$

Thus, by Parseval's equation applied to (7.6), we obtain that

$$
\sum_{k \in \mathbb{Z}} |\langle f, T_k \phi \rangle|^2 = \int_0^1 \left| \sum_{n \in \mathbb{Z}} \hat{f}(\gamma + n) \overline{\hat{\phi}(\gamma + n)} \right|^2 \, d\gamma. \tag{7.7}
$$

Since

$$
||f||^2 = \int_{-\infty}^{\infty} |f(x)|^2 \, dx = \int_{-\infty}^{\infty} |\hat{f}(\gamma)|^2 \, d\gamma = \int_0^1 \sum_{n \in \mathbb{Z}} \left| \hat{f}(\gamma + n) \right|^2 \, d\gamma,
$$

the definition of a Bessel sequence together with (7.7) shows that

$$
\int_0^1 \left| \sum_{n \in \mathbb{Z}} \hat{f}(\gamma + n) \overline{\hat{\phi}(\gamma + n)} \right|^2 \, d\gamma \leq B \int_0^1 \sum_{n \in \mathbb{Z}} \left| \hat{f}(\gamma + n) \right|^2 \, d\gamma. \tag{7.8}
$$

Now, assume that the set

$$
\mathcal{E} := \left\{ \gamma \in [0, 1] \ | \ \sum_{n \in \mathbb{Z}} \left| \hat{\phi}(\gamma + n) \right|^2 > B \right\}
$$

has positive measure. For $\gamma \in [0, 1[$, let $g(\gamma) = \chi_{\mathcal{E}}(\gamma)$, and extend g to a 1-periodic function on $\mathbb{R}$. Define the function $f \in L^2(\mathbb{R})$ via its Fourier transform by

$$
\hat{f}(\gamma) = g(\gamma) \hat{\phi}(\gamma).
$$

Then

$$\sum_{n\in\mathbb{Z}}\left|\hat{f}(\gamma+n)\right|^2 = \sum_{n\in\mathbb{Z}}\left|\chi_{\mathcal{E}}(\gamma)\hat{\phi}(\gamma+n)\right|^2 = \chi_{\mathcal{E}}(\gamma)\sum_{n\in\mathbb{Z}}\left|\hat{\phi}(\gamma+n)\right|^2,$$

and

$$\sum_{n\in\mathbb{Z}}\left|\hat{f}(\gamma+n)\overline{\hat{\phi}(\gamma+n)}\right| = \chi_{\mathcal{E}}(\gamma)\sum_{n\in\mathbb{Z}}\left|\hat{\phi}(\gamma+n)\right|^2.$$

Thus,

$$
\begin{aligned}
\int_0^1\left|\sum_{n\in\mathbb{Z}}\hat{f}(\gamma+n)\overline{\hat{\phi}(\gamma+n)}\right|^2 d\gamma &= \int_{\mathcal{E}}\left(\sum_{n\in\mathbb{Z}}\left|\hat{\phi}(\gamma+n)\right|^2\right)^2 d\gamma \\
&> B\int_{\mathcal{E}}\sum_{n\in\mathbb{Z}}\left|\hat{\phi}(\gamma+n)\right|^2 d\gamma \\
&= B\int_0^1\sum_{n\in\mathbb{Z}}\left|\hat{f}(\gamma+n)\right|^2 d\gamma,
\end{aligned}
$$

which contradicts (7.8). Thus $\mathcal{E}$ is a zero-set, which proves (ii). $\qquad\square$

Associated with a given function $\phi\in L^2(\mathbb{R})$ and motivated by Lemma 7.1.4(ii), we will consider the function

$$\Phi(\gamma) = \sum_{k\in\mathbb{Z}}\left|\hat{\phi}(\gamma+k)\right|^2, \ \gamma\in\mathbb{R}. \tag{7.9}$$

We will now state some of the properties for the function Φ; in particular, we show that $\Phi(\gamma)$ is finite for almost all $\gamma\in\mathbb{R}$.

Lemma 7.1.5 *Given $\phi\in L^2(\mathbb{R})$, the associated function Φ is 1-periodic and belongs to $L^1(0,1)$; in particular, $\Phi(\gamma)$ is finite for almost all $\gamma\in\mathbb{R}$. Its Fourier coefficients are*

$$c_k = \int_{-\infty}^{\infty}\phi(x)\overline{\phi(x+k)}\,dx, \ k\in\mathbb{Z}. \tag{7.10}$$

Proof. The expression for Φ is 1-periodic. Further, by Lemma 7.1.3,

$$\int_0^1\Phi(\gamma)\,d\gamma = \int_0^1\sum_{k\in\mathbb{Z}}\left|\hat{\phi}(\gamma+k)\right|^2 d\gamma = \int_{-\infty}^{\infty}\left|\hat{\phi}(\gamma)\right|^2 d\gamma$$
$$< \infty.$$

Thus $\Phi\in L^1(0,1)$; in particular, $\Phi(\gamma) < \infty$ for a.e. $\gamma\in\mathbb{R}$. Using the 1-periodicity of the function E_k, the Fourier coefficients for Φ can be expressed by

$$c_k = \int_0^1 \Phi(\gamma) e^{-2\pi i k \gamma} \, d\gamma$$

$$= \int_0^1 \sum_{n \in \mathbb{Z}} \left(|\hat{\phi}(\gamma + n)|^2 e^{-2\pi i k (\gamma + n)} \right) d\gamma.$$

Via Lemma 7.1.3, we obtain that

$$c_k = \int_0^1 \sum_{n \in \mathbb{Z}} \left(|\hat{\phi}(\gamma + n)|^2 e^{-2\pi i k (\gamma + n)} \right) d\gamma$$

$$= \int_{-\infty}^{\infty} |\hat{\phi}(\gamma)|^2 e^{-2\pi i k \gamma} d\gamma$$

$$= \langle \hat{\phi}, E_k \hat{\phi} \rangle$$

$$= \langle \mathcal{F}\phi, \mathcal{F}T_{-k}\phi \rangle$$

$$= \langle \phi, T_{-k}\phi \rangle.$$

This concludes the proof. $\qquad\qquad\qquad\qquad\qquad\qquad\square$

The next lemma will be needed repeatedly as a tool to analyze series expansions consisting of translates of a function ϕ.

Lemma 7.1.6 *Let $\phi \in L^2(\mathbb{R})$ and assume that $\{T_k\phi\}_{k \in \mathbb{Z}}$ is a Bessel sequence. Let $\{c_k\}_{k \in \mathbb{Z}} \in \ell^2(\mathbb{Z})$. Then $\sum_{k \in \mathbb{Z}} c_k T_k \phi$ converges in $L^2(\mathbb{R})$ and $\sum_{k \in \mathbb{Z}} c_k E_{-k}$ converges in $L^2(0,1)$, and*

$$\mathcal{F} \sum_{k \in \mathbb{Z}} c_k T_k \phi = \left(\sum_{k \in \mathbb{Z}} c_k E_{-k} \right) \hat{\phi}. \qquad (7.11)$$

Proof. That $\sum_{k \in \mathbb{Z}} c_k T_k \phi$ and $\sum_{k \in \mathbb{Z}} c_k E_{-k}$ converge as described follows from Theorem 3.1.3, so we only have to prove (7.11). We first observe that (Exercise 7.2)

$$\left(\sum_{k \in \mathbb{Z}} c_k E_{-k} \right) \hat{\phi} \in L^2(\mathbb{R}). \qquad (7.12)$$

Also, note that

$$\mathcal{F} \sum_{k \in \mathbb{Z}} c_k T_k \phi = \sum_{k \in \mathbb{Z}} c_k \mathcal{F} T_k \phi = \sum_{k \in \mathbb{Z}} (c_k E_{-k} \hat{\phi}),$$

where the series on the right-hand side converges in $L^2(\mathbb{R})$. We have to prove that

$$\sum_{k \in \mathbb{Z}} (c_k E_{-k} \hat{\phi}) = \left(\sum_{k \in \mathbb{Z}} c_k E_{-k} \right) \hat{\phi},$$

i.e., that

$$\left\| \sum_{|k| \leq N} c_k E_{-k} \hat{\phi} - \left(\sum_{k \in \mathbb{Z}} c_k E_{-k} \right) \hat{\phi} \right\|_{L^2(\mathbb{R})} \to 0 \text{ as } N \to \infty.$$

Because $\sum_{k \in \mathbb{Z}} c_k E_{-k}$ is 1-periodic,

$$\left\| \sum_{|k| \leq N} c_k E_{-k} \hat{\phi} - \left(\sum_{k \in \mathbb{Z}} c_k E_{-k} \right) \hat{\phi} \right\|_{L^2(\mathbb{R})}$$

$$= \left(\int_{-\infty}^{\infty} \left| \sum_{|k| \leq N} c_k E_{-k}(\gamma) \hat{\phi}(\gamma) - \left(\sum_{k \in \mathbb{Z}} c_k E_{-k}(\gamma) \right) \hat{\phi}(\gamma) \right|^2 d\gamma \right)^{1/2}$$

$$= \left(\int_0^1 \left| \sum_{|k| \leq N} c_k E_{-k}(\gamma) - \sum_{k \in \mathbb{Z}} c_k E_{-k}(\gamma) \right|^2 \sum_{k \in \mathbb{Z}} |\hat{\phi}(\gamma + k)|^2 d\gamma \right)^{1/2}$$

$$\leq \sqrt{B} \left\| \sum_{|k| \leq N} c_k E_{-k} - \sum_{k \in \mathbb{Z}} c_k E_{-k} \right\|_{L^2(0,1)} ;$$

here B denotes a Bessel bound for $\{T_k \phi\}_{k \in \mathbb{Z}}$. The last term converges to 0 as $N \to \infty$, and the proof is completed. $\square$

We are now ready for the announced characterization of the frame properties for $\{T_k \phi\}_{k \in \mathbb{Z}}$. They are formulated in terms of the function Φ associated with ϕ.

Theorem 7.1.7 *Let $\phi \in L^2(\mathbb{R})$. For any $A, B > 0$, the following characterizations hold:*

(i) *$\{T_k \phi\}_{k \in \mathbb{Z}}$ is an orthonormal sequence if and only if*

$$\Phi(\gamma) = 1, \text{ a.e. } \gamma \in [0, 1].$$

(ii) *$\{T_k \phi\}_{k \in \mathbb{Z}}$ is a Riesz sequence with bounds A, B if and only if*

$$A \leq \Phi(\gamma) \leq B, \text{ a.e. } \gamma \in [0, 1].$$

(iii) *$\{T_k \phi\}_{k \in \mathbb{Z}}$ is a frame sequence with bounds A, B if and only if*

$$A \leq \Phi(\gamma) \leq B, \text{ a.e. } \gamma \in [0, 1] \setminus N,$$

where $N = \{\gamma \in [0, 1] : \Phi(\gamma) = 0\}$.

Proof. We first prove (iii) via Lemma 5.4.5. Because of Lemma 7.1.4, we will assume that $\{T_k \phi\}_{k \in \mathbb{Z}}$ is a Bessel sequence and concentrate our analysis on the lower bounds. Consequently, the equality (7.4), i.e.,

$$||T\{c_k\}_{k\in\mathbb{Z}}||^2 = \int_0^1 \left|\sum_{k\in\mathbb{Z}} c_k e^{-2\pi i k\gamma}\right|^2 \Phi(\gamma)\,d\gamma, \tag{7.13}$$

now holds for all sequences $\{c_k\}_{k\in\mathbb{Z}} \in \ell^2(\mathbb{Z})$. With the set N defined as in the statement of the theorem, it follows that the kernel of the operator T is

$$\mathcal{N}_T = \left\{\{c_k\}_{k\in\mathbb{Z}} \in \ell^2(\mathbb{Z}) \mid \sum_{k\in\mathbb{Z}} c_k e^{-2\pi i k\gamma} = 0 \text{ for } \gamma \in [0,1]\setminus N\right\}. \tag{7.14}$$

For arbitrary sequences $\{c_k\}_{k\in\mathbb{Z}}, \{d_k\}_{k\in\mathbb{Z}} \in \ell^2(\mathbb{Z})$, the fact that $\{E_k\}_{k\in\mathbb{Z}}$ is an orthonormal basis for $L^2(0,1)$ implies that

$$\langle\{c_k\}_{k\in\mathbb{Z}}, \{d_k\}_{k\in\mathbb{Z}}\rangle_{\ell^2(\mathbb{Z})} = 0 \Leftrightarrow \left\langle\sum_{k\in\mathbb{Z}} c_k E_{-k}, \sum_{k\in\mathbb{Z}} d_k E_{-k}\right\rangle_{L^2(0,1)} = 0;$$

it follows that (Exercise 7.3)

$$\mathcal{N}_T^\perp = \left\{\{c_k\}_{k\in\mathbb{Z}} \in \ell^2(\mathbb{Z}) \mid \sum_{k\in\mathbb{Z}} c_k e^{-2\pi i k\gamma} = 0 \text{ for } \gamma \in N\right\}. \tag{7.15}$$

So for $\{c_k\}_{k\in\mathbb{Z}} \in \mathcal{N}_T^\perp$,

$$\sum_{k\in\mathbb{Z}}|c_k|^2 = \int_{[0,1]}\left|\sum_{k\in\mathbb{Z}} c_k e^{-2\pi i k\gamma}\right|^2 d\gamma = \int_{[0,1]\setminus N}\left|\sum_{k\in\mathbb{Z}} c_k e^{-2\pi i k\gamma}\right|^2 d\gamma;$$

using (7.13), the left-hand condition in (5.19) in Lemma 5.4.5 therefore is equivalent to

$$A\int_{[0,1]\setminus N}\left|\sum_{k\in\mathbb{Z}} c_k e^{-2\pi i k\gamma}\right|^2 d\gamma$$

$$\leq \int_{[0,1]\setminus N}\left|\sum_{k\in\mathbb{Z}} c_k e^{-2\pi i k\gamma}\right|^2 \Phi(\gamma)\,d\gamma, \ \forall\{c_k\}_{k\in\mathbb{Z}} \in \mathcal{N}_T^\perp.$$

This, in turn, is equivalent to (Exercise 7.3)

$$A \leq \Phi(\gamma), \ a.e. \ \gamma \in [0,1]\setminus N. \tag{7.16}$$

This proves (iii). For the rest of the proof, recall from Theorem 5.2.1 that the Riesz bounds and the frame bounds coincide for Riesz sequences. By Theorem 3.3.7, $\{T_k\phi\}_{k\in\mathbb{Z}}$ is a Riesz sequence if and only if the inequality (5.19) holds for all $\{c_k\}_{k\in\mathbb{Z}} \in \ell^2(\mathbb{Z})$. This is the case if and only if $\mathcal{N}_T = \{0\}$, i.e., by (7.14), if and only if N is a null set; this gives (ii). The result in (i) now follows from Proposition 3.2.8. $\square$

As a very important consequence of Theorem 7.1.7, we now prove that the integer-translates of any B-spline forms a Riesz sequence. We formulate the result for the symmetric B-splines B_n defined in (6.11), but the same result holds for the B-splines N_n in (6.2); this remark in fact applies to all results concerning B-splines in the current chapter.

Theorem 7.1.8 *For each $n \in \mathbb{N}$, the sequence $\{T_k B_n\}_{k \in \mathbb{Z}}$ is a Riesz sequence.*

Proof. For $n = 1$, $\{T_k B_1\}_{k \in \mathbb{Z}}$ is an orthonormal system, and therefore a Riesz sequence. In order to prove the result for $n > 1$, we apply Theorem 7.1.7 to B_1; this shows that

$$\sum_{k \in \mathbb{Z}} \left| \widehat{B_1}(\gamma + k) \right|^2 = 1, \quad a.e. \ \gamma \in \mathbb{R}.$$

Since $|\widehat{B_1}(\gamma)| \le 1$ for all $\gamma \in \mathbb{R}$ and $\widehat{B_n}(\gamma) = (\widehat{B_1}(\gamma))^n$ by Corollary 6.2.1, it immediately follows that

$$\sum_{k \in \mathbb{Z}} \left| \widehat{B_n}(\gamma + k) \right|^2 \le \sum_{k \in \mathbb{Z}} \left| \widehat{B_1}(\gamma + k) \right|^2 = 1, \quad a.e. \ \gamma \in \mathbb{R}.$$

Thus $\{T_k B_n\}_{k \in \mathbb{Z}}$ is a Bessel sequence. In order to prove that $\{T_k B_n\}_{k \in \mathbb{Z}}$ satisfies the lower Riesz basis condition, we again use Corollary 6.2.1: it shows that, for a.e. $\gamma \in \mathbb{R}$,

$$\sum_{k \in \mathbb{Z}} \left| \widehat{B_n}(\gamma + k) \right|^2 \ge \inf_{\gamma \in [-\frac{1}{2}, \frac{1}{2}]} \left| \widehat{B_n}(\gamma) \right|^2 = \left(\frac{\sin(\pi/2)}{\pi/2} \right)^{2n} = \left(\frac{2}{\pi} \right)^{2n}. \quad (7.17)$$

The result now follows from Theorem 7.1.7. $\qquad \square$

Example 7.1.9 Define the function $\phi \in L^2(\mathbb{R})$ by $\hat{\phi} = \chi_{[-1/3, 1/3]}$. Then Theorem 7.1.7 shows that $\{T_k \phi\}_{k \in \mathbb{Z}}$ is an overcomplete frame sequence. $\square$

Frames of translates are genuine examples of frame sequences: as we show now, they cannot be frames for all of $L^2(\mathbb{R})$, no matter how $\phi \in L^2(\mathbb{R})$ is chosen. Thus, a more exact name would be to call them *frame sequences* of translates. However, in the literature they are simply called frames of translates, and we will follow this tradition.

Proposition 7.1.10 *A sequence $\{T_k \phi\}_{k \in \mathbb{Z}}$ can at most be a frame for a proper subspace of $L^2(\mathbb{R})$.*

Proof. Let $\phi \in L^2(\mathbb{R})$. Given a number $h \in]0, 1/2[$,

$$\sum_{k \in \mathbb{Z}} |\langle \chi_{]0,h[}, T_k\phi \rangle|^2 = \sum_{k \in \mathbb{Z}} |\langle \chi_{]0,h[}, \chi_{]0,h[} T_k\phi \rangle|^2$$

$$\leq \sum_{k \in \mathbb{Z}} ||\chi_{]0,h[}||^2 \, ||\chi_{]0,h[} T_k\phi||^2$$

$$= ||\chi_{]0,h[}||^2 \sum_{k \in \mathbb{Z}} \int_k^{k+h} |\phi(x)|^2 \, dx.$$

Letting

$$\Delta_h := \bigcup_{k \in \mathbb{Z}}]h, h + k[,$$

it follows that

$$\sum_{k \in \mathbb{Z}} |\langle \chi_{]0,h[}, T_k\phi \rangle|^2 \leq ||\chi_{]0,h[}||^2 \int_{\Delta_h} |\phi(x)|^2 \, dx.$$

By Lebesgue's theorem of dominated convergence,

$$\int_{\Delta_h} |\phi(x)|^2 \, dx \to 0 \text{ as } h \to 0;$$

thus, $\{T_k\phi\}_{k \in \mathbb{Z}}$ does not satisfy the lower frame condition on $L^2(\mathbb{R})$. $\square$

Proposition 7.1.10 shows that $\overline{\text{span}}\{T_k\phi\}_{k \in \mathbb{Z}}$ is a proper subspace of $L^2(\mathbb{R})$ for any frame sequence $\{T_k\phi\}_{k \in \mathbb{Z}}$. Therefore, in order to apply the frame decomposition, it is necessary to be able to characterize membership of $\overline{\text{span}}\{T_k\phi\}_{k \in \mathbb{Z}}$. This can be done in terms of the Fourier transform:

Lemma 7.1.11 *Assume that $\phi \in L^2(\mathbb{R})$ and that $\{T_k\phi\}_{k \in \mathbb{Z}}$ is a frame sequence. Then a function $f \in L^2(\mathbb{R})$ belongs to $\overline{\text{span}}\{T_k\phi\}_{k \in \mathbb{Z}}$ if and only if there exists a 1-periodic function F whose restriction to $[0, 1[$ belongs to $L^2(0, 1)$, such that*

$$\hat{f} = F\hat{\phi}.$$

Proof. A function $f \in L^2(\mathbb{R})$ belongs to $\overline{\text{span}}\{T_k\phi\}_{k \in \mathbb{Z}}$ if and only if there exists a sequence $\{c_k\}_{k \in \mathbb{Z}} \in \ell^2(\mathbb{Z})$ such that

$$f = \sum_{k \in \mathbb{Z}} c_k T_k\phi;$$

via the Fourier transform and Lemma 7.1.6, this is equivalent to

$$\hat{f} = \left(\sum_{k \in \mathbb{Z}} c_k E_{-k} \right) \hat{\phi}. \qquad \square$$

7.2   The canonical dual frame

We now turn the focus to the canonical dual of a frame of translates $\{T_k\phi\}_{k\in\mathbb{Z}}$. Assuming that $\{T_k\phi\}_{k\in\mathbb{Z}}$ is a frame sequence, i.e., a frame for

$$V := \overline{\operatorname{span}}\{T_k\phi\}_{k\in\mathbb{Z}},$$

we know from general frame theory that the frame operator

$$S : V \to V, \ Sf = \sum_{k\in\mathbb{Z}}\langle f, T_k\phi\rangle T_k\phi$$

is invertible. The frame decomposition, see Theorem 5.9, takes the form

$$f = \sum_{k\in\mathbb{Z}}\langle f, S^{-1}T_k\phi\rangle T_k\phi, \ f \in V. \tag{7.18}$$

In order to apply the frame decomposition, we need to be able to calculate the elements $S^{-1}T_k\phi$ in the canonical dual frame. For this purpose, we need the following lemma.

Lemma 7.2.1 *Let $\phi \in L^2(\mathbb{R})$ and assume that $\{T_k\phi\}_{k\in\mathbb{Z}}$ is a frame for its closed linear span V. Then for all $f \in V$,*

$$ST_kf = T_kSf, \ \forall k \in \mathbb{Z},$$

and

$$S^{-1}T_kf = T_kS^{-1}f, \ \forall k \in \mathbb{Z}.$$

Proof.   Given $f \in V$ and $k \in \mathbb{Z}$, we have

$$\begin{aligned}
ST_kf &= \sum_{k'\in\mathbb{Z}}\langle T_kf, T_{k'}\phi\rangle T_{k'}\phi \\
&= \sum_{k'\in\mathbb{Z}}\langle f, T_{k'-k}\phi\rangle T_{k'}\phi.
\end{aligned}$$

Replacing the summation index k' by $k' + k$ gives

$$\begin{aligned}
ST_kf &= \sum_{k'\in\mathbb{Z}}\langle f, T_{k'}\phi\rangle T_{k'+k}\phi \\
&= T_kSf.
\end{aligned}$$

The second part of the result follows from here.   □

By Lemma 7.2.1, the frame decomposition can now be written

$$f = \sum_{k\in\mathbb{Z}}\langle f, T_kS^{-1}\phi\rangle T_k\phi, \ f \in V. \tag{7.19}$$

This result is a great simplification compared with (7.18). In fact, in order to apply (7.18), we would have to compute the action of S^{-1} on the infinite family of functions $\{T_k\phi\}_{k\in\mathbb{Z}}$. On the other hand, application of (7.19) only requires that we find $S^{-1}\phi$; the other functions in the canonical dual frame are obtained by translation of this function. This is certainly a simplification, but we are still left with the question of finding $S^{-1}\phi$. In the current context, we are now able to find $S^{-1}\phi$ in terms of its Fourier transform. Again, the associated function Φ in (7.9) will play a role.

Proposition 7.2.2 *Let $\phi \in L^2(\mathbb{R})$ and assume that $\{T_k\phi\}_{k\in\mathbb{Z}}$ is a frame for its closed linear span V, with frame operator S. Consider the set*

$$\mathcal{E} := \{\gamma \in \mathbb{R} : \Phi(\gamma) \neq 0\},$$

and define the function θ via its Fourier transform by

$$\hat{\theta}(\gamma) := \begin{cases} \frac{\hat{\phi}(\gamma)}{\Phi(\gamma)} & \text{if } \gamma \in \mathcal{E}, \\ 0 & \text{if } \gamma \notin \mathcal{E}. \end{cases} \tag{7.20}$$

Then $\theta = S^{-1}\phi$, and the canonical dual frame of $\{T_k\phi\}_{k\in\mathbb{Z}}$ is given by $\{T_k\theta\}_{k\in\mathbb{Z}}$.

Proof. The function

$$\gamma \mapsto \begin{cases} \frac{1}{\Phi(\gamma)} & \text{if } \gamma \in \mathcal{E}, \\ 0 & \text{if } \gamma \notin \mathcal{E} \end{cases}$$

is 1-periodic, and by Theorem 7.1.7 its restriction to $]0,1[$ belongs to $L^2(0,1)$. Thus, Lemma 7.1.11 shows that the function θ defined by (7.20) belongs to V. Using the definition of the frame operator, properties of the Fourier transform, and Lemma 7.1.6, we have that

$$\begin{aligned} \mathcal{F}S\theta &= \mathcal{F}\sum_{k\in\mathbb{Z}}\langle\theta, T_k\phi\rangle T_k\phi \\ &= \mathcal{F}\sum_{k\in\mathbb{Z}}\langle\hat{\theta}, \mathcal{F}T_k\phi\rangle T_k\phi \\ &= \left(\sum_{k\in\mathbb{Z}}\langle\hat{\theta}, E_{-k}\hat{\phi}\rangle E_{-k}\right)\hat{\phi}. \end{aligned} \tag{7.21}$$

Now, using that the functions $\gamma \mapsto e^{2\pi ik\gamma}$ are 1-periodic, the definition of θ, and Lemma 7.1.3, we arrive at

$$\begin{aligned} \langle\hat{\theta}, E_{-k}\hat{\phi}\rangle &= \int_{-\infty}^{\infty} \hat{\theta}(\gamma)\overline{\hat{\phi}(\gamma)}e^{2\pi ik\gamma}\,d\gamma \\ &= \int_0^1 \sum_{n\in\mathbb{Z}}\left(\hat{\theta}(\gamma+n)\overline{\hat{\phi}(\gamma+n)}e^{2\pi i(k+n)\gamma}\right)d\gamma \\ &= \int_0^1 \sum_{n\in\mathbb{Z}}\frac{|\hat{\phi}(\gamma+n)|^2}{\Phi(\gamma+n)}\chi_{\mathcal{E}}(\gamma+n)e^{2\pi ik\gamma}\,d\gamma. \end{aligned}$$

Note that if $\gamma \in \mathcal{E}$, then $\gamma + n \in \mathcal{E}$ for all $n \in \mathbb{Z}$; thus, for such γ,

$$\sum_{n\in\mathbb{Z}} \frac{|\hat{\phi}(\gamma+n)|^2}{\Phi(\gamma+n)} \chi_{\mathcal{E}}(\gamma+n) = \sum_{n\in\mathbb{Z}} \frac{|\hat{\phi}(\gamma+n)|^2}{\Phi(\gamma)} = 1.$$

Similarly, if $\gamma \notin \mathcal{E}$, then

$$\sum_{n\in\mathbb{Z}} \frac{|\hat{\phi}(\gamma+n)|^2}{\Phi(\gamma+n)} \chi_{\mathcal{E}}(\gamma+n) = 0.$$

Thus, we conclude that

$$\langle \hat{\theta}, E_{-k}\hat{\phi} \rangle = \int_0^1 \chi_{\mathcal{E}}(\gamma)\overline{E_{-k}(\gamma)}\, d\gamma,$$

which is the $-k$-th Fourier coefficient for the function $\chi_{\mathcal{E}}$ in $L^2(0,1)$. Therefore

$$\sum_{k\in\mathbb{Z}} \langle \hat{\theta}, E_{-k}\hat{\phi} \rangle E_{-k} = \chi_{\mathcal{E}}.$$

Noting that $\chi_{\mathcal{E}}(\gamma) = 1$ if $\hat{\phi}(\gamma) \neq 0$, (7.21) now implies that

$$\mathcal{F}S\theta = \chi_{\mathcal{E}}\hat{\phi} = \hat{\phi}.$$

Therefore $S\theta = \phi$; since S is an invertible operator on V, we conclude that $\theta = S^{-1}\phi$. Now Lemma 7.2.1 gives the rest.   □

Example 7.2.3 Consider the B-spline B_n for $n \geq 2$. As we saw in Theorem 7.1.8, $\{T_k B_n\}_{k\in\mathbb{Z}}$ is a Riesz sequence. By Proposition 7.2.2, the canonical dual frame is given by $\{T_k\theta\}_{k\in\mathbb{Z}}$, where

$$\hat{\theta}(\gamma) = \frac{\widehat{B_n}(\gamma)}{\sum_{k\in\mathbb{Z}} |\widehat{B_n}(\gamma+k)|^2}.$$

Denoting the Fourier coefficients for the function $\left(\sum_{k\in\mathbb{Z}} |\widehat{B_n}(\gamma+k)|^2\right)^{-1}$ by $\{c_k\}_{k\in\mathbb{Z}}$, this implies that

$$\hat{\theta}(\gamma) = \sum_{k\in\mathbb{Z}} c_k e^{2\pi i k\gamma}\widehat{B_n}(\gamma) = \mathcal{F}\sum_{k\in\mathbb{Z}} c_k T_{-k} B_n(\gamma),$$

i.e., that

$$\theta = \sum_{k\in\mathbb{Z}} c_k T_{-k} B_n.$$

It follows from Lemma 7.1.5 that $\Phi(\gamma) = \sum_{k\in\mathbb{Z}} |\widehat{B_n}(\gamma+k)|^2$ is a trigonometric polynomial. Because Φ is positive and not constant, the inverse $\Phi(\gamma)^{-1}$ can not be a trigonometric polynomial (Exercise 7.4). Therefore, the sequence $\{c_k\}_{k\in\mathbb{Z}}$ of Fourier coefficients is infinite. In particular, the generator θ does not have compact support. $\qquad\square$

7.3 Compactly supported generators

In applications of systems of the form $\{T_k\phi\}_{k\in\mathbb{Z}}$, it is usually important that the generator ϕ has compact support. This excludes $\{T_k\phi\}_{k\in\mathbb{Z}}$ from being an overcomplete frame sequence:

Proposition 7.3.1 *Assume that $\phi \in L^2(\mathbb{R})$ has compact support. Then the following holds:*

(i) $\{T_k\phi\}_{k\in\mathbb{Z}}$ *is a Bessel sequence.*

(ii) $\{T_k\phi\}_{k\in\mathbb{Z}}$ *cannot be an overcomplete frame sequence.*

Proof. Let $\{c_k\}_{k\in\mathbb{Z}}$ be the Fourier coefficients for Φ. Because of the compact support of ϕ, (7.10) in Lemma 7.1.5 shows that there is an $N \in \mathbb{N}$ such that $c_k = 0$ if $|k| > N$. Thus, the associated function Φ in (7.9) is a trigonometric polynomial and therefore continuous. Now Theorem 7.1.7 shows that $\{T_k\phi\}_{k\in\mathbb{Z}}$ is a Bessel sequence, and that overcompleteness of $\{T_k\phi\}_{k\in\mathbb{Z}}$ is impossible. $\qquad\square$

Thus, if $\phi \in L^2(\mathbb{R})$ has compact support, then $\{T_k\phi\}_{k\in\mathbb{Z}}$ can at most be a Riesz sequence. In cases where the concrete frame bounds are irrelevant, we can check whether this is the case or not via the following consequence of Theorem 7.1.7. We ask the reader to give the proof in Exercise 7.5.

Corollary 7.3.2 *Assume that $\phi \in L^2(\mathbb{R})$ is compactly supported. Then $\{T_k\phi\}_{k\in\mathbb{Z}}$ is a Riesz sequence if and only if there is no $\gamma \in \mathbb{R}$ such that*

$$\hat{\phi}(\gamma + k) = 0, \ \forall k \in \mathbb{Z}.$$

The assumption of $\phi \in L^2(\mathbb{R})$ having compact support has the additional benefit that we can find an explicit expression for the associated function Φ, see (7.9):

Lemma 7.3.3 *Assume that $\phi \in L^2(\mathbb{R})$ has compact support in an interval of length N for some $N \in \mathbb{N}$ and is real–valued. Let $\{c_k\}_{k\in\mathbb{Z}}$ denote the Fourier coefficients for the function Φ in (7.9). Then Φ is a trigonometric polynomial of the form*

$$\Phi(\gamma) = c_0 + 2 \sum_{k=1}^{N} c_k \cos(2\pi k\gamma).$$

Proof. Via Lemma 7.1.5, the assumption that ϕ is real-valued implies that $c_k = c_{-k}$ for all $k \in \mathbb{Z}$; and the compact support implies that $c_k = 0$ if $|k| > N$. Expressing Φ via its Fourier series, we see that

$$
\begin{aligned}
\Phi(\gamma) &= \sum_{|k| \leq N} c_k e^{2\pi i k \gamma} \\
&= c_0 + \sum_{k=1}^{N} c_k (e^{2\pi i k \gamma} + e^{-2\pi i k \gamma}) \\
&= c_0 + 2 \sum_{k=1}^{N} c_k \cos(2\pi k \gamma).
\end{aligned}
$$

$\square$

The concrete expression for Φ in Lemma 7.3.3 makes it easy to check whether $\{T_k \phi\}_{k \in \mathbb{Z}}$ is a Riesz sequence or not:

Example 7.3.4 Let $\phi = \chi_{[-1,2[}$. By Lemma 7.1.5, the Fourier coefficients for Φ are

$$
c_k = \begin{cases}
3 & \text{if } k = 0, \\
2 & \text{if } k = \pm 1, \\
1 & \text{if } k = \pm 2, \\
0 & \text{otherwise.}
\end{cases}
$$

Thus, by Lemma 7.3.3,

$$
\Phi(\gamma) = 3 + 4\cos(2\pi\gamma) + 2\cos(4\pi\gamma).
$$

Note that Φ is continuous and that Φ has two isolated zero's on $[0, 1[$: $\Phi(\gamma) = 0$ for $\gamma = \frac{1}{3}$ and for $\gamma = \frac{2}{3}$. By Theorem 7.1.7, it follows that $\{T_k \phi\}_{k \in \mathbb{Z}}$ is not a frame sequence. $\square$

Overcomplete frames $\{T_k \phi\}_{k \in \mathbb{Z}}$ are not used much in practice. One reason is that, as we have seen, the generator ϕ cannot have compact support. Another reason is that for overcomplete frames of that type, the function ϕ cannot be well-localized in time and frequency simultaneously: either ϕ or its Fourier transform decays slowly. See Exercise 7.8.

7.4 Frames of translates and oblique duals

In the discussion of general frame theory in Section 5.1, we have given arguments that it often is an advantage to search for other dual frames than the canonical dual frame. Assume that $\{T_k \phi\}_{k \in \mathbb{Z}}$ is an overcomplete frame sequence, i.e., a frame for

$$
V = \overline{\text{span}}\{T_k \phi\}_{k \in \mathbb{Z}};
$$

then Theorem 5.2.3 tells us that there exist various choices of sequences of functions $\{g_k\}_{k\in\mathbb{Z}} \subset V$ such that

$$f = \sum_{k\in\mathbb{Z}} \langle f, g_k \rangle T_k\phi, \ \forall f \in V. \tag{7.22}$$

It is natural to insist on the dual frame $\{g_k\}_{k\in\mathbb{Z}}$ having the same structure as $\{T_k\phi\}_{k\in\mathbb{Z}}$, i.e., being translates of a single function. Unfortunately, Corollary 7.4.2 will show us that this removes all the freedom: the canonical dual frame $\{T_k S^{-1}\phi\}_{k\in\mathbb{Z}}$ is the only dual frame, which consists of translates of a single function and belongs to $\overline{\text{span}}\{T_k\phi\}_{k\in\mathbb{Z}}$.

In order to gain flexibility, we will remove the constraint that the elements of the dual frame should belong to $\overline{\text{span}}\{T_k\phi\}_{k\in\mathbb{Z}}$. In fact, we will just search for *some* function $\tilde{\phi} \in L^2(\mathbb{R})$ such that

$$f = \sum_{k\in\mathbb{Z}} \langle f, T_k\tilde{\phi} \rangle T_k\phi, \ \forall f \in V. \tag{7.23}$$

A family $\{T_k\tilde{\phi}\}_{k\in\mathbb{Z}}$ for which (7.23) holds will be called an *oblique dual* of $\{T_k\phi\}_{k\in\mathbb{Z}}$. Note that we do not require $\{T_k\tilde{\phi}\}_{k\in\mathbb{Z}}$ to be a frame sequence; this is the reason that we use the name "oblique dual" rather than "oblique dual frame." The results in this section are taken from [14] and [15].

Given two Bessel sequences $\{T_k\phi\}_{k\in\mathbb{Z}}$ and $\{T_k\tilde{\phi}\}_{k\in\mathbb{Z}}$, the following theorem provides a necessary and sufficient condition on the generators ϕ and $\tilde{\phi}$ such that $\{T_k\tilde{\phi}\}_{k\in\mathbb{Z}}$ is an oblique dual of $\{T_k\phi\}_{k\in\mathbb{Z}}$. Again, the function Φ defined in (7.9) will play a role.

Theorem 7.4.1 *Let* $\phi, \tilde{\phi} \in L^2(\mathbb{R})$, *and assume that* $\{T_k\phi\}_{k\in\mathbb{Z}}$ *and* $\{T_k\tilde{\phi}\}_{k\in\mathbb{Z}}$ *are Bessel sequences. Then the following are equivalent:*

(i) $f = \sum_{k\in\mathbb{Z}} \langle f, T_k\tilde{\phi} \rangle T_k\phi, \ \forall f \in V.$

(ii) $\sum_{k\in\mathbb{Z}} \hat{\phi}(\gamma + k)\overline{\hat{\tilde{\phi}}(\gamma + k)} = 1 \ \ a.e. \ on \ \{\gamma : \ \Phi(\gamma) \neq 0\}.$

In case the equivalent conditions are satisfied, $\{T_k\phi\}_{k\in\mathbb{Z}}$ *is a frame sequence.*

Proof. First, consider an arbitrary function $f \in L^2(\mathbb{R})$ for which the map

$$\gamma \mapsto \sum_{k\in\mathbb{Z}} |\hat{f}(\gamma + k)|^2$$

is bounded. Since we have assumed that $\{T_k\tilde{\phi}\}_{k\in\mathbb{Z}}$ is a Bessel sequence, Lemma 7.1.4 and Cauchy–Schwarz' inequality imply that

$$[\gamma \mapsto \sum_{k\in\mathbb{Z}} \hat{f}(\gamma + k)\overline{\hat{\tilde{\phi}}(\gamma + k)}] \in L^2(0, 1).$$

Now observe that, via Lemma 7.1.3 and Lemma 7.1.6,

$$\mathcal{F}\sum_{k\in\mathbb{Z}}\langle f,T_k\tilde{\phi}\rangle T_k\phi(\gamma)$$

$$= \sum_{k\in\mathbb{Z}}\left(\int_{-\infty}^{\infty}\hat{f}(\mu)\overline{\hat{\tilde{\phi}}(\mu)}e^{2\pi ik\mu}\,d\mu\right)e^{-2\pi ik\gamma}\hat{\phi}(\gamma)$$

$$= \hat{\phi}(\gamma)\sum_{k\in\mathbb{Z}}\left(\int_0^1\sum_{n\in\mathbb{Z}}\hat{f}(\mu+n)\overline{\hat{\tilde{\phi}}(\mu+n)}e^{2\pi ik\mu}\,d\mu\right)e^{-2\pi ik\gamma}$$

$$= \hat{\phi}(\gamma)\sum_{n\in\mathbb{Z}}\hat{f}(\gamma+n)\overline{\hat{\tilde{\phi}}(\gamma+n)}. \qquad (7.24)$$

Assuming that (i) holds and letting $f=\phi$, it follows that

$$\sum_{k\in\mathbb{Z}}\hat{\phi}(\gamma+k)\overline{\hat{\tilde{\phi}}(\gamma+k)}=1 \text{ a.e. on } \{\gamma:\ \hat{\phi}(\gamma)\neq 0\}.$$

Using the above calculation with γ replaced by $\gamma+m$ for some $m\in\mathbb{Z}$ (and using the periodicity of $\gamma\mapsto\sum_{k\in\mathbb{Z}}\hat{\phi}(\gamma+k)\overline{\hat{\tilde{\phi}}(\gamma+k)}$), we even arrive at

$$\sum_{k\in\mathbb{Z}}\hat{\phi}(\gamma+k)\overline{\hat{\tilde{\phi}}(\gamma+k)}=1 \text{ a.e. on } \{\gamma:\ \hat{\phi}(\gamma+m)\neq 0\},\ \forall m\in\mathbb{Z}.$$

This proves (ii). On the other hand, assuming (ii), our calculation (7.24) shows that for $m\in\mathbb{Z}$,

$$\mathcal{F}\sum_{k\in\mathbb{Z}}\langle T_m\phi,T_k\tilde{\phi}\rangle T_k\phi(\gamma) = \hat{\phi}(\gamma)\sum_{n\in\mathbb{Z}}\mathcal{F}T_m\phi(\gamma+n)\overline{\hat{\tilde{\phi}}(\gamma+n)}$$

$$= \hat{\phi}(\gamma)\sum_{n\in\mathbb{Z}}\hat{\phi}(\gamma+n)e^{-2\pi im(\gamma+n)}\overline{\hat{\tilde{\phi}}(\gamma+n)}$$

$$= \hat{\phi}(\gamma)e^{-2\pi im\gamma}$$

$$= \mathcal{F}T_m\phi(\gamma).$$

This shows that (i) holds for all functions $T_m\phi, m\in\mathbb{Z}$, and hence for all functions $f\in\text{span}\{T_k\phi\}_{k\in\mathbb{Z}}$. Now, because $\{T_k\phi\}_{k\in\mathbb{Z}}$ and $\{T_k\tilde{\phi}\}_{k\in\mathbb{Z}}$ are Bessel sequences, the operator

$$f\mapsto\sum_{k\in\mathbb{Z}}\langle f,T_k\tilde{\phi}\rangle T_k\phi$$

is continuous; in fact, it is a composition of the pre-frame operator associated with $\{T_k\phi\}_{k\in\mathbb{Z}}$ and the analysis operator associated with $\{T_k\tilde{\phi}\}_{k\in\mathbb{Z}}$, see page 100. Therefore, (i) holds for all $f\in\overline{\text{span}}\{T_k\phi\}_{k\in\mathbb{Z}}$.

Now assume that the equivalent conditions hold. In order to show that $\{T_k\phi\}_{k\in\mathbb{Z}}$ is a frame sequence, we need to show that the lower frame bound is satisfied. Via (i), for all $f \in V$, we have

$$\|f\|^2 = \sum_{k \in \mathbb{Z}} \langle f, T_k \tilde{\phi} \rangle \langle T_k \phi, f \rangle;$$

that $\{T_k \phi\}_{k \in \mathbb{Z}}$ is a frame sequence now follows from Cauchy–Schwarz' inequality and the assumption that $\{T_k \tilde{\phi}\}_{k \in \mathbb{Z}}$ is a Bessel sequence. □

One important consequence of Theorem 7.4.1 is that the canonical dual frame is the only dual frame that consists of integer–translates of a single function, i.e., the only dual that has same structure as $\{T_k \phi\}_{k \in \mathbb{Z}}$ itself:

Corollary 7.4.2 *Let* $\phi \in L^2(\mathbb{R})$ *and assume that* $\{T_k \phi\}_{k \in \mathbb{Z}}$ *is a frame sequence. Then there is a unique function* $\tilde{\phi} \in \overline{span}\{T_k \phi\}_{k \in \mathbb{Z}}$ *such that*

$$f = \sum_{k \in \mathbb{Z}} \langle f, T_k \tilde{\phi} \rangle T_k \phi, \ \forall f \in \overline{span}\{T_k \phi\}_{k \in \mathbb{Z}}, \qquad (7.25)$$

namely $\tilde{\phi} = S^{-1} \phi$.

Proof. The condition $\tilde{\phi} \in \overline{span}\{T_k \phi\}_{k \in \mathbb{Z}}$ implies by Lemma 7.1.10 that

$$\hat{\tilde{\phi}} = F \hat{\phi}$$

for some 1-periodic function $F \in L^2(0, 1)$. Now, if (7.25) holds, condition (ii) in Theorem 7.4.1 implies that

$$\overline{F(\gamma)} \sum_{k \in \mathbb{Z}} \hat{\phi}(\gamma + k) \overline{\hat{\phi}(\gamma + k)} = 1, \ \text{a.e. on} \ \{\gamma : \ \Phi(\gamma) \neq 0\},$$

i.e., that

$$\overline{F(\gamma)} \sum_{k \in \mathbb{Z}} \left| \hat{\phi}(\gamma + k) \right|^2 = 1, \ \text{a.e. on} \ \{\gamma : \ \Phi(\gamma) \neq 0\}.$$

This defines the function F uniquely, except on the zero-set for Φ. For γ such that $\Phi(\gamma) = 0$, we can define $F(\gamma)$ arbitrarily, but regardless of the choice, we arrive at $\hat{\tilde{\phi}}(\gamma) = F(\gamma) \hat{\phi}(\gamma) = 0$. Thus $\hat{\tilde{\phi}}$ is uniquely defined, and so is $\tilde{\phi}$. □

The role of Theorem 7.4.1 is that it might be used to construct oblique duals with better or more convenient properties than the canonical dual frame. Later in this section, we will show that one might be able to find oblique duals generated by a compactly supported function, even in cases where the canonical dual frame is generated by a function supported on $\mathbb{R}$. For now, we will show how to construct oblique duals $\{T_k \tilde{\phi}\}_{k \in \mathbb{Z}}$ with generators $\tilde{\phi}$ belonging to prescribed subspaces. In fact, the following consequence of Theorem 7.4.1 shows that if $\{T_k \phi\}_{k \in \mathbb{Z}}$ is a frame sequence, then certain conditions imply that we can find an oblique dual $\{T_k \tilde{\phi}\}_{k \in \mathbb{Z}}$ with a generator $\tilde{\phi}$ belonging to a space of the form $\overline{\text{span}}\{T_k\phi_1\}_{k\in\mathbb{Z}}$ for some $\phi_1 \in L^2(\mathbb{R})$. We ask the reader to provide the proof in Exercise 7.9.

Corollary 7.4.3 Let $\phi, \phi_1 \in L^2(\mathbb{R})$, and assume that $\{T_k\phi\}_{k\in\mathbb{Z}}$ and $\{T_k\phi_1\}_{k\in\mathbb{Z}}$ are frame sequences. If there exists a constant $A > 0$ such that

$$\left|\sum_{k\in\mathbb{Z}}\hat{\phi}(\gamma+k)\overline{\hat{\phi}_1(\gamma+k)}\right| \geq A \text{ a.e. on } \{\gamma: \Phi(\gamma) \neq 0\}, \qquad (7.26)$$

then there exists a function $\tilde{\phi} \in \overline{\text{span}}\{T_k\phi_1\}_{k\in\mathbb{Z}}$ such that

$$f = \sum_{k\in\mathbb{Z}}\langle f, T_k\tilde{\phi}\rangle T_k\phi, \ \forall f \in \overline{\text{span}}\{T_k\phi\}_{k\in\mathbb{Z}}; \qquad (7.27)$$

one choice of $\tilde{\phi} \in \overline{\text{span}}\{T_k\phi_1\}_{k\in\mathbb{Z}}$ satisfying (7.27) is given in the Fourier domain by

$$\hat{\tilde{\phi}}(\gamma) = \begin{cases} \left(\sum_{k\in\mathbb{Z}}\hat{\phi}(\gamma+k)\overline{\hat{\phi}_1(\gamma+k)}\right)^{-1}\widehat{\phi_1}(\gamma) & \text{on } \{\gamma: \Phi(\gamma) \neq 0\}, \\ 0 & \text{on } \{\gamma: \Phi(\gamma) = 0\}. \end{cases}$$

Corollary 7.4.3 can be used to "tailor" an oblique dual: that is, if the canonical dual frame does not satisfy the requirements for a specific application, we might search for an oblique dual that does. As an example, we show that one might be able to construct oblique duals of arbitrary smoothness, even if the canonical dual frame consists of non-continuous functions:

Example 7.4.4 Consider the B-spline B_n for some $n \in \mathbb{N}$. By Theorem 7.1.8, we know that $\{T_kB_n\}_{k\in\mathbb{Z}}$ is a Riesz sequence; in particular, $\{T_kB_n\}_{k\in\mathbb{Z}}$ has a unique dual frame consisting of elements in $\overline{\text{span}}\{T_kB_n\}_{k\in\mathbb{Z}}$. Now, fix any $m \in \mathbb{N}$; we will show that there exists an oblique dual $\{T_k\tilde{\phi}\}_{k\in\mathbb{Z}}$ of $\{T_kB_n\}_{k\in\mathbb{Z}}$ belonging to $\overline{\text{span}}\{T_kB_{n+2m}\}_{k\in\mathbb{Z}}$, i.e., such that

$$\tilde{\phi} \in \overline{\text{span}}\{T_kB_{n+2m}\}_{k\in\mathbb{Z}}.$$

For any $\gamma \in \mathbb{R}$, the argument in (7.17) shows that

$$\sum_{k\in\mathbb{Z}}\widehat{B_n}(\gamma+k)\overline{\widehat{B_{n+2m}}(\gamma+k)} = \sum_{k\in\mathbb{Z}}\left(\frac{\sin\pi(\gamma+k)}{\pi(\gamma+k)}\right)^{2(m+n)} \geq \left(\frac{2}{\pi}\right)^{2(m+n)}.$$

By Corollary 7.4.3, there exists a function $\tilde{\phi} \in \overline{\text{span}}\{T_kB_{n+2m}\}_{k\in\mathbb{Z}}$ that generates an oblique dual of $\{T_kB_n\}_{k\in\mathbb{Z}}$. That is, for an arbitrary spline B_n, we can find an oblique dual $\{T_k\tilde{\phi}\}_{k\in\mathbb{Z}}$ for which the generator $\tilde{\phi}$ has prescribed smoothness. In contrast, the canonical dual of $\{T_kB_n\}_{k\in\mathbb{Z}}$ has the same smoothness as B_n itself; for example, the canonical dual frame of $\{T_kB_1\}_{k\in\mathbb{Z}}$ is generated by B_1, which is not even continuous. $\square$

In the rest of this section, we will restrict our attention to frame sequences $\{T_k\phi\}_{k\in\mathbb{Z}}$ generated by compactly supported functions. As we have seen in Example 7.2.3, this does not imply that the canonical dual frame necessarily is generated by a compactly supported function. Nevertheless, we will now show that it often is possible to find oblique duals $\{T_k\tilde{\phi}\}_{k\in\mathbb{Z}}$ for which the function $\tilde{\phi}$ has compact (and small) support. For convenience, we choose the support to be the interval $[0,1]$, but the same considerations work on any other interval.

In Theorem 7.4.1, we characterized the oblique duals $\{T_k\tilde{\phi}\}_{k\in\mathbb{Z}}$ associated with a given frame of translates $\{T_k\phi\}_{k\in\mathbb{Z}}$. We first show that this result has a much simpler version if we assume that the functions ϕ and $\tilde{\phi}$ are compactly supported:

Lemma 7.4.5 *Assume that the functions $\phi, \tilde{\phi} \in L^2(\mathbb{R})$ have compact support. Then the following are equivalent:*

(i) $f = \sum_{k\in\mathbb{Z}}\langle f, T_k\tilde{\phi}\rangle T_k\phi, \ \forall f \in V;$

(ii) $\langle \phi, T_k\tilde{\phi}\rangle = \delta_{k,0}.$

Proof. If (i) holds, then Theorem 7.4.1 shows that $\{T_k\phi\}_{k\in\mathbb{Z}}$ is a frame for V. Because of the compact support of ϕ, this implies by Proposition 7.3.1 that $\{T_k\phi\}_{k\in\mathbb{Z}}$ is a Riesz sequence. Using (i) on $f = \phi$, the statement in (ii) follows because the expansion coefficients in terms of a Riesz basis are unique. On the other hand, if (ii) holds, then (i) holds for $f = \phi$. A change of the summation index proves that then (i) holds for any translate $T_k\phi$, and therefore on $\text{span}\{T_k\phi\}_{k\in\mathbb{Z}}$; finally, by continuity of the operator $f \mapsto \sum_{k\in\mathbb{Z}}\langle f, T_k\tilde{\phi}\rangle T_k\phi$ we obtain that (i) holds for all $f \in V$. $\qquad\square$

Lemma 7.4.5 shows that, with the given assumptions, the question of finding an oblique dual of a frame $\{T_k\phi\}_{k\in\mathbb{Z}}$ can be reformulated as a so-called *moment problem*, see Section 2.6.

The essence of the following result is that, under certain conditions, a Riesz sequence $\{T_k\phi\}_{k\in\mathbb{Z}}$ has an oblique dual $\{T_k\tilde{\phi}\}_{k\in\mathbb{Z}}$, where $\tilde{\phi}$ has the form

$$\tilde{\phi}(x) = \left(\sum_{\ell=0}^{N-1} d_\ell \phi(x+\ell)\right)\chi_{[0,1]}(x). \tag{7.28}$$

For practical reasons, we will formulate a more general version of the result. The reason is that even if ϕ is smooth, the multiplication with the characteristic function $\chi_{[0,1]}$ in the expression for $\tilde{\phi}$ in (7.28) might lead to a function that is discontinuous at $x = 0$ or $x = 1$. On the other hand, multiplying the function in (7.28) with a function of the type $x^p(1-x)^q$ for some $p, q \in \mathbb{N}$ will lead to a continuous function if ϕ is continuous – and we show that functions of that type can be used as generators as well:

Theorem 7.4.6 *Assume that $\phi \in L^2(\mathbb{R})$ is a real-valued function with support on an interval $[0, N]$ for some $N \in \mathbb{N}$, and that $\{T_k\phi\}_{k\in\mathbb{Z}}$ is a Riesz sequence. Assume that*

$$\sum_{k=0}^{N-1} c_k\phi(x + k) = 0, \forall x \in [0, 1] \Rightarrow c_0 = 0. \tag{7.29}$$

Then, for any $p, q \in \{0\} \cup \mathbb{N}$, $\{T_k\phi\}_{k\in\mathbb{Z}}$ has an oblique dual $\{T_k\tilde{\phi}\}_{k\in\mathbb{Z}}$, where $\tilde{\phi}$ has the form

$$\tilde{\phi}(x) = x^p(1 - x)^q \left(\sum_{\ell=0}^{N-1} d_\ell\phi(x + \ell)\right) \chi_{[0,1]}(x) \tag{7.30}$$

for some coefficients $d_0, \ldots, d_{N-1} \in \mathbb{R}$.

Proof. We use Lemma 7.4.5, and search for a function $\tilde{\phi}$ such that $\langle \phi, T_k\tilde{\phi}\rangle = \delta_{k,0}$. First, for any function $\tilde{\phi} \in L^2(\mathbb{R})$ with support on $[0, 1]$, we have that

$$\langle \phi, T_k\tilde{\phi}\rangle = \int_{-\infty}^{\infty} \phi(x)\overline{\tilde{\phi}(x - k)}\, dx = \int_{-\infty}^{\infty} \phi(x + k)\overline{\tilde{\phi}(x)}\, dx$$

$$= \int_0^1 \phi(x + k)\overline{\tilde{\phi}(x)}\, dx.$$

Assuming that ϕ has support on $[0, N]$, this shows that

$$\langle \phi, T_k\tilde{\phi}\rangle = 0 \text{ if } k \notin \{0, 1, \ldots, N - 1\}.$$

Thus the moment problem in Lemma 7.4.5(ii) takes the form

$$\langle T_{-k}\phi, \tilde{\phi}\rangle = \delta_{k,0}, \ k = 0, 1, \ldots, N - 1. \tag{7.31}$$

Because of the assumption (7.29), we know that if

$$\sum_{k=0}^{N-1} c_k\phi(x + k)x^{p/2}(1 - x)^{q/2} = 0 \text{ for all } x \in [0, 1],$$

then $c_0 = 0$. Thus, according to Lemma 2.6.1 with $\mathcal{H} = L^2(0, 1)$ and f_k corresponding to the functions $x \mapsto \phi(x+k)x^{p/2}(1-x)^{q/2}, k = 0, \ldots, N-1$, the moment problem

$$\begin{cases} 1 = \displaystyle\int_0^1 \phi(x)x^{p/2}(1 - x)^{q/2}h(x)\, dx \\[2mm] 0 = \displaystyle\int_0^1 \phi(x + 1)x^{p/2}(1 - x)^{q/2}h(x)\, dx \\[1mm] \ . \\[1mm] \ . \\[1mm] 0 = \displaystyle\int_0^1 \phi(x + N - 1)x^{p/2}(1 - x)^{q/2}h(x)\, dx \end{cases}$$

has a solution h of the form

$$h(x) = \left(\sum_{\ell=0}^{N-1} d_\ell \phi(x+\ell) x^{p/2} (1-x)^{q/2} \right) \chi_{[0,1]}(x).$$

This means that the function

$$\tilde{\phi}(x) := x^{p/2}(1-x)^{q/2} h(x) = \left(\sum_{\ell=0}^{N-1} d_\ell \phi(x+\ell) \right) x^p (1-x)^q \chi_{[0,1]}(x)$$

solves the moment problem (7.31). $\qquad\square$

Note that the coefficients $d_0, \ldots, d_{N-1}$ in (7.30) are determined by the conditions in (7.31), i.e., by the equations

$$\sum_{\ell=0}^{N-1} \int_0^1 d_\ell \phi(x+k)\phi(x+\ell) x^p (1-x)^q \, dx = \delta_{k,0}, \ \ k = 0, \ldots, N-1. \ (7.32)$$

On matrix form, this takes the form

$$M\mathbf{d} = \mathbf{e},$$

where M is the $N \times N$ symmetric matrix with entries

$$M_{k,\ell} = \int_0^1 x^p (1-x)^q \phi(x+k)\phi(x+\ell) \, dx, \ \ k,\ell = 0, \ldots, N-1$$

and

$$\mathbf{d} = \begin{pmatrix} d_0 \\ d_1 \\ \cdot \\ \cdot \\ d_{N-1} \end{pmatrix}, \ \mathbf{e} = \begin{pmatrix} 1 \\ 0 \\ \cdot \\ \cdot \\ 0 \end{pmatrix}.$$

Recall that the parameters p and q in (7.30) were introduced in order to ensure higher-order derivatives of $\tilde{\phi}$ to exist. We see that this only affects the integrals in the entries of matrix M, but not the size of the matrix; thus, the computational complexity does not increase.

Let us apply the results to B-splines. Using Theorem 7.4.6, we will be able to find an oblique dual of any $\{T_k B_n\}_{k\in\mathbb{Z}}$, which is generated by a compactly supported function. In contrast, we saw in Example 7.2.3 that for $n \geq 2$, the generator for the canonical dual frame never has compact support.

Example 7.4.7 For any $n \in \mathbb{N}$, Lemma 6.1.5 shows that the functions $B_n(\cdot + k), k = 0, \ldots n-1,$ are linearly independent on the interval $[0,1]$. Thus, the condition (7.29) is satisfied. Therefore, for any $p, q \in \mathbb{N} \cup \{0\}$,

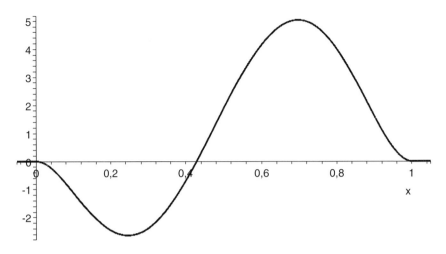

Figure 7.1. The generator $\tilde{\phi}$ in (7.30) corresponding to $\phi = B_2, p = q = 2$.

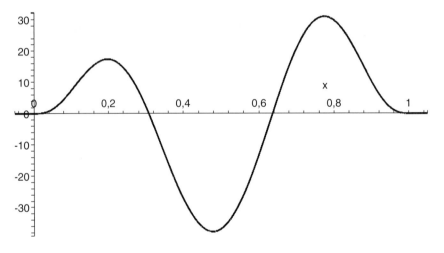

Figure 7.2. The generator $\tilde{\phi}$ in (7.30) corresponding to $\phi = B_3, p = q = 3$.

the frame sequence $\{T_k B_n\}_{k \in \mathbb{Z}}$ has an oblique dual $\{T_k \tilde{\phi}\}_{k \in \mathbb{Z}}$ with a function $\tilde{\phi}$ of the form

$$\tilde{\phi}(x) = \left(\sum_{\ell=0}^{n-1} d_\ell B_n(x + \ell) \right) x^p (1 - x)^q \chi_{[0,1]}(x).$$

Note that on the interval $[0, 1]$, this function is a polynomial of degree $p + q + n - 1$.

Figures 7.1 and 7.2 show some oblique dual generators for Riesz sequences generated by the B-splines B_2 and B_3, for various values of p and q. We ask the reader to do the calculations (Exercise 7.12).   □

7.5   An application to sampling theory

We will now return to the Paley–Wiener space PW, which we introduced in Section 3.8. Already in Theorem 3.8.1 we saw that any continuous function $f \in PW \cap L^1(\mathbb{R})$ can be recovered from its samples $\{f(k)\}_{k \in \mathbb{Z}}$ via

$$f(x) = \sum_{k \in \mathbb{Z}} f(k)\operatorname{sinc}(x - k). \tag{7.33}$$

As discussed in Section 5.9, one will always encounter quantization errors when this result is applied in practice. We will now show that the effect of the quantization error can be reduced via *oversampling*, i.e, by invoking samples $\{f(k/M)\}_{k \in \mathbb{Z}}$ for some $M \in \mathbb{N}, M > 1$. Note that for the sequence $\{f(k/M)\}_{k \in \mathbb{Z}}$, the distance between two consecutive samples is $1/M$.

First, as noticed in Theorem 3.8.1, the expansion (7.33) is really an expansion in terms of the orthonormal basis $\{\operatorname{sinc}(\cdot - k)\}_{k \in \mathbb{Z}}$ for the Paley–Wiener space PW. Given any $M \in \mathbb{N}$, it follows that for each $m = 0, \ldots, M - 1$, the family $\{\operatorname{sinc}(\cdot - k - m/M)\}_{k \in \mathbb{Z}}$ also forms an orthonormal basis for PW. The union of these bases, i.e.,

$$\bigcup_{m=0}^{M-1} \{\operatorname{sinc}(\cdot - k - m/M)\}_{k \in \mathbb{Z}} = \{\operatorname{sinc}(\cdot - k/M)\}_{k \in \mathbb{Z}},$$

therefore forms a tight frame for PW with frame bound $A = M$. Thus, for each $f \in PW$,

$$f = \frac{1}{M} \sum_{k \in \mathbb{Z}} \langle f, \operatorname{sinc}(\cdot - k/M) \rangle \operatorname{sinc}(\cdot - k/M). \tag{7.34}$$

Note that because $\{\operatorname{sinc}(\cdot - k)\}_{k \in \mathbb{Z}}$ forms an orthonormal basis for PW, we know from (7.33) that

$$f(k) = \langle f, T_k \operatorname{sinc} \rangle;$$

this implies that

$$
\begin{aligned}
\langle f, \operatorname{sinc}(\cdot - k/M) \rangle &= \int_{-\infty}^{\infty} f(x)\operatorname{sinc}(x - k/M)\, dx \\
&= \int_{-\infty}^{\infty} T_{-k/M} f(x)\operatorname{sinc}(x)\, dx \\
&= T_{-k/M} f(0) \\
&= f(k/M).
\end{aligned}
$$

Thus, the expansion (7.34) takes the form

$$f = \frac{1}{M} \sum_{k \in \mathbb{Z}} f(k/M) \operatorname{sinc}(\cdot - k/M). \tag{7.35}$$

We now adopt the model for quantization errors discussed in Section 5.9. In particular, Proposition 5.9.1 shows that oversampling with the factor M, i.e., the use of (7.35) instead of (7.33) reduces the energy of the quantization noise by a factor M:

$$E|(Qw)_k|^2 = \frac{\sigma^2}{A} = \frac{\sigma^2}{M}.$$

The relevance of this result is that it usually is easier to increase the redundancy of the frame than to increase the quantization precision.

7.6   Exercises

7.1 Prove Lemma 7.1.3.

7.2 Prove (7.12) in the proof of Lemma 7.1.6.

7.3 In this exercise, we ask the reader to provide some details in the proof of Theorem 7.1.7.

(i) Prove (7.15).

(ii) Prove the equivalence between (7.16) and the statement preceding it.

7.4 Let Φ be a positive trigonometric polynomial. Show that if $\Phi(\cdot)^{-1}$ is a trigonometric polynomial, then Φ is a constant.

7.5 Prove Corollary 7.3.2.

7.6 Show that the sequence $\{T_k \chi_{]0,2[}\}_{k \in \mathbb{Z}}$ does not form a Riesz sequence in $L^2(\mathbb{R})$.

7.7 For which values of $c \in]0, 2[$ does the system $\{T_k \chi_{]0,c[}\}_{k \in \mathbb{Z}}$ form a Riesz sequence in $L^2(\mathbb{R})$?

7.8 Let $\phi \in L^2(\mathbb{R})$ and assume that $\{T_k \phi\}_{k \in \mathbb{Z}}$ is an overcomplete frame sequence. We will show that either $\phi \notin L^1(\mathbb{R})$ or there is no constant $C > 0$ for which

$$|\hat{\phi}(\gamma)| \le C \left(\frac{1}{1 + |\gamma|^2} \right)^{1/2}. \tag{7.36}$$

(i) Assume that an estimate of the type (7.36) is available. Show that for $\gamma \in [0, 1[$ and all $N \in \mathbb{N}$,

$$\left| \Phi(\gamma) - \sum_{|k| \leq N} |\hat{\phi}(\gamma + k)|^2 \right| \leq 2C^2 \sum_{k=N}^{\infty} \frac{1}{1+k^2}.$$

(ii) Argue now that if (7.36) holds, then $\phi \notin L^1(\mathbb{R})$.

7.9 Prove Corollary 7.4.3. (Hint: use the argument from the proof of Corollary 7.4.2.)

7.10 This exercise connects to Exercise 3.7 and Exercise 5.2. Let $\mathcal{H}$ be a separable Hilbert space.

(i) Find a sequence $\{f_k\}_{k=1}^{\infty}$ in $\mathcal{H}$ that satisfies the lower frame condition but not the upper frame condition.

(ii) Find a sequence $\{f_k\}_{k=1}^{\infty}$ of vectors with norm 1 that satisfies the lower frame condition but not the upper frame condition.

7.11 Let $d > 0$, and prove that $\{e^{ikx/d}\}_{k \in \mathbb{Z}}$ is a frame for $L^2(-R, R)$ if and only if $R \in]0, \pi d]$.

7.12 Calculate the oblique duals of $\{T_k B_n\}_{k \in \mathbb{Z}}$ in Theorem 7.4.6 for $n = p = q = 2$ and $n = p = q = 3$ (use Maple or a similar program).

8

Shift-Invariant Systems

Chapter 7 dealt with systems of functions generated by integer-translates of a single function in $L^2(\mathbb{R})$. We will now generalize this setup and consider translates of a given countable family of functions rather than just one function. Such systems of functions are called shift-invariant systems. Our goal is to characterize various frame properties for shift-invariant systems, a subject that was treated first in the paper [58] by Ron and Shen. The presentation is inspired by the approach by Janssen in [44]. The derived results will play an important role in the analysis of Gabor systems in Chapter 9.

The theory for shift-invariant systems is based on two classes of operators on $L^2(\mathbb{R})$, namely,

Translation by $a \in \mathbb{R}$, $T_a : L^2(\mathbb{R}) \to L^2(\mathbb{R})$, $(T_a f)(x) = f(x - a)$;

Modulation by $b \in \mathbb{R}$, $E_b : L^2(\mathbb{R}) \to L^2(\mathbb{R})$, $(E_b f)(x) = e^{2\pi i b x} f(x)$.

Both classes of operators were introduced in Section 2.9; in particular, we will use their interaction with the Fourier transform, a subject that is also treated in Section 2.9.

8.1 Frame-properties of shift-invariant systems

Let $\{g_m\}_{m \in I}$ be a countable collection of functions in $L^2(\mathbb{R})$ and $a > 0$ be a given (shift-) parameter. The *shift-invariant system* generated by $\{g_m\}_{m \in I}$ and a is the collection of functions $\{g_m(\cdot - na)\}_{m \in I, n \in \mathbb{Z}}$. Formulated in

O. Christensen, *Frames and Bases*. DOI: 10.1007/978-0-8176-4678-3_8,
© Springer Science+Business Media, LLC 2008

terms of the translation operator, the system has the form $\{T_{na}g_m\}_{m\in I, n\in\mathbb{Z}}$. Usually, we will let $I = \mathbb{Z}$, in which case we simply write

$$\{g_{nm}\} := \{T_{na}g_m\}_{m,n\in\mathbb{Z}}. \tag{8.1}$$

As already mentioned, our goal is to characterize various frame properties for systems of the form $\{g_{nm}\}$. The Fourier transform will be an important tool; in fact, the characterizations will be formulated in terms of certain conditions on the functions $\widehat{g_m}$.

In particular, we will present equivalent conditions for two systems $\{g_{nm}\}$ and $\{h_{nm}\}$ to form dual frames. Given the two shift-invariant Bessel systems $\{g_{nm}\}$ and $\{h_{nm}\}$, and two functions $e, f \in L^2(\mathbb{R})$, the analysis of the function $\rho(e, f)$ defined by

$$\rho(e, f) : \mathbb{R} \to \mathbb{C}, \ \rho(e, f)(x) = \sum_{m,n\in\mathbb{Z}} \langle T_x e, g_{nm}\rangle \langle h_{nm}, T_x f\rangle \tag{8.2}$$

will play a central role. The reason for considering this function is apparent from our discussion of general dual frame pairs in Section 5.7: in fact, Lemma 5.7.1 shows that two Bessel sequences $\{g_{nm}\}$ and $\{h_{nm}\}$ form dual frames for $L^2(\mathbb{R})$ if and only if

$$\rho(e, f)(0) = \langle e, f\rangle, \ \forall e, f \in L^2(\mathbb{R}).$$

In the first result, we will derive a useful consequence of the Bessel condition.

Lemma 8.1.1 *Assume that $\{g_{nm}\}$ is a Bessel sequence with bound B. Then*

$$\sum_{m\in\mathbb{Z}} |\widehat{g_m}(\nu)|^2 \le aB, \ a.e. \ \nu \in \mathbb{R}. \tag{8.3}$$

Proof. Let $f \in L^2(\mathbb{R})$, and consider the function

$$\rho(f, f)(x) = \sum_{m,n\in\mathbb{Z}} |\langle T_x f, g_{nm}\rangle|^2; \tag{8.4}$$

it corresponds to the general expression in (8.2) in the case $h_m = g_m$. The assumption that $\{g_{nm}\}$ is a Bessel sequence with bound B implies that $\rho(f, f)$ is bounded: in fact,

$$\rho(f, f)(x) \le B\, ||T_x f||^2 = B\, ||f||^2, \ \forall x \in \mathbb{R}. \tag{8.5}$$

The shift-invariance of the system $\{g_{nm}\}$ implies that $\rho(f, f)$ is periodic with period a (Exercise 8.1), so we can consider its Fourier expansion in

$L^2(0, a)$. By definition, the Fourier coefficient c_0 is given by

$$
\begin{aligned}
c_0 &= \frac{1}{a} \int_0^a \sum_{m,n \in \mathbb{Z}} |\langle T_x f, g_{nm} \rangle|^2 \, dx \\
&= \frac{1}{a} \sum_{m,n \in \mathbb{Z}} \int_0^a \left| \int_{-\infty}^{\infty} f(z-x) \overline{g_m(z-na)} \, dz \right|^2 dx \\
&= \frac{1}{a} \sum_{m,n \in \mathbb{Z}} \int_0^a \left| \int_{-\infty}^{\infty} f(z-(x-na)) \overline{g_m(z)} \, dz \right|^2 dx. \qquad (8.6)
\end{aligned}
$$

Introducing the functions $\Phi_m(x)$, $m \in \mathbb{Z}$, by

$$
\Phi_m(x) := \left| \int_{-\infty}^{\infty} f(z-x) \overline{g_m(z)} \, dz \right|^2 = |\langle T_x f, g_m \rangle|^2, \quad x \in \mathbb{R},
$$

this can be written as

$$
\begin{aligned}
c_0 &= \frac{1}{a} \sum_{m \in \mathbb{Z}} \sum_{n \in \mathbb{Z}} \int_0^a \Phi_m(x - na) \, dx \\
&= \frac{1}{a} \sum_{m \in \mathbb{Z}} \int_{-\infty}^{\infty} \Phi_m(x) \, dx \\
&= \frac{1}{a} \sum_{m \in \mathbb{Z}} \int_{-\infty}^{\infty} |\langle T_x f, g_m \rangle|^2 dx.
\end{aligned}
$$

By direct calculation, we see that for arbitrary functions $e, \phi \in L^2(\mathbb{R})$,

$$
\begin{aligned}
\langle T_x e, \phi \rangle &= \langle \mathcal{F} T_x e, \mathcal{F} \phi \rangle \\
&= \langle E_{-x} \hat{e}, \hat{\phi} \rangle \\
&= \int_{-\infty}^{\infty} \hat{e}(\nu) \overline{\hat{\phi}(\nu)} e^{-2\pi i x \nu} \, d\nu \\
&= \mathcal{F} \left(\hat{e} \, \overline{\hat{\phi}} \right)(x). \qquad (8.7)
\end{aligned}
$$

Thus, via Parseval's equation,

$$
\begin{aligned}
c_0 &= \frac{1}{a} \sum_{m \in \mathbb{Z}} \int_{-\infty}^{\infty} \left| \mathcal{F}(\hat{f} \overline{\widehat{g_m}})(\nu) \right|^2 d\nu \\
&= \frac{1}{a} \sum_{m \in \mathbb{Z}} \int_{-\infty}^{\infty} \left| \hat{f}(\nu) \overline{\widehat{g_m}(\nu)} \right|^2 d\nu \\
&= \frac{1}{a} \int_{-\infty}^{\infty} |\hat{f}(\nu)|^2 \sum_{m \in \mathbb{Z}} |\widehat{g_m}(\nu)|^2 d\nu. \qquad (8.8)
\end{aligned}
$$

On the other hand, via the definition of c_0 used in (8.6), an estimate of c_0 can be obtained:

$$c_0 = \frac{1}{a} \int_0^a \sum_{m,n \in \mathbb{Z}} |\langle T_x f, g_{nm} \rangle|^2 \, dx$$

$$\leq B \, \|f\|^2$$

$$= B \int_{-\infty}^{\infty} |\hat{f}(\nu)|^2 d\nu.$$

Thus, via (8.8) we see that

$$\frac{1}{a} \int_{-\infty}^{\infty} |\hat{f}(\nu)|^2 \sum_{m \in \mathbb{Z}} |\widehat{g_m}(\nu)|^2 d\nu \leq B \int_{-\infty}^{\infty} |\hat{f}(\nu)|^2 d\nu.$$

Since this holds for all $f \in L^2(\mathbb{R})$, we conclude that

$$\frac{1}{a} \sum_{m \in \mathbb{Z}} |\widehat{g_m}(\nu)|^2 \leq B, \quad a.e. \ \nu \in \mathbb{R};$$

the desired result now follows. □

We will now analyze the function $\rho(e, f)$ in (8.2).

Lemma 8.1.2 *Assume that two shift-invariant systems $\{g_{nm}\}$ and $\{h_{nm}\}$ are Bessel sequences and let $e, f \in L^2(\mathbb{R})$. Then the function*

$$\rho(e, f) : \mathbb{R} \to \mathbb{C}, \ \rho(e, f)(x) = \sum_{m,n \in \mathbb{Z}} \langle T_x e, g_{nm} \rangle \langle h_{nm}, T_x f \rangle$$

is continuous and has period a. Its Fourier series in $L^2(0, a)$ is

$$\rho(e, f)(x) = \sum_{k \in \mathbb{Z}} c_k e^{2\pi i k x / a}, \tag{8.9}$$

where

$$c_k = \frac{1}{a} \int_{-\infty}^{\infty} \hat{e}(\nu) \overline{\hat{f}(\nu + k/a)} \sum_{m \in \mathbb{Z}} \widehat{g_m}(\nu) \overline{\hat{h}_m(\nu + k/a)} \, d\nu, \ k \in \mathbb{Z}. \tag{8.10}$$

Proof. Assume that $\{g_{nm}\}$ and $\{h_{nm}\}$ are Bessel sequences; then, an application of Cauchy–Schwarz' inequality shows that the series defining $\rho(e, f)(x)$ converges absolutely for all $x \in \mathbb{R}$. In particular, this demonstrates that the function $\rho(e, f)(x)$ is well defined. We will now prove that $\rho(e, f)$ is a continuous function. First, given $x, x_0 \in \mathbb{R}$,

$$|\rho(e, f)(x) - \rho(e, f)(x_0)|$$

$$= \left| \sum_{m,n \in \mathbb{Z}} (\langle T_x e, g_{nm} \rangle \langle h_{nm}, T_x f \rangle - \langle T_{x_0} e, g_{nm} \rangle \langle h_{nm}, T_{x_0} f \rangle) \right|$$

$$\leq \sum_{m,n \in \mathbb{Z}} |\langle T_x e, g_{nm} \rangle \langle h_{nm}, T_x f \rangle - \langle T_{x_0} e, g_{nm} \rangle \langle h_{nm}, T_{x_0} f \rangle|.$$

Writing $T_x e = T_x e - T_{x_0} + T_{x_0} e$, we see that

$$\langle T_x e, g_{nm}\rangle\langle h_{nm}, T_x f\rangle - \langle T_{x_0} e, g_{nm}\rangle\langle h_{nm}, T_{x_0} f\rangle$$
$$= \langle T_x e - T_{x_0} e, g_{nm}\rangle\langle h_{nm}, T_x f\rangle + \langle T_{x_0} e, g_{nm}\rangle\langle h_{nm}, T_x f\rangle$$
$$- \langle T_{x_0} e, g_{nm}\rangle\langle h_{nm}, T_{x_0} f\rangle$$
$$= \langle T_x e - T_{x_0} e, g_{nm}\rangle\langle h_{nm}, T_x f\rangle + \langle T_{x_0} e, g_{nm}\rangle\langle h_{nm}, T_x f - T_{x_0} f\rangle;$$

thus, letting B denote a common Bessel bound for $\{g_{nm}\}$ and $\{h_{nm}\}$,

$$|\rho(e,f)(x) - \rho(e,f)(x_0)|$$
$$\leq \sum_{m,n\in\mathbb{Z}} |\langle T_x e - T_{x_0} e, g_{nm}\rangle| \ |\langle h_{nm}, T_x f\rangle|$$
$$+ \sum_{m,n\in\mathbb{Z}} |\langle T_{x_0} e, g_{nm}\rangle| \ |\langle h_{nm}, T_x f - T_{x_0} f\rangle|$$
$$\leq \left(\sum_{m,n\in\mathbb{Z}} |\langle T_x e - T_{x_0} e, g_{nm}\rangle|^2\right)^{1/2} \left(\sum_{m,n\in\mathbb{Z}} |\langle h_{nm}, T_x f\rangle|^2\right)^{1/2}$$
$$+ \left(\sum_{m,n\in\mathbb{Z}} |\langle T_{x_0} e, g_{nm}\rangle|^2\right)^{1/2} \left(\sum_{m,n\in\mathbb{Z}} |\langle h_{nm}, T_x f - T_{x_0} f\rangle|^2\right)^{1/2}$$
$$\leq B \ ||T_x e - T_{x_0} e|| \ ||T_x f|| + B \ ||T_{x_0} e|| \ ||T_x f - T_{x_0} f||$$
$$= B \ ||T_{x-x_0} e - e|| \ ||f|| + B \ ||e|| \ ||T_{x-x_0} f - f||.$$

The last expression converges to zero for $x \to x_0$ by Lemma 2.9.1; this proves the desired continuity.

The periodicity of $\rho(e,f)$ follows from the structure of the shift-invariant systems $\{g_{nm}\}$ and $\{h_{nm}\}$ (Exercise 8.2). For the computation of the Fourier coefficients, we first assume that e, f are continuous functions with compact support; this will justify the following interchanges of sums and integrals. The coefficients in the Fourier expansion with respect to $\{e^{2\pi ikx/a}\}_{k\in\mathbb{Z}}$ are given by (Exercise 8.2)

$$\begin{aligned} c_k &= \frac{1}{a} \int_0^a \rho(e,f)(x) e^{-2\pi ikx/a} \, dx \\ &= \frac{1}{a} \sum_{m\in\mathbb{Z}} \sum_{n\in\mathbb{Z}} \int_0^a \langle T_x e, g_m(\cdot - na)\rangle\langle h_m(\cdot - na), T_x f\rangle e^{-2\pi ikx/a} \, dx \\ &= \frac{1}{a} \sum_{m\in\mathbb{Z}} \int_{-\infty}^{\infty} \langle T_x e, g_m\rangle\langle h_m, T_x f\rangle e^{-2\pi ikx/a} \, dx \\ &= \frac{1}{a} \sum_{m\in\mathbb{Z}} \int_{-\infty}^{\infty} \langle T_x e, g_m\rangle\overline{\langle T_x f, h_m\rangle} e^{2\pi ikx/a} \, dx. \end{aligned} \tag{8.11}$$

Using (8.7), it follows that

$$\langle T_x f, h_m \rangle e^{2\pi i k x / a} = E_{k/a} \mathcal{F}\left(\hat{f}\, \overline{\widehat{h_m}}\right)(x)$$

$$= \mathcal{F}\left(T_{-k/a}(\hat{f}\, \overline{\widehat{h_m}})\right)(x).$$

Inserting this as well as (8.7) in (8.11) and using Plancherel's equation leads to

$$c_k = \frac{1}{a} \sum_{m \in \mathbb{Z}} \int_{-\infty}^{\infty} \langle T_x e, g_m \rangle \overline{\langle T_x f, h_m \rangle e^{2\pi i k x / a}}\, dx$$

$$= \frac{1}{a} \sum_{m \in \mathbb{Z}} \int_{-\infty}^{\infty} \mathcal{F}\left(\hat{e}\, \overline{\widehat{g_m}}\right)(x) \overline{\mathcal{F}\left(T_{-k/a}(\hat{f}\, \overline{\widehat{h_m}})\right)(x)}\, dx$$

$$= \frac{1}{a} \sum_{m \in \mathbb{Z}} \int_{-\infty}^{\infty} \left(\hat{e}\, \overline{\widehat{g_m}}\right)(\nu) \overline{T_{-k/a}(\hat{f}\, \overline{\widehat{h_m}})(\nu)}\, d\nu$$

$$= \frac{1}{a} \sum_{m \in \mathbb{Z}} \int_{-\infty}^{\infty} \hat{e}(\nu) \overline{\widehat{g_m}(\nu)} \hat{f}(\nu + k/a) \overline{\widehat{h_m}(\nu + k/a)}\, d\nu.$$

The reader can check (Exercise 8.2) that the series

$$\int_{-\infty}^{\infty} |\hat{e}(\nu)|\, |\hat{f}(\nu + k/a)| \sum_{m \in \mathbb{Z}} |\widehat{g_m}(\nu)|\, |\widehat{h_m}(\nu + k/a)|\, d\nu \qquad (8.12)$$

is convergent. Thus, we can interchange the sum and the integral in the above expression for c_0. This proves the result for all functions $e, f \in C_c(\mathbb{R})$. The general case now follows by a density argument (Exercise 8.2).   □

It is very complicated to check the frame conditions for a shift-invariant system directly via the definition. We will now derive equivalent conditions in terms of matrix-valued functions. However, some preparation is needed before we state the main results in Theorem 8.1.6 and Theorem 8.1.7.

We begin with a definition.

Definition 8.1.3 *For $f \in L^2(\mathbb{R})$, the fiber of f at a point $\nu \in \mathbb{R}$ is defined as the sequence*

$$\hat{\mathbf{f}}(\nu) := \{\hat{f}(\nu - k/a)\}_{k \in \mathbb{Z}}.$$

The following lemma shows that the sequence $\hat{\mathbf{f}}(\nu)$ belongs to $\ell^2(\mathbb{Z})$ for a.e. $\nu \in \mathbb{Z}$; for this reason, $||\hat{\mathbf{f}}(\nu)||$ will always denote the ℓ^2-norm. We ask the reader to provide the proof (Exercise 8.3).

Lemma 8.1.4 *Let $f \in L^2(\mathbb{R})$. Then $\mathbf{f}(\nu) \in \ell^2(\mathbb{Z})$ for a.e. $\nu \in \mathbb{Z}$. Furthermore, for any interval I of length $1/a$,*

$$||f||^2 = \int_I \sum_{k \in \mathbb{Z}} |\hat{f}(\nu - k/a)|^2\, d\nu = \int_I ||\hat{\mathbf{f}}(\nu)||^2 d\nu.$$

Given a shift-invariant system $\{g_{nm}\}$ as in (8.1), define the matrix-valued function

$$H(\nu) := (\widehat{g_m}(\nu - k/a))_{k,m\in\mathbb{Z}}, \quad a.e. \ \nu \in \mathbb{R}. \tag{8.13}$$

Note that the columns in the matrix $H(\nu)$ consist of the fibers for the functions g_m. In case $H(\nu)$ defines a bounded operator on $\ell^2(\mathbb{Z})$ for some $\nu \in \mathbb{R}$, the adjoint operator will be denoted by $H(\nu)^*$; it is given by

$$H(\nu)^* = \left(\overline{\widehat{g_m}(\nu - k/a)}\right)_{m,k\in\mathbb{Z}}.$$

For technical reasons, several of the following results will first be proven for functions f belonging to a subspace of the *Schwartz space* $\mathcal{S}$ of rapidly decaying functions. Recall that $\mathcal{S}$ consists of all infinitely often differentiable functions f on $\mathbb{R}$, which decay faster than any inverse polynomial; that is, for any $\alpha, k \in \mathbb{N} \cup \{0\}$,

$$\sup_{x\in\mathbb{R}} \left| x^\alpha \frac{d^k f}{dx^k}(x) \right| < \infty.$$

Let

$$\mathcal{D} := \{f \in \mathcal{S} \mid \hat{f} \text{ is compactly supported}\}. \tag{8.14}$$

One can prove that $\mathcal{D}$ is a dense subspace of $L^2(\mathbb{R})$.

Lemma 8.1.5 *Let $\{g_{nm}\}$ be a shift-invariant system in $L^2(\mathbb{R})$, and assume that for a.e. $\nu \in \mathbb{R}$, the matrix $H(\nu)$ defines a bounded operator on $\ell^2(\mathbb{Z})$. Then, for any interval I of length $1/a$ and any function $f \in \mathcal{D}$,*

$$\int_I \|H(\nu)^*\hat{\mathbf{f}}(\nu)\|^2 d\nu = a \sum_{m,n\in\mathbb{Z}} |\langle f, g_{nm}\rangle|^2.$$

Proof. For the set of $\nu \in \mathbb{R}$ for which the matrix $H(\nu)$ is bounded, its adjoint is bounded, and $\|H(\nu)\| = \|H(\nu)^*\|$. Given any interval I of length $1/a$,

$$\int_I \|H(\nu)^*\hat{\mathbf{f}}(\nu)\|^2 d\nu = \int_I \sum_{m\in\mathbb{Z}} \left| \sum_{k\in\mathbb{Z}} \overline{\widehat{g_m}(\nu - k/a)}\hat{f}(\nu - k/a) \right|^2 d\nu$$

$$= \sum_{m\in\mathbb{Z}} \int_I \left| \sum_{k\in\mathbb{Z}} \overline{\widehat{g_m}(\nu - k/a)}\hat{f}(\nu - k/a) \right|^2 d\nu.$$

Now, for each $m \in \mathbb{Z}$, one can prove (Exercise 8.4) that the mapping

$$\nu \mapsto \sum_{k\in\mathbb{Z}} \overline{\widehat{g_m}(\nu - k/a)}\hat{f}(\nu - k/a) \tag{8.15}$$

is well defined for a.e. $\nu \in \mathbb{R}$ and defines a function in $L^2(I)$ with Fourier series

$$\sum_{k\in\mathbb{Z}} \overline{\widehat{g_m}(\cdot - k/a)} \hat{f}(\cdot - k/a) = \sum_{n\in\mathbb{Z}} a\langle f, g_{nm}\rangle e^{2\pi i n a(\cdot)}. \tag{8.16}$$

Parseval's equation, see (3.24), shows that

$$a \int_I \left| \sum_{k\in\mathbb{Z}} \overline{\widehat{g_m}(\nu - k/a)} \hat{f}(\nu - k/a) \right|^2 d\nu = a^2 \sum_{n\in\mathbb{Z}} |\langle f, g_{nm}\rangle|^2.$$

Thus,

$$a \sum_{m,n\in\mathbb{Z}} |\langle f, g_{nm}\rangle|^2 = \sum_{m\in\mathbb{Z}} \int_I \left| \sum_{k\in\mathbb{Z}} \overline{\widehat{g_m}(\nu - k/a)} \hat{f}(\nu - k/a) \right|^2 d\nu$$

$$= \int_I ||H(\nu)^* \hat{\mathbf{f}}(\nu)||^2 d\nu,$$

as desired.   □

We are now ready to state characterizations of several frame properties for shift-invariant systems. They are formulated in terms of the matrix-valued function $H(\nu)$ in (8.13), i.e., in terms of the fibers associated with the generators g_m. We begin with the upper frame condition.

Theorem 8.1.6 *A shift-invariant system $\{g_{nm}\}$ is a Bessel sequence with bound B if and only if the matrix $H(\nu)$ for a.e. $\nu \in \mathbb{R}$ defines a bounded operator on $\ell^2(\mathbb{Z})$ with norm at most $\sqrt{aB}$.*

Proof. Let us first assume that the matrix $H(\nu)$ for a.e. $\nu \in \mathbb{R}$ defines a bounded operator on $\ell^2(\mathbb{Z})$ with norm at most $\sqrt{aB}$. Fix any interval I of length $1/a$. Then, for any function f belonging to the space $\mathcal{D}$ defined in (8.14), Lemma 8.1.5 shows that

$$\sum_{m,n\in\mathbb{Z}} |\langle f, g_{nm}\rangle|^2 = \frac{1}{a} \int_I ||H(\nu)^* \hat{\mathbf{f}}(\nu)||^2 d\nu \le \frac{aB}{a} \int_I ||\hat{\mathbf{f}}(\nu)||^2 d\nu = B ||f||^2.$$

Since $\mathcal{D}$ is dense in $L^2(\mathbb{R})$, it follows from Lemma 3.1.6 that $\{g_{nm}\}$ is a Bessel sequence with bound B.

Assume now that $\{g_{nm}\}$ is a Bessel sequence with bound B. We have to prove that for almost all $\nu \in \mathbb{R}$, the inequality

$$\sum_{k\in\mathbb{Z}} \left| \sum_{m\in\mathbb{Z}} \widehat{g_m}(\nu - k/a) c_m \right|^2 \le aB \sum_{m\in\mathbb{Z}} |c_m|^2 \tag{8.17}$$

holds for all $\{c_m\}_{m\in\mathbb{Z}} \in \ell^2(\mathbb{Z})$. We first assume that $\{c_m\}_{m\in\mathbb{Z}}$ is a finite sequence. Given another finite sequence $\{d_n\}_{n\in\mathbb{Z}}$, we consider the trigonometric polynomial

$$\varphi(\nu) = \sum_{n \in \mathbb{Z}} d_n e^{-2\pi i n a \nu}.$$

Note that φ has period $1/a$. For any interval I of length $1/a$, Parseval's theorem, see (3.24), shows that

$$\sum_{n \in \mathbb{Z}} |d_n|^2 = a \int_I |\varphi(\nu)|^2 d\nu.$$

The periodicity of φ implies that for any such interval I, we have that

$$\int_I |\varphi(\nu)|^2 \sum_{k \in \mathbb{Z}} \left| \sum_{m \in \mathbb{Z}} \widehat{g_m}(\nu - k/a) c_m \right|^2 d\nu$$

$$= \int_I \sum_{k \in \mathbb{Z}} \left| \sum_{m \in \mathbb{Z}} \varphi(\nu - k/a) \widehat{g_m}(\nu - k/a) c_m \right|^2 d\nu$$

$$= \int_{-\infty}^{\infty} \left| \sum_{m \in \mathbb{Z}} \varphi(\gamma) \widehat{g_m}(\nu) c_m \right|^2 d\nu$$

$$= \int_{-\infty}^{\infty} \left| \sum_{m \in \mathbb{Z}} \sum_{n \in \mathbb{Z}} d_n c_m e^{-2\pi i n a \nu} \widehat{g_m}(\nu) \right|^2 d\nu$$

$$= \int_{-\infty}^{\infty} \left| \mathcal{F} \sum_{m \in \mathbb{Z}} \sum_{n \in \mathbb{Z}} d_n c_m g_m(\nu - na) \right|^2 d\nu$$

$$= \left\| \sum_{m,n \in \mathbb{Z}} d_n c_m g_{nm} \right\|^2.$$

Using that $\{g_{nm}\}$ is a Bessel sequence with bound B, we can estimate this term as follows:

$$\left\| \sum_{m,n \in \mathbb{Z}} d_n c_m g_{nm} \right\|^2 \leq B \sum_{m,n \in \mathbb{Z}} |d_n c_m|^2$$

$$= B \sum_{m \in \mathbb{Z}} |c_m|^2 \sum_{n \in \mathbb{Z}} |d_n|^2$$

$$= aB \sum_{m \in \mathbb{Z}} |c_m|^2 \int_I |\varphi(\nu)|^2 d\nu.$$

Altogether, we arrive at the inequality

$$\int_I |\varphi(\nu)|^2 \sum_{k \in \mathbb{Z}} \left| \sum_{m \in \mathbb{Z}} \widehat{g_m}(\nu - k/a) c_m \right|^2 d\nu \leq aB \sum_{m \in \mathbb{Z}} |c_m|^2 \int_I |\varphi(\nu)|^2 d\nu.$$

Since this holds for all trigonometric polynomials φ with period $1/a$, we conclude that (8.17) holds for a.e. $\nu \in I$, for any given finite sequence $\{c_m\}_{m\in\mathbb{Z}}$; for reasons of periodicity, it therefore holds for a.e. $\nu \in \mathbb{R}$ for such sequences.

However, we have to prove that there is a null set $N \subset \mathbb{R}$ such that (8.17) holds for *all* $\{c_m\}_{m\in\mathbb{Z}} \in \ell^2(\mathbb{Z})$ if $\nu \in \mathbb{R} \setminus N$. In order to do so, let $V \subset \ell^2(\mathbb{Z})$ be a subset formed by a countable and dense collection of finite sequences $\{c_m\}_{m\in\mathbb{Z}}$. Note that a countable union of null sets again is a null set; this implies that there exists a null set $N \subset J$ such that (8.17) holds for all $\nu \in \mathbb{R} \setminus N$ and all $\{c_m\}_{m\in\mathbb{Z}} \in V$. Via Lemma 8.1.1, we might also assume that the inequality (8.3) holds for all $\nu \in \mathbb{R} \setminus N$. We now prove that for $\nu \in \mathbb{R} \setminus N$, the result in (8.17) actually holds for all $\{c_m\}_{m\in\mathbb{Z}} \in \ell^2(\mathbb{Z})$. In order to do so, let $\{c_m\}_{m\in\mathbb{Z}} \in \ell^2(\mathbb{Z})$ be given, and take a sequence of finite sequences $\{c_m^n\}_{m\in\mathbb{Z}} \in V$, $n \in \mathbb{N}$, such that

$$\{c_m^n\}_{m\in\mathbb{Z}} \to \{c_m\}_{m\in\mathbb{Z}} \text{ in } \ell^2(\mathbb{Z}) \text{ as } n \to \infty.$$

Now let $\nu \in \mathbb{R} \setminus N$. For all $k \in \mathbb{Z}$, we know that $\{\widehat{g_m}(\nu - k/a)\}_{m\in\mathbb{Z}} \in \ell^2(\mathbb{Z})$; thus,

$$
\begin{aligned}
\sum_{m\in\mathbb{Z}} \widehat{g_m}(\nu - k/a)c_m &= \langle \{\widehat{g_m}(\nu - k/a)\}_{m\in\mathbb{Z}}, \{\overline{c_m}\}_{m\in\mathbb{Z}} \rangle_{\ell^2(\mathbb{Z})} \\
&= \lim_{n\to\infty} \langle \{\widehat{g_m}(\nu - k/a)\}_{m\in\mathbb{Z}}, \{\overline{c_m^n}\}_{m\in\mathbb{Z}} \rangle_{\ell^2(\mathbb{Z})} \\
&= \liminf_{n\to\infty} \langle \{\widehat{g_m}(\nu - k/a)\}_{m\in\mathbb{Z}}, \{\overline{c_m^n}\}_{m\in\mathbb{Z}} \rangle_{\ell^2(\mathbb{Z})} \\
&= \liminf_{n\to\infty} \sum_{m\in\mathbb{Z}} \widehat{g_m}(\nu - k/a)c_m^n.
\end{aligned}
$$

By Fatou's lemma, see Lemma 2.7.4, this implies that

$$
\begin{aligned}
\sum_{k\in\mathbb{Z}} \left| \sum_{m\in\mathbb{Z}} \widehat{g_m}(\nu - k/a)c_m \right|^2 &= \sum_{k\in\mathbb{Z}} \liminf_{n\to\infty} \left| \sum_{m\in\mathbb{Z}} \widehat{g_m}(\nu - k/a)c_m^n \right|^2 \\
&\leq \liminf_{n\to\infty} \sum_{k\in\mathbb{Z}} \left| \sum_{m\in\mathbb{Z}} \widehat{g_m}(\nu - k/a)c_m^n \right|^2 \\
&\leq aB \liminf_{n\to\infty} \sum_{m\in\mathbb{Z}} |c_m^n|^2 \\
&= aB \sum_{m\in\mathbb{Z}} |c_m|^2.
\end{aligned}
$$

This shows that for $\nu \in \mathbb{R} \setminus N$, the inequality (8.17) indeed holds for all $\{c_m\}_{m\in\mathbb{Z}} \in \ell^2(\mathbb{Z})$. This completes the proof. $\qquad\square$

We now state characterizations of various frame properties for the system $\{g_{nm}\}$ in terms of the matrices $H(\nu)$ and their adjoints. The first of these is formulated as a matrix inequality, involving certain positive bi-infinite matrices indexed by $\mathbb{Z} \times \mathbb{Z}$. Recall that for two such matrices M and N, the inequality $N \leq M$ means that for all sequences $\{c_k\} \in \ell^2(\mathbb{Z})$,

$$\langle N\{c_k\}, \{c_k\}\rangle \leq \langle M\{c_k\}, \{c_k\}\rangle. \tag{8.18}$$

Theorem 8.1.7 *The following characterizations hold:*

(i) *A Bessel sequence $\{g_{nm}\}$ is a frame for $L^2(\mathbb{R})$ with lower frame bound A if and only if*

$$aAI \leq H(\nu)H(\nu)^*, \text{ a.e. } \nu \in \mathbb{R}. \tag{8.19}$$

(ii) *$\{g_{nm}\}$ is a tight frame for $L^2(\mathbb{R})$ if and only if there is a constant $c > 0$ such that*

$$\sum_{m \in \mathbb{Z}} \overline{\widehat{g_m}(\nu)} \widehat{g_m}(\nu + k/a) = c\delta_{k,0}, \ k \in \mathbb{Z}, \text{ a.e. } \nu \in \mathbb{R}. \tag{8.20}$$

In case (8.20) is satisfied, the frame bound is $A = c/a$.

(iii) *Two shift-invariant systems $\{g_{nm}\}$ and $\{h_{nm}\}$, which form Bessel sequences, are dual frames if and only if*

$$\sum_{m \in \mathbb{Z}} \overline{\widehat{g_m}(\nu)} \widehat{h_m}(\nu + k/a) = a\delta_{k,0}, \ k \in \mathbb{Z}, \text{ a.e. } \nu \in \mathbb{R}. \tag{8.21}$$

Proof. To prove (i), first assume that (8.19) holds. According to (8.18), this means that

$$aA \, ||\{c_k\}_{k \in \mathbb{Z}}||^2 \leq \langle H(\nu)H(\nu)^*\{c_k\}_{k \in \mathbb{Z}}, \{c_k\}_{k \in \mathbb{Z}}\rangle$$

for all $\{c_k\}_{k \in \mathbb{Z}} \in \ell^2(\mathbb{Z})$; or, equivalently, that

$$||H(\nu)^*\{c_k\}_{k \in \mathbb{Z}}||^2 \geq aA \, ||\{c_k\}_{k \in \mathbb{Z}}||^2 \tag{8.22}$$

holds for all $\{c_k\}_{k \in \mathbb{Z}} \in \ell^2(\mathbb{Z})$. Let $I \subset \mathbb{R}$ be an arbitrary interval of length $1/a$. Then, for any function $f \in \mathcal{D}$, Lemma 8.1.5 shows that

$$\sum_{m,n \in \mathbb{Z}} |\langle f, g_{nm}\rangle|^2 = \frac{1}{a} \int_I ||H(\nu)^*\hat{f}(\nu)||^2 d\nu \geq A \int_I ||\hat{f}(\nu)||^2 d\nu = A \, ||f||^2;$$

by Lemma 5.1.2, this implies that the number A is a lower frame bound for $\{g_{nm}\}$.

The second part of the proof of (i) is more technical. Assume that A is a lower frame bound for $\{g_{nm}\}$; we want to prove that the inequality in (8.22) holds for all $\{c_k\}_{k \in \mathbb{Z}} \in \ell^2(\mathbb{Z})$. Like in the proof of Theorem 8.1.6, we first consider a finite sequence $\{c_k\}_{k \in \mathbb{Z}}$. For technical reasons that will become clear soon, we furthermore consider a function φ for which $\hat{\varphi}$ is supported on an interval I of length $1/a$. We can associate the function φ to a fiber of a certain function f. In fact, define the function f in terms of its Fourier transform by the following: given $\nu \in \mathbb{R}$, choose the unique number $k \in \mathbb{Z}$ such that $\nu + k/a \in I$, and let

$$\hat{f}(\nu) = c_k \hat{\varphi}(\nu + k/a).$$

By definition, the ℓ-th coordinate in the fiber for a function f is $\hat{f}(\nu - \ell/a)$. Thus, directly by the definition of the function f, we see that for $\nu \in I$,

$$\hat{f}(\nu - \ell/a) = c_\ell \hat{\varphi}(\nu).$$

Thus,

$$\hat{\mathbf{f}}(\nu) = \hat{\varphi}(\nu)\{c_k\}_{k\in\mathbb{Z}}, \ \nu \in I.$$

By construction and Lemma 8.1.4,

$$||f||^2 = \int_I ||\hat{\mathbf{f}}(\nu)||^2 d\nu = \int_I |\hat{\varphi}(\nu)|^2 d\nu \sum_{k\in\mathbb{Z}} |c_k|^2;$$

thus, Lemma 8.1.5 shows that

$$
\begin{aligned}
\int_I |\hat{\varphi}(\gamma)|^2 ||H^*(\nu)\{c_k\}_{k\in\mathbb{Z}}||^2 d\nu &= \int_I ||H^*(\nu)\hat{\mathbf{f}}(\nu)||^2 d\nu \\
&\geq a \sum_{m,n\in\mathbb{Z}} |\langle f, g_{nm}\rangle|^2 \\
&\geq aA\,||f||^2 \\
&= aA \int_I |\hat{\varphi}(\nu)|^2 d\nu \sum_{k\in\mathbb{Z}} |c_k|^2.
\end{aligned}
$$

Since this holds for all functions φ for which $\hat{\varphi}$ is supported on an interval I of length $1/a$, we conclude that

$$||H^*(\nu)\{c_k\}_{k\in\mathbb{Z}}||^2 \geq aA \sum_{k\in\mathbb{Z}} |c_k|^2, \text{ a.e. } \nu \in I. \qquad (8.23)$$

The null set depends on the sequence $\{c_k\}_{k\in\mathbb{Z}}$. It remains to show that there is a null set N such that (8.23) holds for all $\nu \in \mathbb{R} \setminus N$ and all $\{c_k\}_{k\in\mathbb{Z}} \in \ell^2(\mathbb{Z})$; this part is left to the reader (Exercise 8.5).

We now prove (iii). The proof is based on the functions $\rho(e, f)$ from Lemma 8.1.2 and the derived expression for their Fourier coefficients. Recall from Lemma 5.7.1 that two Bessel sequences $\{g_{nm}\}$ and $\{h_{nm}\}$ are dual frames if and only if

$$\langle e, f\rangle = \sum_{m,n\in\mathbb{Z}} \langle e, g_{nm}\rangle\langle h_{nm}, f\rangle, \ \forall e, f \in L^2(\mathbb{R}). \qquad (8.24)$$

If we assume that $\{g_{nm}\}, \{h_{nm}\}$ are dual frames, it follows from this identity that for all $e, f \in L^2(\mathbb{R})$,

$$\rho(e, f)(x) = \sum_{m,n\in\mathbb{Z}} \langle T_x e, g_{nm}\rangle\langle h_{nm}, T_x f\rangle = \langle T_x e, T_x f\rangle = \langle e, f\rangle, \ x \in \mathbb{R}.$$

Hence the function $\rho(e, f)(x)$ and the constant $\langle e, f \rangle$ have the same Fourier coefficients in $L^2(0, a)$. Via Lemma 8.1.2, this implies that for all $k \in \mathbb{Z}$,

$$\frac{1}{a} \int_{-\infty}^{\infty} \hat{e}(\nu) \overline{\hat{f}(\nu + k/a)} \sum_{m \in \mathbb{Z}} \overline{\widehat{g_m}(\nu)} \widehat{h_m}(\nu + k/a) \, d\nu = \delta_{k,0} \langle e, f \rangle$$

$$= \delta_{k,0} \int_{-\infty}^{\infty} \hat{e}(\nu) \overline{\hat{f}(\nu)} \, d\nu.$$

Since this holds for all $e \in L^2(\mathbb{R})$,

$$\overline{\hat{f}(\nu + k/a)} \sum_{m \in \mathbb{Z}} \overline{\widehat{g_m}(\nu)} \widehat{h_m}(\nu + k/a) = a \delta_{k,0} \overline{\hat{f}(\nu)}, \ a.e. \ \nu \in \mathbb{R}, \ \forall f \in L^2(\mathbb{R}).$$

For $k = 0$ this implies that

$$\sum_{m \in \mathbb{Z}} \overline{\widehat{g_m}(\nu)} \widehat{h_m}(\nu) = a, \ a.e. \ \nu \in \mathbb{R};$$

furthermore, the choices $\hat{f} = \chi_{[\ell/a, (\ell+1)/a[}, \ell \in \mathbb{Z}$, lead to

$$\sum_{m \in \mathbb{Z}} \overline{\widehat{g_m}(\nu)} \widehat{h_m}(\nu + k/a) = 0, \ a.e. \ \nu \in \mathbb{R} \text{ when } k \neq 0.$$

The opposite implication can be obtained by reversing the above steps: assuming that (8.21) is satisfied, it follows that the function $\rho(e, f)(x)$ and the constant $\langle e, f \rangle$ have the same Fourier coefficients, so by continuity

$$\rho(e, f)(x) = \langle e, f \rangle, \ \forall x \in \mathbb{R}.$$

Now take $x = 0$, and we obtain (8.24).

We now prove (ii). Via Exercise 5.7, the equivalence in (ii) follows directly from (iii); thus, we only need to prove the statement about the frame bound. Now, if (8.20) is satisfied, the entry in the k-th row and ℓ-th column of the matrix $H(\nu) H(\nu)^*$ is

$$\sum_{m \in \mathbb{Z}} \widehat{g_m}(\nu - k/a) \overline{\widehat{g_m}(\nu - \ell/a)} = c \delta_{k,\ell};$$

by the result in (i), this implies that the frame bound is $A = c/a$.   □

Theorem 8.1.7 characterizes frames of the type $\{g_{nm}\}$ in terms of the Fourier transforms $\widehat{g_m}$. One speaks about characterizations in the *frequency-domain*, as opposed to *time-domain* characterizations directly in terms of the functions g_m.

8.2   Representations of the frame operator

As we have seen, the frame operator plays a very prominent role in frame theory. In order to facilitate manipulations on the frame operator, it is important to have various ways of representing it. In the context of shift-invariant systems, we will now provide a representation in the Fourier-domain in terms of fibers.

Theorem 8.2.1 *Assume that the shift-invariant system $\{g_{nm}\}$ is a Bessel sequence. Define the functions d_k, $k \in \mathbb{Z}$, by*

$$d_k(\nu) := \sum_{m \in \mathbb{Z}} \widehat{g_m}(\nu)\overline{\widehat{g_m}(\nu - k/a)}, \ \nu \in \mathbb{R}.$$

Then the frame operator S associated with $\{g_{nm}\}$ has a representation in the Fourier-domain, given by

$$\widehat{Sf}(\nu) = \frac{1}{a} \sum_{k \in \mathbb{Z}} d_k(\nu)\hat{f}(\nu - k/a), \tag{8.25}$$

with absolute convergence for a.e. $\nu \in \mathbb{R}$.

Proof. We ask the reader (Exercise 8.6) to verify that the series defining $d_k(\nu)$ is convergent for a.e. $\nu \in \mathbb{R}$ and that

$$\sum_{k \in \mathbb{Z}} |d_k(\nu)|^2 \leq (aB)^2, \ a.e. \ \nu \in \mathbb{R}. \tag{8.26}$$

For any $f \in L^2(\mathbb{R})$, Lemma 8.1.4, shows that

$$\sum_{k \in \mathbb{Z}} |\hat{f}(\nu - k/a)|^2 < \infty, \ a.e. \ \nu \in \mathbb{R};$$

via Cauchy–Schwarz' inequality, this implies that the series on the right-hand side of (8.25) is absolutely convergent for a.e. $\gamma \in \mathbb{R}$. In order to show that it represents the frame operator S in the Fourier-domain, we first note that for $f, h \in L^2(\mathbb{R})$,

$$\langle Sf, h \rangle = \langle \sum_{m,n \in \mathbb{Z}} \langle f, g_{nm} \rangle g_{nm}, h \rangle = \sum_{m,n \in \mathbb{Z}} \langle f, g_{nm} \rangle \langle g_{nm}, h \rangle.$$

By Lemma 8.1.2, the function

$$\rho(f, h)(x) := \sum_{m,n \in \mathbb{Z}} \langle T_x f, g_{nm} \rangle \langle g_{nm}, T_x h \rangle$$

is continuous and has period a; its Fourier series is

$$\rho(f, h)(x) = \sum_{k \in \mathbb{Z}} c_k e^{2\pi i k x/a},$$

where

$$
\begin{aligned}
c_k &= \frac{1}{a} \int_{-\infty}^{\infty} \hat{f}(\nu)\overline{\hat{h}(\nu + k/a)} \sum_{m \in \mathbb{Z}} \widehat{g_m}(\nu)\overline{\widehat{g_m}(\nu + k/a)} \, d\nu \\
&= \frac{1}{a} \int_{-\infty}^{\infty} \hat{f}(\nu)\overline{\hat{h}(\nu + k/a)} d_k(\nu + k/a) \, d\nu, \ k \in \mathbb{Z}.
\end{aligned}
$$

Note that $\rho(f, h)(0) = \langle Sf, h \rangle$. We will use the Fourier expansion of $\rho(f, h)$ to calculate $\langle Sf, h \rangle$ for functions $h \in \mathcal{D}$, see (8.14). In order to do so, we first need to show that the Fourier expansion for $\rho(f, h)$ holds pointwise; according to Theorem 3.5.1, this is the case if we can show that the sequence of Fourier coefficients belongs to $\ell^1(\mathbb{Z})$. Now,

$$
\begin{aligned}
\sum_{k \in \mathbb{Z}} |c_k| &= \frac{1}{a} \sum_{k \in \mathbb{Z}} \left| \int_{-\infty}^{\infty} \hat{f}(\nu)\overline{\hat{h}(\nu + k/a)} d_k(\nu + k/a)\, d\nu \right| \\
&\leq \frac{1}{a} \sum_{k \in \mathbb{Z}} \int_{-\infty}^{\infty} |\hat{f}(\nu)|\, |\hat{h}(\nu + k/a)|\, |d_k(\nu + k/a)|\, d\nu \\
&= \frac{1}{a} \int_{-\infty}^{\infty} \sum_{k \in \mathbb{Z}} |\hat{f}(\nu - k/a)|\, |\hat{h}(\nu)|\, |d_k(\nu)|\, d\nu.
\end{aligned}
$$

Using Cauchy–Schwarz' inequality on the sum leads to

$$
\begin{aligned}
\sum_{k \in \mathbb{Z}} |c_k| &\leq \frac{1}{a} \int_{-\infty}^{\infty} \left(\sum_{k \in \mathbb{Z}} |d_k(\nu)|^2 \right)^{1/2} \left(\sum_{k \in \mathbb{Z}} |\hat{f}(\nu - k/a)|^2 \right)^{1/2} |\hat{h}(\nu)|\, d\nu \\
&\leq B \int_{-\infty}^{\infty} \left(\sum_{k \in \mathbb{Z}} |\hat{f}(\nu - k/a)|^2 \right)^{1/2} |\hat{h}(\nu)|\, d\nu.
\end{aligned}
$$

Applying Cauchy–Schwarz' inequality on the integral now shows that

$$
\begin{aligned}
\sum_{k \in \mathbb{Z}} |c_k| &\leq B \left(\int_{\operatorname{supp} \hat{h}} \sum_{k \in \mathbb{Z}} |\hat{f}(\nu - k/a)|^2 d\nu \right)^{1/2} \left(\int_{\operatorname{supp} \hat{h}} |\hat{h}(\nu)|^2 d\nu \right)^{1/2} \\
&\leq B \, ||h||^2 \left(\int_{\operatorname{supp} \hat{h}} \sum_{k \in \mathbb{Z}} |\hat{f}(\nu - k/a)|^2 d\nu \right)^{1/2}.
\end{aligned}
$$

The compact set $\operatorname{supp} \hat{h}$ can be covered by a finite number of intervals of length $1/a$, so Lemma 8.1.4 implies that $\{c_k\}_{k \in \mathbb{Z}} \in \ell^1(\mathbb{Z})$. Thus, we conclude that the Fourier series for $\rho(f, h)$ is absolutely convergent for all $x \in \mathbb{R}$. In particular,

$$
\begin{aligned}
\langle Sf, h \rangle = \rho(f, h)(0) &= \sum_{k \in \mathbb{Z}} c_k \\
&= \frac{1}{a} \sum_{k \in \mathbb{Z}} \int_{-\infty}^{\infty} \hat{f}(\nu)\overline{\hat{h}(\nu + k/a)} d_k(\nu + k/a)\, d\nu \\
&= \frac{1}{a} \sum_{k \in \mathbb{Z}} \int_{-\infty}^{\infty} d_k(\nu)\hat{f}(\nu - k/a)\overline{\hat{h}(\nu)}\, d\nu \\
&= \left\langle \frac{1}{a} \sum_{k \in \mathbb{Z}} d_k(\cdot)\hat{f}(\cdot - k/a),\, \hat{h}(\cdot) \right\rangle.
\end{aligned}
$$

Since $\langle Sf, h \rangle = \langle \widehat{Sf}, \hat{h} \rangle$, we conclude that

$$\langle \frac{1}{a} \sum_{k \in \mathbb{Z}} d_k(\cdot) \hat{f}(\cdot - k/a), \hat{h}(\cdot) \rangle = \langle \widehat{Sf}, \hat{h} \rangle.$$

This holds for all $h \in \mathcal{D}$; thus,

$$\widehat{Sf} = \frac{1}{a} \sum_{k \in \mathbb{Z}} d_k(\cdot) \hat{f}(\cdot - k/a),$$

as desired.   □

8.3   Exercises

8.1 Show that the function $\rho(f, f)$ in (8.4) has period a.

8.2 In this exercise, we ask the reader to provide some details in the proof of Lemma 8.1.2.

(i) Verify that the function $\rho(e, f)$ has period a.

(ii) Justify the calculations leading to (8.11).

(iii) Prove that the series in (8.12) is convergent.

(iv) Provide the density argument at the end of the proof.

8.3 Prove Lemma 8.1.4.

8.4 Complete the proof of Lemma 8.1.5 by showing that the infinite series in (8.15) converges absolutely for a.e. $\nu \in \mathbb{R}$ and that the resulting function has the Fourier expansion stated in (8.16). (Hint: use the periodicity of the function in (8.15) followed by an application of Fourier's inversion theorem.)

8.5 Complete the proof of Theorem 8.1.7(i) by showing that there is a null set N such that (8.23) holds for all $\nu \in \mathbb{R} \setminus N$ and all $\{c_k\}_{k \in \mathbb{Z}} \in \ell^2(\mathbb{Z})$.

8.6 Prove that the series defining $d_k(\nu)$ in Theorem 8.2.1 converges for a.e. $\nu \in \mathbb{R}$, and that (8.26) holds. (Hint: use Lemma 8.1.1.)

9
Gabor Frames in $L^2(\mathbb{R})$

The mathematical theory for Gabor analysis in $L^2(\mathbb{R})$ is based on two classes of operators on $L^2(\mathbb{R})$, namely

Translation by $a \in \mathbb{R}$, $T_a : L^2(\mathbb{R}) \to L^2(\mathbb{R}), (T_a f)(x) = f(x - a)$,

Modulation by $b \in \mathbb{R}$, $E_b : L^2(\mathbb{R}) \to L^2(\mathbb{R}), (E_b f)(x) = e^{2\pi i b x} f(x)$.

Gabor analysis aims at representing functions $f \in L^2(\mathbb{R})$ as superpositions of translated and modulated versions of a fixed function $g \in L^2(\mathbb{R})$. There are two ways one can think about this. The first is to restrict the translation and modulation parameters to a lattice $\{(na, mb)\}_{m,n \in \mathbb{Z}} \subset \mathbb{R}^2$ and ask for series expansions of f in terms of the functions

$$\{e^{2\pi i m b x} g(x - na)\}_{m,n \in \mathbb{Z}}. \tag{9.1}$$

This naturally fits into the discrete frame theory, which has been our main concern so far. The second approach is to ask for *integral representations* involving all possible translations and modulations, i.e., representations like

$$f(x) = \int_{-\infty}^{\infty} \int_{-\infty}^{\infty} c_f(a, b) e^{2\pi i b x} g(x - a)\, db\, da; \tag{9.2}$$

here we have to search for a function c_f of two variables making this true, either pointwise, or in some weak sense. Formulated this way, the continuous frames discussed in Section 5.8 become a natural tool.

Concerning the first approach, the natural question is how we can choose $g \in L^2(\mathbb{R})$ and the parameters $a, b > 0$ such that the functions in (9.1) constitute a frame for $L^2(\mathbb{R})$. We actually saw the first example of such

O. Christensen, *Frames and Bases*. DOI: 10.1007/978-0-8176-4678-3_9,
© Springer Science+Business Media, LLC 2008

a frame in Example 3.5.3, where we proved that $\{E_m T_n \chi_{0,1]}\}_{m,n \in \mathbb{Z}}$ is an orthonormal basis for $L^2(\mathbb{R})$.

The basic idea goes back to Gabor, who considered a sequence of functions of the form $\{E_{mb} T_{na} g\}_{m,n \in \mathbb{Z}}$, where $ab = 1$ and g is the Gaussian, $g(x) = e^{-x^2/2}$. The mathematical analysis of Gabor systems was initiated by Janssen in a series of papers around 1980. Gabor analysis took an entirely new direction with the fundamental paper [27] by Daubechies, Grossmann, and Meyer from 1986. Here one finds for the first time the idea of combining Gabor analysis with frame theory. The authors constructed tight frames for $L^2(\mathbb{R})$ having the form $\{E_{mb} T_{na} g\}_{m,n \in \mathbb{Z}}$ at hand, and this contribution was the beginning of an intense activity that is still ongoing.

We begin in Section 9.1 by introducing Gabor frames in $L^2(\mathbb{R})$ and some of their fundamental properties. In Section 9.2, we characterize tight Gabor frames. In Section 9.3, we prove that the canonical dual frame of a Gabor frame again has Gabor structure. We discuss some of the properties of the canonical dual frame and how other duals can be found. These results are applied in Section 9.4, where explicitly given pairs of dual frames with compactly supported generators are constructed. Section 9.5 introduces some Banach spaces and technical conditions that appear in Gabor analysis, and Section 9.6 presents various representations of the Gabor frame operator. The Zak transform and its connections to Gabor systems are discussed in Section 9.7. Section 9.8 deals with a very important aspect in applications of Gabor systems, namely, how time–frequency localization of a signal affects its representation via a Gabor frame. Finally, we consider continuous representations in Section 9.9.

The commutator relations (2.23) show that modulation in the time domain corresponds to translation in the Fourier domain; for this reason, the functions $E_b T_a g$ are called *time–frequency shifts* of g. For more information about the role of Gabor frames in time–frequency analysis, we refer to the book [37] by Gröchenig. For a broader perspective on Gabor frames and their use in many different fields, the reader should consult the two books [32], [33] edited by Feichtinger and Strohmer, which contain surveys and research articles covering several aspects.

9.1 Basic Gabor frame theory

A collection of functions on the form $\{E_{mb} T_{na} g\}_{m,n \in \mathbb{Z}}$ is called a *Gabor system*. Explicitly, these functions have the form

$$E_{mb} T_{na} g(x) = e^{2\pi i m b x} g(x - na).$$

We will consider frames of that particular form:

Definition 9.1.1 *A Gabor frame is a frame for $L^2(\mathbb{R})$ of the form $\{E_{mb}T_{na}g\}_{m,n\in\mathbb{Z}}$, where $a,b > 0$ are given and $g \in L^2(\mathbb{R})$ is a fixed function.*

Frames of this type are also called *Weyl–Heisenberg frames*. The function g is called the *window function* or the *generator*. Note the convention, which is implicit in our definition: when speaking about a Gabor frame, it is understood that it is a frame for all of $L^2(\mathbb{R})$, i.e., we will not deal with frames for subspaces in the current chapter.

The Gabor system $\{E_{mb}T_{na}g\}_{m,n\in\mathbb{Z}}$ only involves translates with parameters na, $n \in \mathbb{Z}$ and modulations with parameters mb, $m \in \mathbb{Z}$. The points $\{(na, mb)\}_{m,n\in\mathbb{Z}}$ form a *lattice* in $\mathbb{R}^2$, and for this reason one frequently calls $\{E_{mb}T_{na}g\}_{m,n\in\mathbb{Z}}$ a *regular Gabor system*.

We begin with a proposition that gives a necessary condition for $\{E_{mb}T_{na}g\}_{m,n\in\mathbb{Z}}$ to be a frame for $L^2(\mathbb{R})$. It depends on the interplay between the function g and the translation parameter a and is expressed in terms of the function

$$G(x) := \sum_{n\in\mathbb{Z}} |g(x - na)|^2, \ x \in \mathbb{R}. \tag{9.3}$$

Proposition 9.1.2 *Let $g \in L^2(\mathbb{R})$ and $a,b > 0$ be given, and assume that $\{E_{mb}T_{na}g\}_{m,n\in\mathbb{Z}}$ is a frame with bounds A, B. Then*

$$bA \leq \sum_{n\in\mathbb{Z}} |g(x - na)|^2 \leq bB, \ a.e. \ x \in \mathbb{R}. \tag{9.4}$$

More precisely: if the upper bound in (9.4) is violated, then $\{E_{mb}T_{na}g\}_{m,n\in\mathbb{Z}}$ is not a Bessel sequence; if the lower bound is violated, then $\{E_{mb}T_{na}g\}_{m,n\in\mathbb{Z}}$ does not satisfy the lower frame condition.

Proof. The proof is by contradiction. Assume that the upper condition in (9.4) is violated. Then there exists a measurable set $\Delta \subseteq \mathbb{R}$ with positive measure such that $G(x) = \sum_{n\in\mathbb{Z}} |g(x - na)|^2 > bB$ on Δ. We can assume that Δ is contained in an interval of length $\frac{1}{b}$. Letting

$$\Delta_0 := \{x \in \Delta \mid G(x) \geq 1 + bB\},$$
$$\Delta_k := \{x \in \Delta \mid \frac{1}{k+1} + bB \leq G(x) < \frac{1}{k} + bB\}, \ k \in \mathbb{N},$$

we obtain a partition of Δ into disjoint measurable sets. At least one of them, say, $\Delta_{k'}$, has positive measure. Now consider the function $f = \chi_{\Delta_{k'}}$, and note that $||f||^2 = |\Delta_{k'}|$. For $n \in \mathbb{Z}$, the function $f \, \overline{T_{na}g}$ has support in $\Delta_{k'}$; since $\Delta_{k'}$ is contained in an interval of length $1/b$ and the functions $\{\sqrt{b}E_{mb}\}_{m\in\mathbb{Z}} = \{\sqrt{b}e^{2\pi imbx}\}_{m\in\mathbb{Z}}$ constitute an orthonormal basis for $L^2(I)$ for every interval I of length $1/b$, we have

$$\sum_{m\in\mathbb{Z}} |\langle f, E_{mb}T_{na}g\rangle|^2 = \sum_{m\in\mathbb{Z}} |\langle f\overline{T_{na}g}, E_{mb}\rangle|^2 = \frac{1}{b}\int_{-\infty}^{\infty} |f(x)|^2 \, |g(x-na)|^2 \, dx.$$

Thus, for our particular choice of f,

$$
\begin{aligned}
\sum_{m,n\in\mathbb{Z}} |\langle f, E_{mb}T_{na}g\rangle|^2 &= \frac{1}{b}\sum_{n\in\mathbb{Z}}\int_{-\infty}^{\infty} |f(x)|^2\, |g(x-na)|^2\, dx \\
&= \frac{1}{b}\int_{\Delta_{k'}} G(x)\, dx \\
&\geq \frac{1}{b}\left(\frac{1}{k'+1} + bB\right) ||f||^2 \\
&= \left(B + \frac{1}{b(k'+1)}\right) ||f||^2.
\end{aligned}
$$

But then B cannot be an upper frame bound for $\{E_{mb}T_{na}g\}_{m,n\in\mathbb{Z}}$. A similar proof shows that if the lower condition in (9.4) is violated, then A cannot be a lower frame bound for $\{E_{mb}T_{na}g\}_{m,n\in\mathbb{Z}}$.   □

Proposition 9.1.2 implies that a function g generating a Gabor frame $\{E_{mb}T_{na}g\}_{m,n\in\mathbb{Z}}$ necessarily is bounded. Note also that Proposition 9.1.2 gives a relationship between the frame bounds and the lower and upper bounds for the function G in (9.3). In Corollary 9.1.7, we will see that under certain circumstances, the necessary condition (9.4) is also sufficient for $\{E_{mb}T_{na}g\}_{m,n\in\mathbb{Z}}$ to be a frame for $L^2(\mathbb{R})$.

Proposition 9.1.2 also implies that the function G is constant if $\{E_{mb}T_{na}g\}_{m,n\in\mathbb{Z}}$ is a tight frame. An extension of this observation actually leads to a characterization of tight frames, see Theorem 9.2.1.

We now discuss various conditions for $\{E_{mb}T_{na}g\}_{m,n\in\mathbb{Z}}$ to be a frame for $L^2(\mathbb{R})$. We begin with two lemmas, which will be needed repeatedly.

Lemma 9.1.3 *Let $f, g \in L^2(\mathbb{R})$ and $a, b > 0$ be given. Then, for any $n \in \mathbb{N}$ the following hold:*

(i) *The series*

$$
\sum_{k\in\mathbb{Z}} f(x-k/b)\overline{g(x-na-k/b)}, \quad x \in \mathbb{R}, \tag{9.5}
$$

converges absolutely for a.e. $x \in \mathbb{R}$.

(ii) *The mapping $x \mapsto \sum_{k\in\mathbb{Z}} |f(x-k/b)\overline{g(x-na-k/b)}|$ belongs to $L^1(0, 1/b)$.*

(iii) *The $1/b$-periodic function $F_n \in L^1(0, 1/b)$ defined by*

$$
F_n(x) = \sum_{k\in\mathbb{Z}} f(x-k/b)\overline{g(x-na-k/b)} \tag{9.6}
$$

has the Fourier coefficients

$$
c_m = b\,\langle f, E_{mb}T_{na}g\rangle, \quad m \in \mathbb{Z}.
$$

Proof. Since $f, T_{na}g \in L^2(\mathbb{R})$, we have $f\overline{T_{na}g} \in L^1(\mathbb{R})$ for all $n \in \mathbb{Z}$. Thus

$$\int_0^{1/b} \sum_{n \in \mathbb{Z}} |f(x - k/b)\overline{g(x - na - k/b)}| \, dx \;=\; \int_{-\infty}^{\infty} \left| f(x)\overline{g(x - na)} \right| dx$$
$$< \infty.$$

This proves (ii), and also implies that $\sum_{n \in \mathbb{Z}} |f(x - k/b)\overline{g(x - na - k/b)}|$ converges for a.e. $x \in [0, 1/b]$; for reasons of periodicity, it therefore converges for a.e. $x \in \mathbb{R}$. As a consequence, the series in (9.5) converges for a.e. $x \in \mathbb{R}$ and defines a function with period $1/b$. For part (iii) concerning the Fourier coefficients for F_n, note that

$$\langle f, E_{mb}T_{na}g \rangle \;=\; \int_{-\infty}^{\infty} f(x)\overline{g(x - na)}e^{-2\pi imbx} \, dx$$
$$= \sum_{k \in \mathbb{Z}} \int_0^{1/b} f(x - k/b)\overline{g(x - na - k/b)}e^{-2\pi imbx} \, dx$$
$$= \int_0^{1/b} \left(\sum_{k \in \mathbb{Z}} f(x - k/b)\overline{g(x - na - k/b)} \right) e^{-2\pi imbx} \, dx.$$

We leave it to the reader to justify the manipulations. $\qquad\square$

If we want to check that a Gabor system $\{E_{mb}T_{na}g\}_{m,n \in \mathbb{Z}}$ forms a frame by hand, we need to be able to estimate the expression $\sum_{m,n \in \mathbb{Z}} |\langle f, E_{mb}T_{na}g \rangle|^2$ for all functions f belonging to $L^2(\mathbb{R})$ (or at least a dense subset hereof). Under certain conditions on the functions f and g, we can find an explicit expression for this infinite sum:

Lemma 9.1.4 *Suppose that f is a bounded measurable function with compact support and that the function G defined by (9.3) is bounded. Then*

$$\sum_{m,n \in \mathbb{Z}} |\langle f, E_{mb}T_{na}g \rangle|^2$$
$$= \frac{1}{b} \int_{-\infty}^{\infty} |f(x)|^2 \sum_{n \in \mathbb{Z}} |g(x - na)|^2 \, dx$$
$$+ \frac{1}{b} \sum_{k \neq 0} \int_{-\infty}^{\infty} \overline{f(x)}f(x - k/b) \sum_{n \in \mathbb{Z}} g(x - na)\overline{g(x - na - k/b)} \, dx.$$

Proof. Let $n \in \mathbb{Z}$, and consider the $\frac{1}{b}$-periodic function F_n defined in (9.6). We have already given a general argument for F_n being well defined pointwise almost everywhere, but our present assumptions give more; in fact, for a given $x \in \mathbb{R}$, the compact support of f implies that $f(x - k/b)$ only can be non-zero for finitely many k-values. The number of k-values for which $f(x - k/b) \neq 0$ is uniformly bounded, i.e., there is a constant C

such that at most C k-values appear, independently of the chosen x. It follows that F_n is bounded, so $F_n \in L^1(0, 1/b) \cap L^2(0, 1/b)$; in fact, even

$$\left[x \mapsto \sum_{k \in \mathbb{Z}} \left| f(x - k/b)\overline{g(x - na - k/b)} \right| \right] \in L^1(0, 1/b) \cap L^2(0, 1/b).$$

By Lemma 9.1.3, for all $m, n \in \mathbb{Z}$,

$$\langle f, E_{mb}T_{na}g \rangle = \int_0^{1/b} F_n(x)e^{-2\pi imbx}\, dx. \tag{9.7}$$

Parseval's theorem, see (3.24), gives that

$$\sum_{m \in \mathbb{Z}} \left| \int_0^{1/b} F_n(x)e^{-2\pi imbx}\, dx \right|^2 = \frac{1}{b} \int_0^{1/b} |F_n(x)|^2\, dx. \tag{9.8}$$

The assumption on f being a bounded measurable function with compact support will justify all interchanges of integration and summation in the final calculation. This follows from the observation that

$$\sum_{k \in \mathbb{Z}} \int_{-\infty}^{\infty} |\overline{f(x)}f(x - k/b)| \sum_{n \in \mathbb{Z}} |g(x - na)\overline{g(x - na - k/b)}|\, dx < \infty. \tag{9.9}$$

The verification of (9.9) and the proof that this is exactly what we need is left to the reader (Exercise 9.2). Now, via (9.7) and (9.8),

$$
\begin{aligned}
\sum_{n \in \mathbb{Z}} \sum_{m \in \mathbb{Z}} |\langle f, E_{mb}T_{na}g \rangle|^2 &= \sum_{n \in \mathbb{Z}} \sum_{m \in \mathbb{Z}} \left| \int_0^{1/b} F_n(x)e^{-2\pi imbx}\, dx \right|^2 \\
&= \frac{1}{b} \sum_{n \in \mathbb{Z}} \int_0^{1/b} |F_n(x)|^2\, dx.
\end{aligned}
$$

Writing

$$|F_n(x)|^2 = \overline{F_n(x)}F_n(x) = \sum_{\ell \in \mathbb{Z}} \overline{f(x - \ell/b)}g(x - na - \ell/b)F_n(x),$$

and using that F_n is $1/b$-periodic, Lemma 7.1.3 finally implies that

$$\sum_{n\in\mathbb{Z}}\sum_{m\in\mathbb{Z}}|\langle f, E_{mb}T_{na}g\rangle|^2$$

$$= \frac{1}{b}\sum_{n\in\mathbb{Z}}\int_0^{1/b}\sum_{\ell\in\mathbb{Z}}\overline{f(x-\ell/b)}g(x-na-\ell/b)F_n(x)\,dx$$

$$= \frac{1}{b}\sum_{n\in\mathbb{Z}}\int_{-\infty}^{\infty}\overline{f(x)}g(x-na)F_n(x)\,dx$$

$$= \frac{1}{b}\sum_{n\in\mathbb{Z}}\int_{-\infty}^{\infty}\overline{f(x)}g(x-na)\sum_{k\in\mathbb{Z}}f(x-k/b)\overline{g(x-na-k/b)}\,dx \quad (9.10)$$

$$= \frac{1}{b}\int_{-\infty}^{\infty}|f(x)|^2\sum_{n\in\mathbb{Z}}|g(x-na)|^2\,dx$$

$$+\frac{1}{b}\sum_{k\neq0}\int_{-\infty}^{\infty}\overline{f(x)}f(x-k/b)\sum_{n\in\mathbb{Z}}g(x-na)\overline{g(x-na-k/b)}\,dx. \quad \square$$

Lemma 9.1.4 has several important consequences. For example, it leads to a sufficient condition for $\{E_{mb}T_{na}g\}_{m,n\in\mathbb{Z}}$ to form a frame:

Theorem 9.1.5 *Let $g \in L^2(\mathbb{R})$, $a,b > 0$ and suppose that*

$$B := \frac{1}{b}\sup_{x\in[0,a]}\sum_{k\in\mathbb{Z}}\left|\sum_{n\in\mathbb{Z}}g(x-na)\overline{g(x-na-k/b)}\right| < \infty. \quad (9.11)$$

Then $\{E_{mb}T_{na}g\}_{m,n\in\mathbb{Z}}$ is a Bessel sequence with bound B. If also

$$A := \quad (9.12)$$

$$\frac{1}{b}\inf_{x\in[0,a]}\left[\sum_{n\in\mathbb{Z}}|g(x-na)|^2 - \sum_{k\neq0}\left|\sum_{n\in\mathbb{Z}}g(x-na)\overline{g(x-na-k/b)}\right|\right] > 0,$$

then $\{E_{mb}T_{na}g\}_{m,n\in\mathbb{Z}}$ is a frame for $L^2(\mathbb{R})$ with bounds A,B.

Proof. Consider a function $f \in L^2(\mathbb{R})$ that is continuous and has compact support. By Lemma 9.1.4,

$$\sum_{m,n\in\mathbb{Z}}|\langle f, E_{mb}T_{na}g\rangle|^2 \quad (9.13)$$

$$= \frac{1}{b}\int_{-\infty}^{\infty}|f(x)|^2\sum_{n\in\mathbb{Z}}|g(x-na)|^2\,dx \quad (9.14)$$

$$+ \frac{1}{b}\sum_{k\neq0}\int_{-\infty}^{\infty}\overline{f(x)}f(x-k/b)\sum_{n\in\mathbb{Z}}g(x-na)\overline{g(x-na-k/b)}\,dx. \quad (9.15)$$

We want to estimate (9.15). For $k \in \mathbb{Z}$, let

$$H_k(x) := \sum_{n \in \mathbb{Z}} T_{na}g(x)\overline{T_{na+k/b}g(x)}; \tag{9.16}$$

we observe that H_k is well defined a.e. by Lemma 9.1.3. Now,

$$
\begin{aligned}
\sum_{k \neq 0} |T_{-k/b}H_k(x)| &= \sum_{k \neq 0} \left| T_{-k/b} \sum_{n \in \mathbb{Z}} T_{na}g(x)\overline{T_{na+k/b}g(x)} \right| \\
&= \sum_{k \neq 0} \left| \sum_{n \in \mathbb{Z}} T_{na-k/b}g(x)\overline{T_{na}g(x)} \right|.
\end{aligned}
$$

Replacing k with $-k$ (which is allowed because we sum over all $k \neq 0$) and complex conjugating all terms, we arrive at

$$
\begin{aligned}
\sum_{k \neq 0} |T_{-k/b}H_k(x)| &= \sum_{k \neq 0} \left| \sum_{n \in \mathbb{Z}} T_{na+k/b}g(x)\overline{T_{na}g(x)} \right| \\
&= \sum_{k \neq 0} \left| \sum_{n \in \mathbb{Z}} \overline{T_{na+k/b}g(x)}T_{na}g(x) \right| \\
&= \sum_{k \neq 0} |H_k(x)|.
\end{aligned}
$$

Now,

$$
\begin{aligned}
&\left| \sum_{k \neq 0} \int_{-\infty}^{\infty} \overline{f(x)}f(x-k/b) \sum_{n \in \mathbb{Z}} g(x-na)\overline{g(x-na-k/b)}\, dx \right| \\
&\leq \sum_{k \neq 0} \int_{-\infty}^{\infty} |f(x)|\, |T_{k/b}f(x)|\, |H_k(x)|\, dx \\
&= \sum_{k \neq 0} \int_{-\infty}^{\infty} |f(x)|\sqrt{|H_k(x)|}\, |T_{k/b}f(x)|\sqrt{|H_k(x)|}\, dx \\
&= (*).
\end{aligned}
$$

Using Cauchy–Schwarz' inequality twice, first on the integral, and then on the sum over $k \neq 0$,

$$(*) \quad \leq \quad \sum_{k \neq 0} \left(\int_{-\infty}^{\infty} |f(x)|^2 |H_k(x)| \, dx \right)^{1/2} \left(\int_{-\infty}^{\infty} |T_{k/b}f(x)|^2 |H_k(x)| \, dx \right)^{1/2}$$

$$\leq \quad \left(\sum_{k \neq 0} \int_{-\infty}^{\infty} |f(x)|^2 |H_k(x)| \, dx \right)^{1/2}$$

$$\times \left(\sum_{k \neq 0} \int_{-\infty}^{\infty} |T_{k/b}f(x)|^2 |H_k(x)| \, dx \right)^{1/2}$$

$$= \quad \left(\int_{-\infty}^{\infty} |f(x)|^2 \sum_{k \neq 0} |H_k(x)| \, dx \right)^{1/2}$$

$$\times \left(\int_{-\infty}^{\infty} |f(x)|^2 \sum_{k \neq 0} |T_{-k/b}H_k(x)| \, dx \right)^{1/2}$$

$$= \quad \int_{-\infty}^{\infty} |f(x)|^2 \sum_{k \neq 0} |H_k(x)| \, dx.$$

Note that the expression

$$\sum_{k \neq 0} |H_k(x)| = \sum_{k \neq 0} \left| \sum_{n \in \mathbb{Z}} T_{na}g(x) \overline{T_{na+k/b}g(x)} \right|$$

defines a periodic function with period a. By (9.13) and the condition (9.11), we now have

$$\sum_{m,n \in \mathbb{Z}} |\langle f, E_{mb}T_{na}g \rangle|^2$$

$$\leq \quad \frac{1}{b} \int_{-\infty}^{\infty} \left(|f(x)|^2 \right.$$

$$\times \left[\sum_{n \in \mathbb{Z}} |g(x - na)|^2 + \sum_{k \neq 0} \left| \sum_{n \in \mathbb{Z}} g(x - na) \overline{g(x - na - k/b)} \right| \right] \right) dx$$

$$= \quad \frac{1}{b} \int_{-\infty}^{\infty} |f(x)|^2 \sum_{k \in \mathbb{Z}} \left| \sum_{n \in \mathbb{Z}} g(x - na) \overline{g(x - na - k/b)} \right| dx$$

$$\leq \quad B \, \|f\|^2.$$

Because this estimate holds on a dense subset of $L^2(\mathbb{R})$, it holds on $L^2(\mathbb{R})$ by Lemma 3.1.6. This proves the first part. If also (9.12) is satisfied, we again consider a continuous function f with compact support and obtain that

$$\sum_{m,n\in\mathbb{Z}} |\langle f, E_{mb}T_{na}g\rangle|^2$$

$$\geq \frac{1}{b}\int_{-\infty}^{\infty} |f(x)|^2$$

$$\times\left[\sum_{n\in\mathbb{Z}} |g(x-na)|^2 - \sum_{k\neq 0}\left|\sum_{n\in\mathbb{Z}} g(x-na)\overline{g(x-na-k/b)}\right|\right] dx$$

$$\geq A\,||f||^2.$$

By Lemma 5.1.2 the lower frame condition actually holds for all $f \in L^2(\mathbb{R})$. This completes the proof. $\qquad\square$

We note that more general versions of Theorem 9.1.5 and Lemma 9.1.4 hold. In fact, the proof of Lemma 9.1.4 did not use that we were summing over all $n \in \mathbb{Z}$: for *all* index sets $I \subseteq \mathbb{Z}$, the assumptions imply that

$$\sum_{n\in I}\sum_{m\in\mathbb{Z}} |\langle f, E_{mb}T_{na}g\rangle|^2$$

$$= \frac{1}{b}\sum_{n\in I}\int_{-\infty}^{\infty}\overline{f(x)}g(x-na)\sum_{k\in\mathbb{Z}} f(x-k/b)\overline{g(x-na-k/b)}\,dx. \quad (9.17)$$

This, in turn, lead to a more general version of Theorem 9.1.5.

The condition (9.11) is sometimes called *condition (CC)*. It leads to an easy sufficient condition for $\{E_{mb}T_{na}g\}_{m,n\in\mathbb{Z}}$ to be a Bessel sequence (Exercise 9.5):

Corollary 9.1.6 *Let $g \in L^2(\mathbb{R})$ be bounded and compactly supported. Then $\{E_{mb}T_{na}g\}_{m,n\in\mathbb{Z}}$ is a Bessel sequence for any choice of $a, b > 0$.*

The condition that the function g is bounded and compactly supported is not sufficient for $\{E_{mb}T_{na}g\}_{m,n\in\mathbb{Z}}$ to be a frame: in fact, as shown in Proposition 9.1.2, the associated function G in (9.3) needs to be bounded below and above (see Exercise 9.6). On the other hand, for a function g with compact support, the condition that the function G is bounded below and above for some $a > 0$ is enough for $\{E_{mb}T_{na}g\}_{m,n\in\mathbb{Z}}$ to be a frame for sufficiently small values of b. We also obtain expressions for the frame operator and its inverse in this case:

Corollary 9.1.7 *Let $a, b > 0$ be given. Suppose that $g \in L^2(\mathbb{R})$ has support in an interval of length $\frac{1}{b}$ and that the function G satisfies (9.4) for some $A, B > 0$. Then $\{E_{mb}T_{na}g\}_{m,n\in\mathbb{Z}}$ is a frame for $L^2(\mathbb{R})$ with bounds A, B. The frame operator S and its inverse S^{-1} are given by*

$$Sf = \frac{G}{b}f, \quad S^{-1}f = \frac{b}{G}f, \quad f \in L^2(\mathbb{R}).$$

Proof. That $\{E_{mb}T_{na}g\}_{m,n\in\mathbb{Z}}$ is a frame follows directly from Lemma 9.1.4 or Theorem 9.1.5 because

$$\sum_{n\in\mathbb{Z}} g(x-na)\overline{g(x-na-k/b)} = 0 \text{ for all } k \neq 0.$$

Given a continuous function f with compact support, Lemma 9.1.4 implies that

$$\begin{aligned}
\langle Sf, f\rangle &= \sum_{m,n\in\mathbb{Z}} |\langle f, E_{mb}T_{na}g\rangle|^2 \\
&= \frac{1}{b}\int_{-\infty}^{\infty} |f(x)|^2 G(x)\,dx \\
&= \langle \frac{G}{b}f, f\rangle.
\end{aligned}$$

By continuity of S, this expression even holds for all $f \in L^2(\mathbb{R})$. Via Lemma 2.4.3, it follows that S acts by multiplication with the function $\frac{G}{b}$. □

For a continuous function g, we can be even more explicit. We leave the proof of the following result to the reader (Exercise 9.7).

Corollary 9.1.8 *Suppose that $g \in L^2(\mathbb{R})$ is a continuous function with support on an interval I with length $|I|$ and that $g(x) > 0$ on the interior of I. Then $\{E_{mb}T_{na}g\}_{m,n\in\mathbb{Z}}$ is a frame for all $(a,b) \in]0, |I|[\times]0, \frac{1}{|I|}]$.*

In particular, this result applies to the B-splines and introduced in Chapter 6. In order to avoid a conflict with our notation for a Gabor system, we will denote the splines by B_ℓ and N_ℓ, $\ell \in \mathbb{N}$, instead of B_n and N_n.

Corollary 9.1.9 *For $\ell \in \mathbb{N}$, the B-splines B_ℓ and N_ℓ generate Gabor frames for all $(a,b) \in]0, \ell[\times]0, 1/\ell]$.*

One might wonder whether the Gabor system $\{E_{mb}T_{na}B_\ell\}_{m,n\in\mathbb{Z}}$ is a frame for $(a,b) \notin]0, \ell[\times]0, 1/\ell]$. The behavior of this Gabor system is actually quite strange, and a complete answer is not known. For the B-spline B_2, it is proved in [39] that if $b \in \mathbb{N}\setminus\{1\}$, this system cannot form a frame for any $a > 0$; see also Exercise 9.6.

For a given function $g \in L^2(\mathbb{R})$, it is usually difficult to find the exact range of parameters $a, b > 0$ for which $\{E_{mb}T_{na}g\}_{m,n\in\mathbb{Z}}$ is a frame. The next examples illustrate this.

Example 9.1.10 For the Gaussian $g(x) = e^{-x^2}$, it is known that the Gabor system $\{E_{mb}T_{na}g\}_{m,n\in\mathbb{Z}}$ is a frame if and only if $ab < 1$; this was proved in 1991 by Lyubarski and independently by Seip and Wallsten. The proof is complicated and requires a deep knowledge of complex analysis. □

Example 9.1.11 Let us now consider characteristic functions,

$$g := \chi_{[0,c[}, \ c > 0.$$

It turns out to be surprisingly complicated to find the exact range of $c > 0$ and parameters $a, b > 0$ for which $\{E_{mb}T_{na}g\}_{m,n\in\mathbb{Z}}$ is a frame. A complete answer has not been obtained yet. Via a scaling (see Proposition 9.1.14 below), we can assume that $b = 1$. A detailed analysis performed by Janssen [49] shows that

(i) $\{E_m T_{na}g\}_{m,n\in\mathbb{Z}}$ is not a frame if $c < a$ or $a > 1$.

(ii) $\{E_m T_{na}g\}_{m,n\in\mathbb{Z}}$ is a frame if $1 \geq c \geq a$.

(iii) $\{E_m T_{na}g\}_{m,n\in\mathbb{Z}}$ is not a frame if $a = 1$ and $c > 1$.

Assuming now that $a < 1, c > 1$, we further have

(iv) $\{E_m T_{na}g\}_{m,n\in\mathbb{Z}}$ is a frame if $a \notin \mathbb{Q}$ and $c \in]1, 2[$.

(v) $\{E_m T_{na}g\}_{m,n\in\mathbb{Z}}$ is not a frame if $a = p/q \in \mathbb{Q}$, $gcd(p,q) = 1$, and $2 - \frac{1}{q} < c < 2$.

(vi) $\{E_m T_{na}g\}_{m,n\in\mathbb{Z}}$ is not a frame if $a > \frac{3}{4}$ and $c = L - 1 + L(1 - a)$ with $L \in \mathbb{N}, L \geq 3$.

The graphical illustration of this result is known as *Janssen's tie*. The example indicates how complicated it is to find the exact range of parameters a, b that generate a frame for a given function g.   □

The results discussed so far concentrate on the interplay between the function g and the parameters a, b. The next result is of a different type. It shows that, regardless of the choice of generator $g \in L^2(\mathbb{R})$, the choice of the parameters a and b puts certain restrictions on the possible frame properties for $\{E_{mb}T_{na}g\}_{m,n\in\mathbb{Z}}$:

Theorem 9.1.12 *Let $g \in L^2(\mathbb{R})$ and $a, b > 0$ be given. Then the following holds:*

(i) *If $ab > 1$, then $\{E_{mb}T_{na}g\}_{m,n\in\mathbb{Z}}$ can not be a frame for $L^2(\mathbb{R})$.*

(ii) *If $\{E_{mb}T_{na}g\}_{m,n\in\mathbb{Z}}$ is a frame, then*

$$ab = 1 \ \Leftrightarrow \ \{E_{mb}T_{na}g\}_{m,n\in\mathbb{Z}} \ \text{is a Riesz basis.} \qquad (9.18)$$

Proof. Recall from the general frame theory that we to any frame $\{f_k\}_{k=1}^\infty$ can associate a canonical tight frame, see Theorem 5.3.4. Now, assume that $\{E_{mb}T_{na}g\}_{m,n\in\mathbb{Z}}$ is a frame, and denote the frame operator by S. We will use that the canonical tight frame associated with $\{E_{mb}T_{na}g\}_{m,n\in\mathbb{Z}}$ can be rewritten as

$$\{S^{-1/2}E_{mb}T_{na}g\}_{m,n\in\mathbb{Z}} = \{E_{mb}T_{na}S^{-1/2}g\}_{m,n\in\mathbb{Z}}, \qquad (9.19)$$

a result that we prove in Theorem 9.3.2.

We first derive an equation, which will play a crucial role in the proof. Proposition 9.1.2 applied to the function $S^{-1/2}g$ implies that

$$\sum_{n\in\mathbb{Z}} |S^{-1/2}g(x-na)|^2 = b \text{ for a.e. } x \in \mathbb{R}. \tag{9.20}$$

Since

$$||S^{-1/2}g||^2 = \int_{-\infty}^{\infty} |S^{-1/2}g(x)|^2 dx = \int_0^a \sum_{n\in\mathbb{Z}} |S^{-1/2}g(x-na)|^2 dx,$$

we conclude that

$$||S^{-1/2}g||^2 = ab. \tag{9.21}$$

Note that (9.21) is a general result: we only used that $\{E_{mb}T_{na}g\}_{m,n\in\mathbb{Z}}$ is a frame in the proof.

In order to prove (i), we will show that $ab \leq 1$ for the arbitrary given frame $\{E_{mb}T_{na}g\}_{m,n\in\mathbb{Z}}$. Now, since $\{E_{mb}T_{na}S^{-1/2}g\}_{m,n\in\mathbb{Z}}$ is a tight frame with frame bounds equal to 1, Exercise 3.1 implies that $||S^{-1/2}g|| \leq 1$. Combining with the equation (9.21), we obtain that $ab \leq 1$ as desired.

For the proof of (ii), assume first that $\{E_{mb}T_{na}g\}_{m,n\in\mathbb{Z}}$ is a Riesz basis. Then by definition $\{S^{-1/2}E_{mb}T_{na}g\}_{m,n\in\mathbb{Z}}$ is also a Riesz basis, i.e., $\{E_{mb}T_{na}S^{-1/2}g\}_{m,n\in\mathbb{Z}}$ is a Riesz basis. By construction, this family is also a tight frame with frame bound 1, so the Riesz bounds are $A = B = 1$; in particular this implies by Theorem 3.3.7 that $||S^{-1/2}g|| = 1$. Again via the equation (9.21), we conclude that $ab = 1$ as desired.

For the other implication in (ii) we now assume that $ab = 1$. Then, via (9.21),

$$||S^{-1/2}g||^2 = ab = 1,$$

and therefore $||E_{mb}T_{na}S^{-1/2}g|| = 1$ for all $m, n \in \mathbb{Z}$. Using Exercise 3.1, we conclude that $\{E_{mb}T_{na}S^{-1/2}g\}_{m,n\in\mathbb{Z}}$ is an orthonormal basis for $\mathcal{H}$, and therefore the family

$$\{E_{mb}T_{na}g\}_{m,n\in\mathbb{Z}} = \{S^{1/2}E_{mb}T_{na}S^{-1/2}g\}_{m,n\in\mathbb{Z}}$$

is a Riesz basis by definition. $\square$

Differently formulated, Theorem 9.1.12 says that $\{E_{mb}T_{na}g\}_{m,n\in\mathbb{Z}}$ can only be a frame if $ab \leq 1$; and if $\{E_{mb}T_{na}g\}_{m,n\in\mathbb{Z}}$ is a frame and $ab < 1$, then the frame is overcomplete.

Example 9.1.13 The following examples relate to the conditions in Theorem 9.1.12.

(i) The assumptions $ab \leq 1$ and $g \neq 0$ are not sufficient for $\{E_{mb}T_{na}g\}_{m,n\in\mathbb{Z}}$ to be a frame. For example, if $a \in]1/2, 1[$, the functions $\{E_m T_{na}\chi_{[0,\frac{1}{2}]}\}_{m,n\in\mathbb{Z}}$ are not complete in $L^2(\mathbb{R})$ and cannot form a frame.

(ii) The assumption that $\{E_{mb}T_{na}g\}_{m,n\in\mathbb{Z}}$ is a frame is necessary for the equivalence in (9.18) to hold. For example, as noted already in Example 9.1.10, the Gaussian $g(x) = e^{-x^2}$ does not generate a frame if $a = b = 1$. □

In many cases, it is convenient to assume that either the translation parameter or the modulation parameter in a Gabor frame is equal to 1. Given an arbitrary Gabor frame $\{E_{mb}T_{na}g\}_{m,n\in\mathbb{Z}}$, this can be obtained by a scaling of g, i.e., by replacing g with a function of the type

$$D_c g(x) = \frac{1}{c^{1/2}} g(x/c)$$

for some $c > 0$.

Proposition 9.1.14 *Let $g \in L^2(\mathbb{R})$ and $a, b, c > 0$ be given, and assume that $\{E_{mb}T_{na}g\}_{m,n\in\mathbb{Z}}$ is a Gabor frame. Then, with $g_c := D_c g$, the Gabor family $\{E_{mb/c}T_{nac}g_c\}_{m,n\in\mathbb{Z}}$ is a frame with the same frame bounds as $\{E_{mb}T_{na}g\}_{m,n\in\mathbb{Z}}$.*

Proof. Operators of the type D_c are studied in Section 2.9, and they are unitary. By Lemma 5.3.3, it follows that $\{D_c E_{mb}T_{na}g\}_{m,n\in\mathbb{Z}}$ is a frame with the same frame bounds as $\{E_{mb}T_{na}g\}_{m,n\in\mathbb{Z}}$. Using the commutator relations in Section 2.9, we see that

$$D_c E_{mb}T_{na} = E_{mb/c}D_c T_{na} = E_{mb/c}T_{nac}D_c,$$

and the proposition follows.   □

In the rest of this section we present some of the more advanced results that are central in Gabor analysis. In order to do so, we first notice that there is a close connection between Gabor systems and the shift-invariant systems considered in Chapter 8. In fact,

$$T_{na}E_{mb}g(x) = e^{-2\pi imnab}e^{2\pi imbx}g(x - na) = e^{-2\pi imnab}E_{mb}T_{na}g(x); \quad (9.22)$$

thus, the functions in the shift-invariant system $\{T_{na}E_{mb}g\}_{m,n\in\mathbb{Z}}$ only differ from the functions in the Gabor system $\{E_{mb}T_{na}g\}_{m,n\in\mathbb{Z}}$ by some complex factors of absolute value one. This implies that the systems $\{T_{na}E_{mb}g\}_{m,n\in\mathbb{Z}}$ and $\{E_{mb}T_{na}g\}_{m,n\in\mathbb{Z}}$ have the same frame properties.

We now present a characterization of Gabor frames, which follows immediately from our discussion of shift-invariant systems in Chapter 8. As for the shift-invariant case, the characterization is formulated in terms of a matrix inequality, which should be interpreted as explained in (8.18). Given a function $g \in L^2(\mathbb{R})$ and two numbers $a, b > 0$, consider the matrix-valued function

$$M(x) := (g(x - na - m/b))_{m,n\in\mathbb{Z}}, \quad x \in \mathbb{R}. \quad (9.23)$$

Theorem 9.1.15 *Let $A, B > 0$ and the Gabor system $\{E_{mb}T_{na}g\}_{m,n\in\mathbb{Z}}$ be given. Then the following holds:*

(i) *$\{E_{mb}T_{na}g\}_{m,n\in\mathbb{Z}}$ is a Bessel sequence with bound B if and only if $M(x)$ for a.e. $x \in \mathbb{R}$ defines a bounded operator on $\ell^2(\mathbb{Z})$ with norm at most $\sqrt{bB}$.*

(ii) *Assuming that $\{E_{mb}T_{na}g\}_{m,n\in\mathbb{Z}}$ is a Bessel sequence, it is a frame for $L^2(\mathbb{R})$ with lower frame bound A if and only if*

$$bAI \le M(x)M(x)^* \ a.e. \ x \in \mathbb{R}, \tag{9.24}$$

where I is the identity operator on $\ell^2(\mathbb{Z})$.

Proof. We derive the result from Theorem 8.1.7. First we note that the Fourier transform $\mathcal{F}$ is unitary; thus, Lemma 5.3.3 shows that $\{E_{mb}T_{na}g\}_{m,n\in\mathbb{Z}}$ is a frame if and only if $\{\mathcal{F}^{-1}E_{mb}T_{na}g\}_{m,n\in\mathbb{Z}}$ is a frame. The commutator relations (2.23) imply that

$$\mathcal{F}^{-1}E_{mb}T_{na}g = T_{-mb}E_{na}\mathcal{F}^{-1}g,$$

which is a shift-invariant system based on the translation parameter b and the functions $E_{na}\mathcal{F}^{-1}g$, $n \in \mathbb{Z}$. Consider the matrix H in (8.13) corresponding to this system; denoting the variable by x rather than ν, the k, n-th entry is

$$\mathcal{F}E_{na}\mathcal{F}^{-1}g(x - k/b) = T_{na}g(x - k/b) = g(x - na - k/b).$$

That is, $H(x)$ equals the matrix $M(x)$ in (9.23), and the result follows from Theorem 8.1.6 and Theorem 8.1.7. $\qquad\square$

Finally, we now state one of the most fundamental and important results in Gabor analysis. It is known as the *duality principle* and was discovered almost simultaneously by three groups of researchers: Janssen [47], Daubechies, Landau, and Landau [29], and Ron and Shen [54]. The duality principle concerns the relationship between frame properties for a function g with respect to the lattice $\{(na, mb)\}_{m,n\mathbb{Z}}$ and with respect to the so-called *dual lattice* $\{(n/b, m/a)\}_{m,n\mathbb{Z}}$. A unified treatment of duality results is given by Gröchenig in [37].

Theorem 9.1.16 *Let $g \in L^2(\mathbb{R})$ and $a, b > 0$ be given. Then the Gabor system $\{E_{mb}T_{na}g\}_{m,n\in\mathbb{Z}}$ is a frame for $L^2(\mathbb{R})$ with bounds A, B if and only if $\{E_{m/a}T_{n/b}g\}_{m,n\in\mathbb{Z}}$ is a Riesz sequence with bounds abA, abB.*

The importance of Theorem 9.1.16 lies in the fact that it often is easier to prove that $\{E_{m/a}T_{n/b}g\}_{m,n\in\mathbb{Z}}$ is a Riesz sequence than to prove directly that $\{E_{mb}T_{na}g\}_{m,n\in\mathbb{Z}}$ is a frame.

9.2   Tight Gabor frames

In applications of frames, it is inconvenient that the frame decomposition, stated in Theorem 5.1.7, requires inversion of the frame operator. As we have seen in the discussion of general frame theory, one way of avoiding the problem is to consider tight frames. We will now characterize tight Gabor frames; the result follows from the characterization in Theorem 8.1.7 of shift-invariant systems generating tight frames (Exercise 9.8), but we include a direct proof.

Theorem 9.2.1 *Let $g \in L^2(\mathbb{R})$ and $a, b > 0$ be given. Then the following are equivalent:*

(i) *$\{E_{mb}T_{na}g\}_{m,n\in\mathbb{Z}}$ is a tight frame for $L^2(\mathbb{R})$ with frame bound $A = 1$.*

(ii) *For a.e. $x \in \mathbb{R}$ the following conditions hold:*
  (a) *$G(x) := \sum_{n\in\mathbb{Z}} |g(x - na)|^2 = b$;*
  (b) *$G_k(x) := \sum_{n\in\mathbb{Z}} g(x - na)\overline{g(x - na - k/b)} = 0$ for all $k \neq 0$.*

Moreover, when the equivalent conditions hold, $\{E_{mb}T_{na}g\}_{m,n\in\mathbb{Z}}$ is an orthonormal basis for $L^2(\mathbb{R})$ if and only if $\|g\| = 1$.

Proof. (i)$\Rightarrow$ (ii): Assume that $\{E_{mb}T_{na}g\}_{m,n\in\mathbb{Z}}$ is a tight frame for $L^2(\mathbb{R})$ with frame bound $A = 1$. Then Proposition 9.1.2 shows that $G(x) = b$ for a.e. $x \in \mathbb{R}$. Therefore

$$\sum_{m,n\in\mathbb{Z}} |\langle f, E_{mb}T_{na}g\rangle|^2 = \frac{1}{b} \int_{-\infty}^{\infty} |f(x)|^2 G(x)\, dx$$

for all functions $f \in L^2(\mathbb{R})$. Using Lemma 9.1.4, we conclude that for all bounded, compactly supported $f \in L^2(\mathbb{R})$,

$$\frac{1}{b}\sum_{k\neq 0} \int_{-\infty}^{\infty} \overline{f(x)} f(x - k/b) \sum_{n\in\mathbb{Z}} g(x - na)\overline{g(x - na - k/b)}\, dx = 0.$$

A change of variable shows that the contribution in the above sum arising from any value of $k \in \mathbb{Z}$ is the complex conjugate of the contribution from the value $-k$. Therefore

$$\sum_{k=1}^{\infty} \mathrm{Re}\left(\int_{-\infty}^{\infty} \overline{f(x)} f(x - k/b) \sum_{n\in\mathbb{Z}} g(x - na)\overline{g(x - na - k/b)}\, dx \right) = 0. \quad (9.25)$$

Now fix $k_0 \geq 1$ and let I be any interval in $\mathbb{R}$ of length at most $1/b$. Define a function $f \in L^2(\mathbb{R})$ by

$$f(x) = \begin{cases} e^{-i \arg(G_{k_0}(x))} & \text{for } x \in I, \\ 1 & \text{for } x \in I + k_0/b, \\ 0 & \text{otherwise.} \end{cases}$$

Then, by (9.25),

$$
\begin{aligned}
0 &= \sum_{k=1}^{\infty} \operatorname{Re} \left(\int_{-\infty}^{\infty} \overline{f(x)} f(x - k/b) \sum_{n \in \mathbb{Z}} g(x - na) \overline{g(x - na - k/b)} \, dx \right) \\
&= \operatorname{Re} \left(\int_{-\infty}^{\infty} \overline{f(x)} f(x - k_0/b) G_{k_0}(x) \, dx \right) = \int_I |G_{k_0}(x)| \, dx.
\end{aligned}
$$

It follows that $G_{k_0}(x) = 0$ for a.e. $x \in I$. Since I was an arbitrary interval of length at most $1/b$, we conclude that $G_{k_0} = 0$. In order to deal with G_k for $k < 0$, a direct computation shows that

$$
G_{-k_0}(x) = \overline{G_{k_0}(x + k_0/b)} = 0;
$$

this shows that statement (b) in (ii) indeed holds for all $k \neq 0$.

(ii)$\Rightarrow$ (i): The assumptions in (ii) imply, again by Lemma 9.1.4, that for all bounded, compactly supported functions $f \in L^2(\mathbb{R})$,

$$
\begin{aligned}
\sum_{m,n \in \mathbb{Z}} |\langle f, E_{mb} T_{na} g \rangle|^2 &= \frac{1}{b} \int_{-\infty}^{\infty} |f(x)|^2 \sum_{n \in \mathbb{Z}} |g(x - na)|^2 \, dx \\
&= \|f\|^2.
\end{aligned}
$$

Since the bounded compactly supported functions are dense in $L^2(\mathbb{R})$, Lemma 5.1.2 implies that $\{E_{mb} T_{na} g\}_{m,n \in \mathbb{Z}}$ is a tight frame with frame bound $A = 1$, as desired. The final part of the theorem is a consequence of Exercise 3.1. $\qquad \square$

In general, it is not easy to construct functions g such that the conditions in Theorem 9.2.1(ii) are satisfied for some given $a, b > 0$. A simplification occurs if we assume that g has compact support: in that case, the condition (b) in (ii) is automatically satisfied for sufficiently small values of the parameter b. In particular, we obtain the following very useful sufficient condition for $\{E_{mb} T_{na} g\}_{m,n \in \mathbb{Z}}$ being a tight Gabor frame. We ask the reader to provide the proof in Exercise 9.9.

Corollary 9.2.2 *Let $a, b > 0$ be given. Assume that $\varphi \in L^2(\mathbb{R})$ is a real-valued non-negative function with support in an interval of length $1/b$, and that*

$$
\sum_{n \in \mathbb{Z}} \varphi(x + na) = 1, \quad a.e. \ x \in \mathbb{R}. \tag{9.26}
$$

Then the function

$$
g(x) := \sqrt{b\varphi(x)}
$$

generates a tight Gabor frame $\{E_{mb} T_{na} g\}_{m,n \in \mathbb{Z}}$ with frame bound $A = 1$.

If (9.26) is satisfied, we say that the functions $\{T_{na}\varphi\}_{n \in \mathbb{Z}}$ form a *partition of unity*. Readers with knowledge of multiresolution analysis will notice that the associated scaling function satisfies this condition for $a = 1$; see, e.g., [64]. In particular, we can apply the result to B-splines:

Example 9.2.3 For any $\ell \in \mathbb{N}$, the B-spline $\varphi = N_\ell$ defined in (6.1) satisfies the requirements in Corollary 9.2.2 with $a = 1$ and any $b \in]0, 1/\ell]$. Thus, for any $b \in]0, 1/\ell]$, the function

$$g(x) = \sqrt{b N_\ell(x)}$$

generates a tight Gabor frame $\{E_{mb}T_n g\}_{m,n\in\mathbb{Z}}$ with frame bound $A = 1.\square$

We note that the frame generators in Example 9.2.3 are very suitable for time–frequency analysis: they are given by an explicit formula, have compact support, and can be chosen with polynomial decay of any desired order in the frequency domain, simply by taking the parameter n sufficiently large.

9.3   The duals of a Gabor frame

For a Gabor frame $\{E_{mb}T_{na}g\}_{m,n\in\mathbb{Z}}$ with associated frame operator S, the frame decomposition, see Theorem 5.1.7, shows that

$$f = \sum_{m,n\in\mathbb{Z}} \langle f, S^{-1}E_{mb}T_{na}g\rangle E_{mb}T_{na}g, \ \forall f \in L^2(\mathbb{R}). \tag{9.27}$$

In order to use the frame decomposition, we need to be able to calculate the canonical dual frame $\{S^{-1}E_{mb}T_{na}g\}_{m,n\in\mathbb{Z}}$. This is usually difficult. Via the following lemma, we will be able to obtain a simplification.

Lemma 9.3.1 *Let $g \in L^2(\mathbb{R})$ and $a, b > 0$ be given, and assume that $\{E_{mb}T_{na}g\}_{m,n\in\mathbb{Z}}$ is a Bessel sequence with frame operator S. Then the following holds:*

(i) $SE_{mb}T_{na} = E_{mb}T_{na}S$ for all $m, n \in \mathbb{Z}$.

(ii) If $\{E_{mb}T_{na}g\}_{m,n\in\mathbb{Z}}$ is a frame, then

$$S^{-1}E_{mb}T_{na} = E_{mb}T_{na}S^{-1}, \ \forall m, n \in \mathbb{Z}.$$

Proof. Let $f \in L^2(\mathbb{R})$, and assume that $\{E_{mb}T_{na}g\}_{m,n\in\mathbb{Z}}$ is a Bessel sequence. Using the commutator relations (2.19),

$$
\begin{aligned}
SE_{mb}T_{na}f &= \sum_{m',n'\in\mathbb{Z}} \langle E_{mb}T_{na}f, E_{m'b}T_{n'a}g\rangle E_{m'b}T_{n'a}g \\
&= \sum_{m',n'\in\mathbb{Z}} \langle f, T_{-na}E_{(m'-m)b}T_{n'a}g\rangle E_{m'b}T_{n'a}g \\
&= \sum_{m',n'\in\mathbb{Z}} \langle f, e^{2\pi ina(m'-m)b}E_{(m'-m)b}T_{(n'-n)a}g\rangle E_{m'b}T_{n'a}g.
\end{aligned}
$$

Performing the change of variables $m' \to m' + m, n' \to n' + n$ and using the commutator relations again,

$$
\begin{aligned}
& SE_{mb}T_{na}f \\
={}& \sum_{m',n'\in\mathbb{Z}} e^{-2\pi inam'b}\langle f, E_{m'b}T_{n'a}g\rangle E_{(m'+m)b}T_{(n'+n)a}g \\
={}& \sum_{m',n'\in\mathbb{Z}} e^{-2\pi inam'b}\langle f, E_{m'b}T_{n'a}g\rangle e^{2\pi inam'b}E_{mb}T_{na}E_{m'b}T_{n'a}g \\
={}& E_{mb}T_{na}Sf.
\end{aligned}
$$

This proves (i). The result in (ii) follows by applying the operator S^{-1} to both sides of the equality in (i). $\qquad\square$

Lemma 9.3.1 has important consequences for the structure of the canonical dual frame of a Gabor frame:

Theorem 9.3.2 *Let $g \in L^2(\mathbb{R})$ and $a, b > 0$ be given, and assume that $\{E_{mb}T_{na}g\}_{m,n\in\mathbb{Z}}$ is a Gabor frame. Then the following holds:*

(i) *The canonical dual frame also has the Gabor structure and is given by $\{E_{mb}T_{na}S^{-1}g\}_{m,n\in\mathbb{Z}}$.*

(ii) *The canonical tight frame associated with $\{E_{mb}T_{na}g\}_{m,n\in\mathbb{Z}}$ is given by $\{E_{mb}T_{na}S^{-1/2}g\}_{m,n\in\mathbb{Z}}$.*

Proof. The assumption that $\{E_{mb}T_{na}g\}_{m,n\in\mathbb{Z}}$ is a frame implies that the frame operator S is invertible, so (i) is consequence of Lemma 9.3.1. Furthermore, Lemma 2.4.4 shows that $S^{-1/2}$ is a limit of polynomials in S^{-1} in the strong operator topology; therefore $S^{-1/2}$ commutes with $E_{mb}T_{na}$. Thus, according to the definition, the canonical tight frame associated with $\{E_{mb}T_{na}g\}_{m,n\in\mathbb{Z}}$ is given by

$$
\{S^{-1/2}E_{mb}T_{na}g\}_{m,n\in\mathbb{Z}} = \{E_{mb}T_{na}S^{-1/2}g\}_{m,n\in\mathbb{Z}};
$$

this proves (ii). $\qquad\square$

Via Theorem 9.3.2, the frame decomposition (9.27) associated with a Gabor frame $\{E_{mb}T_{na}g\}_{m,n\in\mathbb{Z}}$ takes the form

$$f = \sum_{m,n\in\mathbb{Z}} \langle f, E_{mb}T_{na}S^{-1}g\rangle E_{mb}T_{na}g, \ \forall f \in L^2(\mathbb{R}). \tag{9.28}$$

In practice, this version of the frame decomposition is much more convenient than (9.27): instead of calculating the *double infinite* family $\{S^{-1}E_{mb}T_{na}g\}_{m,n\in\mathbb{Z}}$, it is enough to find $S^{-1}g$ and then apply the modulation and translation operators. The function $S^{-1}g$ is called the *dual window function* or the *dual generator*.

Bölcskei and Janssen showed in [7] that the canonical dual frame has many pleasant properties. For example, if $\{E_{mb}T_{na}g\}_{m,n\in\mathbb{Z}}$ is an overcomplete frame and the window function g decays exponentially, i.e., there exist constants $C, \lambda > 0$ such that

$$|g(x)| \le Ce^{-\lambda|x|}, \ a.e. \ x \in \mathbb{R},$$

then also $S^{-1}g$ decays exponentially. Still assuming that g generates an overcomplete frame, it is also proved that exponential decay of $\hat{g}$ implies exponential decay of $\mathcal{F}(S^{-1}g)$.

The same results hold with $S^{-1}g$ replaced by $S^{-1/2}g$. In particular, this leads to the existence of Gabor frames with (theoretically) perfect properties:

Proposition 9.3.3 *Let $g \in L^2(\mathbb{R})$, and assume that g as well as $\hat{g}$ decay exponentially. Let $a, b > 0$ be given and assume that $\{E_{mb}T_{na}g\}_{m,n\in\mathbb{Z}}$ is a frame. Then $\{E_{mb}T_{na}S^{-1/2}g\}_{m,n\in\mathbb{Z}}$ is a tight frame, for which $S^{-1/2}g$ as well as $\mathcal{F}(S^{-1/2}g)$ decay exponentially.*

Applying this result to the Gaussian $g(x) = e^{-x^2/2}$ and any $a, b > 0$ with $ab < 1$ leads to a tight Gabor frame $\{E_{mb}T_{na}S^{-1/2}g\}_{m,n\in\mathbb{Z}}$ with perfect time–frequency localization. The only shortcoming of this example (and all the other known constructions obtained via Proposition 9.3.3) is that we do not have an explicit expression for the window function $S^{-1/2}g$.

We will now leave the discussion of the canonical dual frame and examine the question of how general dual frames of a given Gabor frame $\{E_{mb}T_{na}g\}_{m,n\in\mathbb{Z}}$ can be found; in Section 9.4, we will use the obtained results to construct pairs of explicitly given Gabor frames. The general characterization of all dual frames in Theorem 5.7.4 also applies to Gabor frames, but if $\{E_{mb}T_{na}g\}_{m,n\in\mathbb{Z}}$ is an overcomplete frame, not all of these duals have the Gabor structure (Exercise 9.10). The duals with Gabor structure are characterized in the famous *Wexler–Raz Theorem*. Several proofs of this result exist: we will derive it as a consequence of Theorem 8.1.7.

Theorem 9.3.4 *Let $g, h \in L^2(\mathbb{R})$ and $a, b > 0$ be given. Then, if the two Gabor systems $\{E_{mb}T_{na}g\}_{m,n\in\mathbb{Z}}$ and $\{E_{mb}T_{na}h\}_{m,n\in\mathbb{Z}}$ are Bessel sequences, they are dual frames if and only if*

$$\langle h, E_{m/a}T_{n/b}g \rangle = 0 \text{ for all } (m, n) \neq (0, 0) \text{ and } \langle h, g \rangle = ab. \qquad (9.29)$$

Proof. The commutation relations in (9.22) show that the Bessel sequences $\{E_{mb}T_{na}g\}_{m,n\in\mathbb{Z}}$ and $\{E_{mb}T_{na}h\}_{m,n\in\mathbb{Z}}$ are dual frames if and only if the shift-invariant systems $\{T_{na}E_{mb}g\}_{m,n\in\mathbb{Z}}$ and $\{T_{na}E_{mb}h\}_{m,n\in\mathbb{Z}}$ are dual frames. The generators for the two latter systems are $g_m = E_{mb}g$ and $h_m = E_{mb}h$; by Theorem 8.1.7, they generate dual frames if and only if

$$\sum_{m\in\mathbb{Z}} \overline{\widehat{g_m}(\nu)}\widehat{h_m}(\nu + k/a) = a\delta_{k,0}, \ k \in \mathbb{Z}, \ a.e. \ \nu \in \mathbb{R}.$$

In terms of the functions g and h, this is equivalent to

$$\sum_{m\in\mathbb{Z}} \overline{\hat{g}(\nu - mb)}\hat{h}(\nu + k/a - mb) = a\delta_{k,0}, \ k \in \mathbb{Z}, \ a.e. \ \nu \in \mathbb{R}. \qquad (9.30)$$

We can express this condition in terms of the coefficients in the Fourier expansion with respect to $\{e^{2\pi i n\nu/b}\}_{n\in\mathbb{Z}}$ for the b-periodic functions

$$\phi_k(\nu) := \sum_{m\in\mathbb{Z}} \overline{\hat{g}(\nu - mb)}\hat{h}(\nu + k/a - mb), \ k \in \mathbb{Z}:$$

in fact, (9.30) is equivalent to all coefficients for $\phi_k, k \neq 0$, being zero and the coefficients for ϕ_0 being zero for $k \neq 0$ and equal to a for $k = 0$. Wexler–Raz' theorem is now a consequence of the following computation, which yields the n-th Fourier coefficient for the function ϕ_k. The computation is based on Lemma 7.1.3 and the commutator relations for the Fourier transform and the operators T_a, E_b:

$$\frac{1}{b}\int_0^b \phi_k(\nu)e^{-2\pi i n\nu/b}\, d\nu$$

$$= \frac{1}{b}\int_0^b \sum_{m\in\mathbb{Z}} \overline{\hat{g}(\nu - mb)}\hat{h}(\nu + k/a - mb)e^{-2\pi i n\nu/b}\, d\nu$$

$$= \frac{1}{b}\int_{-\infty}^{\infty} \overline{\hat{g}(\nu)}\hat{h}(\nu + k/a)e^{-2\pi i n\nu/b}d\nu$$

$$= \frac{1}{b}\langle T_{-k/a}\hat{h}, E_{n/b}\hat{g}\rangle$$

$$= \frac{1}{b}\langle \hat{h}, T_{k/a}E_{n/b}\hat{g}\rangle$$

$$= \frac{1}{b}\langle \mathcal{F}h, \mathcal{F}E_{k/a}T_{-n/b}g\rangle.$$

The result now follows from the Fourier transform being unitary. $\quad\square$

For practical purposes, it is not sufficient to characterize the duals of a frame $\{E_{mb}T_{na}g\}_{m,n\in\mathbb{Z}}$: we also need to know how to find them. A constructive procedure to find all dual frames having the Gabor structure is described in Exercise 9.11. In Section 9.4, we will provide constructions of pairs of dual Gabor frames, for which the generators are given explicitly; they are based on the following consequence of Theorem 8.1.7.

Theorem 9.3.5 *Let $g, h \in L^2(\mathbb{R})$ and $a, b > 0$ be given. Two Bessel sequences $\{E_{mb}T_{na}g\}_{m,n\in\mathbb{Z}}$ and $\{E_{mb}T_{na}h\}_{m,n\in\mathbb{Z}}$ form dual frames if and only if*

$$\sum_{k\in\mathbb{Z}} \overline{g(x - n/b - ka)}h(x - ka) = b\delta_{n,0}, \quad a.e. \ x \in [0, a]. \tag{9.31}$$

We leave the proof to the reader (Exercise 9.12).

9.4 Explicit construction of dual frame pairs

So far, we have only seen few examples of Gabor frames and their dual frames. After the preparation in Section 9.3, we are now ready to provide explicit constructions of certain Gabor frames and some particularly convenient duals. The assumptions are tailored to the properties of the B-splines. The results presented here first appeared in [13]. For convenience, we begin with the case where the translation parameter is $a = 1$.

Theorem 9.4.1 *Let $N \in \mathbb{N}$. Let $g \in L^2(\mathbb{R})$ be a real-valued bounded function with supp $g \subseteq [0, N]$, for which*

$$\sum_{k\in\mathbb{Z}} g(x - k) = 1, \ x \in \mathbb{R}. \tag{9.32}$$

Let $b \in]0, \frac{1}{2N-1}]$. Then the function g and the function h defined by

$$h(x) = bg(x) + 2b\sum_{k=1}^{N-1} g(x + k) \tag{9.33}$$

generate dual frames $\{E_{mb}T_n g\}_{m,n\in\mathbb{Z}}$ and $\{E_{mb}T_n h\}_{m,n\in\mathbb{Z}}$ for $L^2(\mathbb{R})$.

Proof. By assumption, the function g has compact support and is bounded; by the definition (9.33), the function h shares these properties. It now follows from Corollary 9.1.6 that $\{E_{mb}T_n g\}_{m,n\in\mathbb{Z}}$ and $\{E_{mb}T_n h\}_{m,n\in\mathbb{Z}}$ are Bessel sequences. In order to verify that these sequences form dual frames, we use Theorem 9.3.5: according to (9.31), we need to check that for $x \in [0, 1]$,

$$\sum_{k \in \mathbb{Z}} g(x - n/b - k)h(x - k) = b\delta_{n,0}. \tag{9.34}$$

The function g has support in $[0, N]$, so by construction h has support in $[-N + 1, N]$; thus (9.34) is satisfied for $n \neq 0$ whenever $1/b \geq 2N - 1$, i.e., if $b \in]0, \frac{1}{2N-1}]$. For $n = 0$, the condition (9.34) means that

$$\sum_{k \in \mathbb{Z}} g(x - k)h(x - k) = b, \ x \in [0, 1];$$

because of the compact support of g, this is equivalent to

$$\sum_{k=0}^{N-1} g(x + k)h(x + k) = b, \ x \in [0, 1]. \tag{9.35}$$

The condition (9.35) is indeed satisfied in our setting. To see this, we use that for $x \in [0, 1]$,

$$1 = \sum_{k=0}^{N-1} g(x + k).$$

This implies that, again for $x \in [0, 1]$,

$$
\begin{aligned}
1 &= \left(\sum_{k=0}^{N-1} g(x + k) \right)^2 \\
&= (g(x) + g(x + 1) + \cdots + g(x + N - 1)) \times \\
&\qquad\qquad\qquad (g(x) + g(x + 1) + \cdots + g(x + N - 1)) \\
&= g(x) \left[g(x) + 2g(x + 1) + 2g(x + 2) + \cdots + 2g(x + N - 1) \right] \\
&\quad + g(x + 1) \left[g(x + 1) + 2g(x + 2) + 2g(x + 3) + \cdots + 2g(x + N - 1) \right] \\
&\quad + g(x + 2) \left[g(x + 2) + 2g(x + 3) + 2g(x + 4) + \cdots + 2g(x + N - 1) \right] \\
&\quad + \cdots \\
&\quad + \cdots \\
&\quad + g(x + N - 2) \left[g(x + N - 2) + 2g(x + N - 1) \right] \\
&\quad + g(x + N - 1) \left[g(x + N - 1) \right] \\
&= \frac{1}{b} \sum_{k=0}^{N-1} g(x + k)h(x + k).
\end{aligned}
$$

Thus the condition (9.35) is satisfied. $\square$

The assumptions in Theorem 9.4.1 are tailored to the properties of the B-splines N_ℓ defined in (6.1):

Corollary 9.4.2 *For any $\ell \in \mathbb{N}$ and $b \in]0, \frac{1}{2\ell-1}]$, the functions N_ℓ and*

$$h_\ell(x) := bN_\ell(x) + 2b \sum_{k=1}^{N-1} N_\ell(x+k) \tag{9.36}$$

generate dual frames $\{E_{mb}T_nN_\ell\}_{m,n\in\mathbb{Z}}$ and $\{E_{mb}T_nh_\ell\}_{m,n\in\mathbb{Z}}$ for $L^2(\mathbb{R})$.

Some of the important features of the dual pair of frame generators (N_ℓ, h_ℓ) in Corollary 9.4.2 are as follows:

- The functions N_ℓ and h_ℓ are splines for all choices of $\ell \in \mathbb{N}$;

- N_ℓ and h_ℓ are explicitly given functions with compact support, i.e., they have perfect time-localization;

- By choosing $\ell \in \mathbb{N}$ sufficiently large, polynomial decay of $\widehat{N_\ell}$ and $\widehat{h_\ell}$ of any desired order can be obtained.

We note that the crucial partition of unity condition (9.32) in Theorem 9.4.1 is satisfied for many other functions than B-splines. For example, it holds for any scaling function, i.e., any function $\phi \in L^2(\mathbb{R})$ satisfying an equation of the form

$$\hat{\phi}(2\gamma) = H_0(\gamma)\hat{\phi}(\gamma)$$

for some bounded 1-periodic function H_0. See Section 3.6 and Chapter 11 for a discussion of scaling functions in the context of wavelet theory.

Example 9.4.3 For the B-spline

$$N_2(x) = \begin{cases} x & x \in [0, 1[, \\ 2 - x & x \in [1, 2[, \\ 0 & x \notin [0, 2[, \end{cases}$$

we can use Corollary 9.4.2 for $b \in]0, 1/3]$. For $b = 1/3$, we obtain the dual generator

$$h_2(x) = \frac{1}{3}N_2(x) + \frac{2}{3}N_2(x+1) = \begin{cases} \frac{2}{3}(x+1) & x \in [-1, 0[, \\ \frac{1}{3}(2-x) & x \in [0, 2[, \\ 0 & x \notin [-1, 2[. \end{cases} \tag{9.37}$$

See Figure 9.1. Figure 9.2 shows a similar construction based on N_3. □

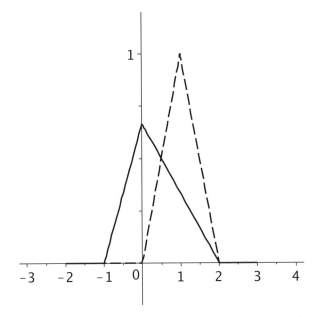

Figure 9.1. The B-spline N_2 and the dual generator h_2 in (9.37).

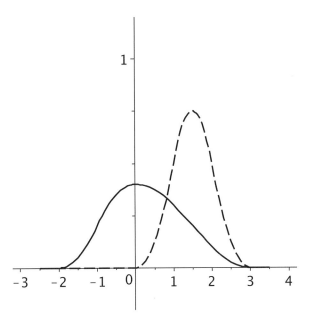

Figure 9.2. The B-spline N_3 and the dual generator h_3 in (9.33) with $b = 1/5$.

Via a scaling, we obtain a version of Theorem 9.4.1 that is valid for any translation parameter $a > 0$:

Theorem 9.4.4 *Let $N \in \mathbb{N}$. Let $g \in L^2(\mathbb{R})$ be a real-valued bounded function with supp $g \subseteq [0, N]$, for which*

$$\sum_{k \in \mathbb{Z}} g(x - k) = 1, \ x \in \mathbb{R}.$$

Let $a, b > 0$ be given such that $ab \in]0, \frac{1}{2N-1}]$, and let

$$h(x) = abg(x) + 2ab \sum_{k=1}^{N-1} g(x + k). \tag{9.38}$$

Then the functions $D_a g$ and $D_a h$ generate dual frames $\{E_{mb} T_{na} D_a g\}_{m,n \in \mathbb{Z}}$ and $\{E_{mb} T_{na} D_a h\}_{m,n \in \mathbb{Z}}$ for $L^2(\mathbb{R})$.

Proof. Via Theorem 9.4.1, the assumptions imply that $\{E_{mab} T_n g\}_{m,n \in \mathbb{Z}}$ and $\{E_{mab} T_n h\}_{m,n \in \mathbb{Z}}$ form dual Gabor frames. Since

$$D_a E_{mab} T_n = E_{mb} T_{na} D_a,$$

the result now follows from D_a being unitary and Exercise 5.32. $\square$

Observe that for large support sizes, Theorem 9.4.4 forces the product ab to be small. For larger values of ab, one can still make a frame construction like in Theorem 9.4.4, however, with more than one generator. See [13] for details. We also note that it recently has been shown how to construct pairs of dual Gabor frames where both generators are symmetric; see [16].

9.5 Popular Gabor conditions

In this section, we shortly discuss some conditions on the generator g for a Gabor frame, which are often used in the literature.

Given a positive number a, the *Wiener space* is defined by

$$W := \left\{ g : \mathbb{R} \to \mathbb{C} \mid g \text{ measurable and } \sum_{k \in \mathbb{Z}} ||g\chi_{[ka,(k+1)a[}||_\infty < \infty \right\}. \tag{9.39}$$

One can prove that W is a Banach space with respect to the norm

$$||g||_{W,a} = \sum_{k \in \mathbb{Z}} ||g\chi_{[ka,(k+1)a[}||_\infty.$$

The space W is independent of the choice of a, and different choices give equivalent norms; both statements follow from the fact that if $0 < a \leq b$

and $\lfloor \frac{b}{a} \rfloor$ denotes the smallest integer that is larger than or equal to b/a, then (Exercise 9.13)

$$\sum_{k\in\mathbb{Z}} ||g\chi_{[kb,(k+1)b[}||_\infty \;\leq\; 2\sum_{k\in\mathbb{Z}} ||g\chi_{[ka,(k+1)a[}||_\infty \qquad (9.40)$$

$$\leq\; 2\left(\lfloor \frac{b}{a}\rfloor + 2\right)\sum_{k\in\mathbb{Z}} ||g\chi_{[kb,(k+1)b[}||_\infty.$$

A function $g \in W$ is automatically bounded and decays so fast that the "local maximum function" $k \mapsto ||g\chi_{[ka,(k+1)a[}||_\infty$ belongs to $\ell^1(\mathbb{Z})$. In Exercise 9.13, we ask the reader to prove that $W \subset L^1(\mathbb{R}) \cap L^2(\mathbb{R})$. The condition for being in W is strong enough to exclude many of the pathological functions, which play a role for the understanding of functions in $L^2(\mathbb{R})$ but are of little practical interest (like functions in $L^2(\mathbb{R})$, which do not decay to zero whenever $|x| \to \infty$).

Lemma 9.5.1 *Let $g \in W$ and $a > 0$ be given. Then*

$$\sum_{n\in\mathbb{Z}} |g(x - na)| \leq ||g||_{W,a}, \quad a.e.\ x \in \mathbb{R}.$$

If also $h \in W$ and $b \in]0, \frac{1}{a}]$, then

$$\sum_{k\in\mathbb{Z}}\left|\sum_{n\in\mathbb{Z}} g(x-na)\overline{h(x-na-k/b)}\right| \leq 2\,||g||_{W,a}||h||_{W,a}, \quad a.e.\ x \in \mathbb{R}.$$

Proof. For the first part, fix $x \in \mathbb{R}$, and observe that for any given $n \in \mathbb{Z}$, there exists exactly one value of $k \in \mathbb{Z}$ such that

$$x - na \in [ka, (k+1)a[, k \in \mathbb{Z};$$

furthermore, different values of n lead to different values for k. Therefore,

$$\sum_{n\in\mathbb{Z}} |g(x-na)| \leq \sum_{k\in\mathbb{Z}} ||g\chi_{[ka,(k+1)a[}||_\infty = ||g||_{W,a}, \quad a.e.\ x \in \mathbb{R}.$$

For the second part, we have

$$\sum_{k\in\mathbb{Z}}\left|\sum_{n\in\mathbb{Z}} g(x-na)\overline{h(x-na-k/b)}\right| \leq \sum_{n\in\mathbb{Z}} |g(x-na)| \sum_{k\in\mathbb{Z}} |h(x-na-k/b)|.$$

The first part of the lemma (applied to the function h and the translation parameter $\frac{1}{b}$) combined with (9.40) gives that

$$\sum_{k\in\mathbb{Z}} |h(x-na-k/b)| \leq ||h||_{W,\frac{1}{b}} \leq 2\,||h||_{W,a},$$

and the lemma follows. $\qquad\square$

If g belongs to the Wiener space, then $\{E_{mb}T_{na}g\}_{m,n\in\mathbb{Z}}$ is a Bessel sequence for all $a, b > 0$:

Proposition 9.5.2 *If $g \in W$ and $a, b > 0$, then $\{E_{mb}T_{na}g\}_{m,n\in\mathbb{Z}}$ is a Bessel sequence. If $ab \leq 1$, then $B := \frac{2}{b}\|g\|_{W,a}^2$ is an upper frame bound.*

Proof. The case $ab \leq 1$ follows immediately from Lemma 9.5.1 combined with Theorem 9.1.5. In case $ab > 1$, we can choose $N \in \mathbb{N}$ such that $\frac{a}{N}b \leq 1$; this implies that $\{E_{mb}T_{na/N}g\}_{m,n\in\mathbb{Z}}$ is a Bessel sequence, and therefore the subsequence $\{E_{mb}T_{na}g\}_{m,n\in\mathbb{Z}}$ is also a Bessel sequence. $\square$

Another important subspace of $L^2(\mathbb{R})$ is known as the *Feichtinger algebra* $\mathcal{S}_0$. In order to define this vector space, we first introduce the *short-time Fourier transform*, also called the *continuous Gabor transform*:

Definition 9.5.3 *Fix a function $g \in L^2(\mathbb{R}) \setminus \{0\}$. The short-time Fourier transform of a function $f \in L^2(\mathbb{R})$, with respect to the window function g is the function $\Psi_g(f)$ of two variables given by*

$$\Psi_g(f)(y, \gamma) = \int_{-\infty}^{\infty} f(x)\overline{g(x-y)}e^{-2\pi i x\gamma}\, dx, \quad y, \gamma \in \mathbb{R}.$$

Note that in terms of the modulation operators and translation operators,

$$\Psi_g(f)(y, \gamma) = \langle f, E_\gamma T_y g\rangle.$$

Now, consider the Gaussian, $g(x) := e^{-x^2}$. The Feichtinger algebra $\mathcal{S}_0$ is defined as the vector space consisting of all functions $f \in L^2(\mathbb{R})$ for which

$$\int_{-\infty}^{\infty}\int_{-\infty}^{\infty} |\langle E_x T_y f, g\rangle|\, dx dy < \infty. \tag{9.41}$$

In terms of the short-time Fourier transform, the Feichtinger algebra consists of the functions $f \in L^2(\mathbb{R})$ for which $\Psi_g(f) \in L^1(\mathbb{R}^2)$.

Feichtinger introduced the space $\mathcal{S}_0$ in 1980. It is a Banach space with respect to the norm

$$\|f\|_{\mathcal{S}_0} = \|\Psi_g(f)\|_{L^1(\mathbb{R}^2)},$$

and it is dense in $L^2(\mathbb{R})$. Several characterizations of $\mathcal{S}_0$ can be found in [31] and [37]; in particular, $\mathcal{S}_0$ consists of all *countable* superpositions of time–frequency shifts of the Gaussian with ℓ^1-coefficients:

$$\mathcal{S}_0 = \left\{f = \sum c_k E_{y_k}T_{x_k}g \; : \; \{(x_k, y_k)\} \subset \mathbb{R}^2, \{c_k\} \in \ell^1\right\}.$$

The infimum of all ℓ^1-norms $\sum |c_k|$, taken over coefficients representing a given f, gives an equivalent norm on $\mathcal{S}_0$.

Nothing is special about g being the Gaussian in the definition of $\mathcal{S}_0$: it can be replaced by any non-zero function in $\mathcal{S}_0$, and (9.41) still characterizes all functions in $\mathcal{S}_0$. It can be proved that $\mathcal{S}_0 \subset W$, see [37].

The space $\mathcal{S}_0$ is, in fact, just one out of a range of function spaces, called *modulation spaces*. A detailed analysis of these spaces can be found in [31] and [37].

In [38], Gröchenig introduced the concept *localization* for frames with particular emphasis on Gabor frames. One of the important consequences of the definition is that a localized Gabor frame not only yields frame expansions in $L^2(\mathbb{R})$: there is, in fact, a whole class of Banach spaces associated with the frame, and within each of these spaces one obtain series expansions that are very similar to the frame expansion in $L^2(\mathbb{R})$. This provides a very elegant way to extend the Hilbert space results to Banach spaces.

Tolimieri and Orr introduced a special condition in Gabor analysis in 1995. A Gabor system $\{E_{mb}T_{na}g\}_{m,n\in\mathbb{Z}}$ is said to satisfy *condition (A)* if

$$\sum_{m,n\in\mathbb{Z}} |\langle g, E_{m/a}T_{n/b}g\rangle| < \infty.$$

Condition (A) is often needed in order to guarantee certain convergence properties of infinite series appearing in Gabor analysis. However, as observed by Gröchenig [37], it is preferable to avoid the condition if possible. For example, condition (A) is very sensitive to the choice of the lattice parameters: even for a simple function like $g = \chi_{[0,1]}$ and an arbitrary translation parameter $a > 0$, it is only satisfied for irrational parameters b of the form $b = 1/q, q \in \mathbb{N}$! A natural question is whether such "irregular behavior" can be avoided if we only consider functions g in certain subspaces of $L^2(\mathbb{R})$. Since $\chi_{[0,1]} \in W$, the Wiener spaces is not suitable. On the other hand, Feichtinger's algebra yields such functions: in [37], it is proved that condition (A) is satisfied for all $a, b > 0$ if $g \in \mathcal{S}_0$.

Janssen introduced another condition in [48], which is frequently used in Gabor analysis. In contrast with condition (A), it only involves the function g and not the actual parameters a, b. We say that a function $g \in L^2(\mathbb{R})$ satisfies *condition (R)* if

$$\lim_{\epsilon\to 0^+} \sum_{k\in\mathbb{Z}} \frac{1}{\epsilon} \int_{-\frac{1}{2}\epsilon}^{\frac{1}{2}\epsilon} |g(k+x) - g(k)|^2 \, dx = 0.$$

Condition (R) might look restrictive, but it is actually satisfied for a dense class of functions in $L^2(\mathbb{R})$ (see Exercise 9.15 and page 250). The condition will play a role in the context of sampling of Gabor frames, discussed in Chapter 10.

9.6 Representations of the Gabor frame operator and duality

The structure of a Gabor frame turns out to have important implications for its frame operator, which can be rewritten in several ways. Many central frame results are based on the obtained representations of the frame operator.

Walnut was the first to rewrite the frame operator S associated with a Gabor frame $\{E_{mb}T_{na}g\}_{m,n\in\mathbb{Z}}$. He obtained what is now known as the *Walnut representation:* it expresses Sf in terms of the functions

$$G_k(x) = \sum_{n\in\mathbb{Z}} g(x - na)\overline{g(x - na - k/b)}, \ k \in \mathbb{Z}. \tag{9.42}$$

By Lemma 9.1.3, the series defining $G_k(x)$ converges absolutely for a.e. $x \in \mathbb{R}$.

Theorem 9.6.1 *Assume that $\{E_{mb}T_{na}g\}_{m,n\in\mathbb{Z}}$ is a Bessel sequence. Then the associated frame operator S has the representation*

$$Sf(x) = \frac{1}{b}\sum_{k\in\mathbb{Z}}(T_{k/b}f)(x)G_k(x) \tag{9.43}$$

with absolute convergence of the series for a.e. $x \in \mathbb{R}$.

Proof. We will derive the result as a consequence of Theorem 8.2.1. For $f \in L^2(\mathbb{R})$, let $\check{f} := \mathcal{F}^{-1}f$. First, we note that

$$\mathcal{F}T_{nb}E_{ma}\check{g} = E_{-nb}T_{ma}g;$$

since the Fourier transform is unitary, this implies that $\{T_{nb}E_{ma}\check{g}\}_{m,n\in\mathbb{Z}}$ is a Bessel sequence. Now let the associated frame operator, which we denote by $\tilde{S}$, act on the inverse Fourier transform of some $f \in L^2(\mathbb{R})$, and apply the Fourier transform to the outcome:

$$
\begin{aligned}
\widehat{\tilde{S}\check{f}} &= \mathcal{F}\sum_{m,n\in\mathbb{Z}} \langle\check{f}, T_{nb}E_{ma}\check{g}\rangle T_{nb}E_{ma}\check{g} \\
&= \sum_{m,n\in\mathbb{Z}} \langle f, \mathcal{F}T_{nb}E_{ma}\check{g}\rangle \mathcal{F}T_{nb}E_{ma}\check{g} \\
&= \sum_{m,n\in\mathbb{Z}} \langle f, E_{-nb}T_{ma}g\rangle E_{-nb}T_{ma}g \\
&= Sf.
\end{aligned}
$$

Clearly, $\{T_{nb}E_{ma}\check{g}\}_{m,n\in\mathbb{Z}}$ has the form of a shift-invariant system $\{g_{nm}\}$ with $g_m = E_{ma}\check{g}$ and shift-parameter b. Thus, Theorem 8.2.1 leads to the representation

$$Sf(x) = \widehat{\check{S}\check{f}}(x) = \frac{1}{b}\sum_{k\in\mathbb{Z}}\left(\sum_{m\in\mathbb{Z}}\mathcal{F}E_{ma}\check{g}(x)\overline{\mathcal{F}E_{ma}\check{g}(x-k/b)}\right)f(x-k/b)$$

$$= \frac{1}{b}\sum_{k\in\mathbb{Z}}\left(\sum_{m\in\mathbb{Z}}T_{ma}g(x)\overline{T_{ma}g(x-k/b)}\right)f(x-k/b)$$

$$= \sum_{k\in\mathbb{Z}}(T_{k/b}f)(x)G_k(x),$$

with absolute convergence for a.e. $x \in \mathbb{R}$. $\qquad\square$

We note that if the function g belongs to the Wiener space W, the representation (9.43) for the frame operator converges in $L^2(\mathbb{R})$; see [37] for a proof.

We already mentioned that there are close relationships between properties of a Gabor system $\{E_{mb}T_{na}g\}_{m,n\in\mathbb{Z}}$ and the Gabor system $\{E_{m/a}T_{n/b}g\}_{m,n\in\mathbb{Z}}$ with respect to the dual lattice. For Gabor systems, they were investigated almost at the same time by three groups of researchers, namely Daubechies, Landau, Landau [29]; Janssen [47]; and Ron, Shen [58]. There is a large overlap between their results, but their methods are quite different. We will now state a result from [29] and sketch its proof: our main purpose is to clarify how the dual lattice comes into play.

In the statement of the result, we will need the pre-frame operators associated with several Gabor systems with respect to different generators and different parameters. For this reason, we will denote the pre-frame operator for $\{E_{mb}T_{na}g\}_{m,n\in\mathbb{Z}}$ by $T_{g;a,b}$ instead of just T.

Proposition 9.6.2 Let $f, g, h \in L^2(\mathbb{R})$ and $a, b > 0$ be given. If $\{E_{mb}T_{na}g\}_{m,n\in\mathbb{Z}}, \{E_{mb}T_{na}f\}_{m,n\in\mathbb{Z}}$ and $\{E_{mb}T_{na}h\}_{m,n\in\mathbb{Z}}$ are Bessel sequences, then

$$T_{f;a,b}T_{g;a,b}^*h = \frac{1}{ab}T_{h;1/b,1/a}T_{g;1/b,1/a}^*f. \tag{9.44}$$

Proof. The complete proof in [29] is technical, and we will prove the result under the additional assumptions that the functions f and h are compactly supported and bounded; this makes all needed interchanges of summations and integrals legal. First, let $\phi \in L^2(\mathbb{R})$. Then

$$T_{f;a,b}^*\phi = \{\langle\phi, E_{mb}T_{na}f\rangle\}_{m,n\in\mathbb{Z}}.$$

By Lemma 9.1.3,

$$\langle\phi, E_{mb}T_{na}f\rangle = \int_0^{1/b}\left(\sum_{k\in\mathbb{Z}}\phi(x-k/b)\overline{f(x-na-k/b)}\right)e^{-2\pi imbx}dx.$$

The interpretation of this equation in Lemma 9.1.3 in terms of Fourier coefficients together with Lemma 3.5.2 now gives that

$$
\begin{aligned}
&\langle T_{f;a,b}T^*_{g;a,b}h, \phi\rangle \\
= {}& \langle T^*_{g;a,b}h, T^*_{f;a,b}\phi\rangle \\
= {}& \sum_{n\in\mathbb{Z}}\sum_{m\in\mathbb{Z}} \langle h, E_{mb}T_{na}g\rangle \overline{\langle \phi, E_{mb}T_{na}f\rangle} \\
= {}& \frac{1}{b}\sum_{n\in\mathbb{Z}} \left\langle \sum_{\ell\in\mathbb{Z}} h(\cdot - \ell/b)\overline{g(\cdot - na - \ell/b)}, \sum_{k\in\mathbb{Z}} \phi(\cdot - k/b)\overline{f(\cdot - na - k/b)}\right\rangle,
\end{aligned}
$$

where the inner product in the last line is in $L^2(0, 1/b)$. When we write it out, we arrive at

$$
\begin{aligned}
\langle T_{f;a,b}T^*_{g;a,b}h, \phi\rangle = {}& \frac{1}{b}\sum_{n\in\mathbb{Z}}\int_0^{1/b}\left(\sum_{\ell\in\mathbb{Z}} h(x-\ell/b)\overline{g(x-na-\ell/b)}\right. \\
& \left. \times \sum_{k\in\mathbb{Z}} \overline{\phi(x-k/b)}f(x-na-k/b)\right)dx \\
= {}& \frac{1}{b}\sum_{n\in\mathbb{Z}}\sum_{\ell\in\mathbb{Z}}\int_{-\infty}^{\infty} h(x-\ell/b)\overline{g(x-na-\ell/b)\phi(x)}f(x-na)\,dx.
\end{aligned}
$$

If we apply this calculation with other choices of the generators and the parameters $1/b, 1/a$ instead of a, b, we obtain that

$$
\begin{aligned}
&\langle T_{h;1/b,1/a}T^*_{g;1/b,1/a}f, \phi\rangle \\
= {}& a\sum_{k\in\mathbb{Z}}\sum_{m\in\mathbb{Z}}\int_{-\infty}^{\infty} h(x-m/b)\overline{g(x-ka-m/b)\phi(x)}f(x-ka)\,dx.
\end{aligned}
$$

This shows that

$$
\langle T_{f;a,b}T^*_{g;a,b}h, \phi\rangle = \frac{1}{ab}\langle T_{h;1/b,1/a}T^*_{g;1/b,1/a}f, \phi\rangle;
$$

since this holds for all $\phi \in L^2(\mathbb{R})$, the conclusion follows.   □

Written in terms of the involved sequences, (9.44) says that

$$
\sum_{m,n\in\mathbb{Z}} \langle h, E_{mb}T_{na}g\rangle E_{mb}T_{na}f = \frac{1}{ab}\sum_{m,n\in\mathbb{Z}} \langle f, E_{m/a}T_{n/b}g\rangle E_{m/a}T_{n/b}h. \quad (9.45)
$$

The right-hand side of (9.45) converges unconditionally in $L^2(\mathbb{R})$ because $\{\langle f, E_{m/a}T_{n/b}g\rangle\}_{m,n\in\mathbb{Z}} \in \ell^2(\mathbb{Z}^2)$ and $\{E_{m/a}T_{n/b}h\}_{m,n\in\mathbb{Z}}$ is a Bessel sequence, see Corollary 3.1.5. In particular, the result leads to a representation of the Gabor frame operator for $\{E_{mb}T_{na}g\}_{m,n\in\mathbb{Z}}$ in terms of the function g and the dual lattice associated with $\{(na, mb)\}_{m,n\in\mathbb{Z}}$:

Corollary 9.6.3 *Let $g \in L^2(\mathbb{R})$ and $a, b > 0$ be given, and assume that $\{E_{mb}T_{na}g\}_{m,n\in\mathbb{Z}}$ is a frame with frame operator S. Consider any $h \in L^2(\mathbb{R})$ for which $\{E_{mb}T_{na}h\}_{m,n\in\mathbb{Z}}$ is a Bessel sequence; then*

$$Sh = \frac{1}{ab} \sum_{m,n\in\mathbb{Z}} \langle g, E_{m/a}T_{n/b}g\rangle E_{m/a}T_{n/b}h.$$

Proof. The result follows from (9.45): just let $f = g$. $\square$

9.7   The Zak transform

The *Zak transform* is a very useful tool to analyze Gabor systems $\{E_{mb}T_{na}g\}_{m,n\in\mathbb{Z}}$ in the case where $ab \in \mathbb{Q}$. For a fixed parameter $\lambda > 0$, the Zak transform $Z_\lambda f$ of $f \in L^2(\mathbb{R})$ is formally defined as a function of two real variables:

$$(Z_\lambda f)(t, \nu) = \lambda^{1/2} \sum_{k\in\mathbb{Z}} f(\lambda(t - k))e^{2\pi ik\nu}, \qquad t, \nu \in \mathbb{R}. \tag{9.46}$$

In the case $\lambda = 1$, we simply write

$$(Zf)(t, \nu) = \sum_{k\in\mathbb{Z}} f(t - k)e^{2\pi ik\nu}, \qquad t, \nu \in \mathbb{R}. \tag{9.47}$$

For functions $f \in C_c(\mathbb{R})$, the Zak transform is defined pointwise, but for general functions in $L^2(\mathbb{R})$ we have to be more precise about how to interpret the definition. Letting $Q := [0, 1[\times[0, 1[$, we now prove that the series defining $Z_\lambda f$ in fact converges in $L^2(Q)$ for all $f \in L^2(\mathbb{R})$:

Lemma 9.7.1 *Given $\lambda > 0$, the Zak transform Z_λ is a unitary map of $L^2(\mathbb{R})$ onto $L^2(Q)$.*

Proof. We first consider the case $\lambda = 1$. Let $f \in L^2(\mathbb{R})$ be given. In order to show that Zf is well defined as a function in $L^2(Q)$, we consider the functions $F_k : Q \to \mathbb{C}, k \in \mathbb{Z}$, defined by

$$F_k(t, \nu) := f(t - k)e^{2\pi ik\nu}.$$

These functions belong to $L^2(Q)$. Denoting their norm by $||F_k||_{L^2(Q)}$, we observe that

$$\sum_{k\in\mathbb{Z}} ||F_k||^2_{L^2(Q)} = \sum_{k\in\mathbb{Z}} \int_0^1 \int_0^1 |F_k(t, \nu)|^2 \, d\nu dt$$

$$= \sum_{k\in\mathbb{Z}} \int_0^1 |f(t - k)|^2 \, dt$$

$$= ||f||^2.$$

Furthermore, for $j \neq k$,

$$\langle F_k, F_j \rangle_{L^2(Q)} = \int_0^1 f(t-k)\overline{f(t-j)} \left(\int_0^1 e^{2\pi i(k-j)\nu} d\nu \right) dt = 0. \quad (9.48)$$

Combining the obtained results shows that $\sum_{k\in\mathbb{Z}} F_k$ in fact converges in $L^2(Q)$; thus, the Zak transform Zf is well defined. Furthermore, we see that

$$\|Zf\|_{L^2(Q)}^2 = \left\| \sum_{k\in\mathbb{Z}} F_k \right\|_{L^2(Q)}^2 = \sum_{k\in\mathbb{Z}} \|F_k\|_{L^2(Q)}^2 = \|f\|^2;$$

thus Z is an isometry from $L^2(\mathbb{R})$ into $L^2(Q)$.

For the rest of the proof, we use the Gabor basis $\{E_m T_n \chi_{[0,1]}\}_{m,n\in\mathbb{Z}}$ for $L^2(\mathbb{R})$, see Example 3.5.3. By direct computation for $(t,\nu) \in Q$,

$$
\begin{aligned}
(Z E_m T_n \chi_{[0,1]})(t,\nu) &= \sum_{k\in\mathbb{Z}} e^{2\pi i m(t-k)} \chi_{[0,1]}(t-n-k) e^{2\pi i k\nu} \\
&= e^{2\pi i m t} e^{-2\pi i n\nu} \sum_{k\in\mathbb{Z}} \chi_{[0,1]}(t-k) e^{2\pi i k\nu} \\
&= e^{2\pi i m t} e^{-2\pi i n\nu}. \quad (9.49)
\end{aligned}
$$

That is, the Zak transform maps the orthonormal basis $\{E_m T_n \chi_{[0,1]}\}_{m,n\in\mathbb{Z}}$ for $L^2(\mathbb{R})$ onto the orthonormal basis $\{e^{-2\pi i n\nu} e^{2\pi i m t}\}_{m,n\in\mathbb{Z}}$ for $L^2(Q)$. This implies that Z is a unitary mapping of $L^2(\mathbb{R})$ onto $L^2(Q)$.

For the general case, we note that in terms of the unitary dilation operator $D_{\lambda^{-1}}$ defined in Section 2.9,

$$Z_\lambda f = Z(D_{\lambda^{-1}} f).$$

As a composition of unitary operators, Z_λ is itself unitary. $\square$

Now where the Zak transform is proved to be well defined almost everywhere on Q, an inspection of the expression (9.46) reveals that $Z_\lambda f(t,\nu)$ even is defined for a.e. $(t,\nu) \in \mathbb{R}^2$ and that the *quasi-periodicity* in Lemma 9.7.2(i) below holds. We collect some more properties of the Zak transform:

Lemma 9.7.2 *Consider the Zak transform Z_λ, $\lambda > 0$, and $f \in L^2(\mathbb{R})$. Then the following holds:*

(i) $Z_\lambda f(t+1,\nu) = e^{2\pi i \nu} Z_\lambda f(t,\nu)$, $Z_\lambda f(t,\nu+1) = Z_\lambda f(t,\nu).$

(ii) *If f is continuous and, for some $C > 0$,*

$$|f(x)| \le \frac{C}{1+|x|^2}, \quad \forall x \in \mathbb{R},$$

then $Z_\lambda f$ is continuous on $\mathbb{R}^2$.

(iii) If $Z_\lambda f$ is continuous on $\mathbb{R}^2$, then there exists $(t, \nu) \in \mathbb{R}^2$ such that $Z_\lambda f(t, \nu) = 0$.

The proof of (ii) is similar to the proof of Proposition 7.8 (Exercise 9.16). A proof of (iii) can be found in [40]. Note that the quasi-periodicity in (i) often leads to jump-discontinuities on the lines $t = k, k \in \mathbb{Z}$: even if $Z_\lambda f$ is continuous on Q, it might not be continuous on $\mathbb{R}^2$. For a concrete example, take the function f whose Zak transform is equal to 1 on Q: in this case, $Z_\lambda f$ is continuous on Q but not on $\mathbb{R}^2$.

If $g \in L^2(\mathbb{R})$ and $ab = 1$, a computation as in (9.49) shows that

$$Z_a E_{mb} T_{na} g(t, \nu) = e^{2\pi i m t} e^{-2\pi i n \nu} Z_a g(t, \nu). \tag{9.50}$$

The family $\{e^{2\pi i m t} e^{-2\pi i n \nu}\}_{m,n \in \mathbb{Z}}$ is an orthonormal basis for $L^2(Q)$, which we denote by $\{E_{(m,n)}\}_{m,n \in \mathbb{Z}}$. The equation (9.50) shows that $\{E_{mb} T_{na} g\}_{m,n \in \mathbb{Z}}$ is complete in $L^2(\mathbb{R})$ (respectively, an orthonormal basis for $L^2(\mathbb{R})$ or a Riesz basis) if and only if $\{E_{(m,n)} Z_a g\}_{m,n \in \mathbb{Z}}$ has the same property in $L^2(Q)$. This observation will be used in the following theorem, which expresses properties for a Gabor system $\{E_{mb} T_{na} g\}_{m,n \in \mathbb{Z}}$ with $ab = 1$ in terms of the Zak transform $Z_a g$. Remember from Theorem 9.1.12 that a Gabor system with $ab = 1$ is a frame if and only if it is a Riesz basis.

Proposition 9.7.3 *Let $g \in L^2(\mathbb{R})$ and $a, b > 0$ with $ab = 1$ be given. Then the following holds:*

(i) $\{E_{mb} T_{na} g\}_{m,n \in \mathbb{Z}}$ is complete in $L^2(\mathbb{R})$ if and only if $Z_a g \neq 0$, a.e.

(ii) $\{E_{mb} T_{na} g\}_{m,n \in \mathbb{Z}}$ is a Bessel sequence with bound B if and only if $|Z_a g|^2 \leq B$, a.e.

(iii) $\{E_{mb} T_{na} g\}_{m,n \in \mathbb{Z}}$ is a Riesz basis for $L^2(\mathbb{R})$ with bounds A, B if and only if $A \leq |Z_a g|^2 \leq B$, a.e.

(iv) $\{E_{mb} T_{na} g\}_{m,n \in \mathbb{Z}}$ is an orthonormal basis for $L^2(\mathbb{R})$ if and only if $|Z_a g|^2 = 1$, a.e.

Proof. To prove (i), consider the subspace $V \subset L^2(\mathbb{R})$ given by

$$V = \{f \in L^2(\mathbb{R}) : \ Z_a f \text{ is bounded}\}.$$

The bounded functions are dense in $L^2(Q)$, so V is dense in $L^2(\mathbb{R})$ by Lemma 9.7.1. Now let $f \in V$. Then

$$\begin{aligned} \langle f, E_{mb} T_{na} g \rangle_{L^2(\mathbb{R})} &= \langle Z_a f, E_{(m,n)} Z_a g \rangle_{L^2(Q)} \\ &= \langle Z_a f \overline{Z_a g}, E_{(m,n)} \rangle_{L^2(Q)}. \end{aligned} \tag{9.51}$$

First assume that $Z_a g \neq 0$ a.e. If $f \neq 0$, then $Z_a f \overline{Z_a g}$ is *not* the zero-function, and there exists $(m, n) \in \mathbb{Z}^2$ such that

$$\langle Z_a f \overline{Z_a g}, E_{(m,n)} \rangle_{L^2(Q)} \neq 0.$$

Therefore, (9.51) shows that $\{E_{mb}T_{na}g\}_{m,n\in\mathbb{Z}}$ is complete. For the other implication, assume that $Z_a g = 0$ on a measurable set $\Delta \subseteq Q$ with positive measure. We leave the slight modifications in the case $\Delta = Q$ to the reader and assume that $Q \setminus \Delta \neq \emptyset$. By choosing $f \in L^2(\mathbb{R})$ such that $Z_a f = \chi_{Q\setminus\Delta}$, it follows that $\langle f, E_{mb}T_{na}g \rangle = 0$ for all $m, n \in \mathbb{Z}$, so $\{E_{mb}T_{na}g\}_{m,n\in\mathbb{Z}}$ is incomplete in $L^2(\mathbb{R})$.

For the rest of the proof, we note that for any $F \in L^2(Q)$ we have $F Z_a g \in L^1(Q)$. Since $\{E_{(m,n)}\}_{m,n}$ is an orthonormal basis for $L^2(Q)$,

$$\sum_{m,n\in\mathbb{Z}} \left| \langle F, E_{(m,n)} Z_a g \rangle_{L^2(Q)} \right|^2 = \sum_{m,n\in\mathbb{Z}} \left| \int_Q (F\overline{Z_a g})\, \overline{E_{(m,n)}} \right|^2$$

$$= \int_Q \left| F\overline{Z_a g} \right|^2. \tag{9.52}$$

(ii)–(iv) now follows by a standard argument (Exercise 9.17), yielding e.g., that

$$\int_Q \left| F\overline{Z_a g} \right|^2 \leq B \, \|F\|^2_{L^2(Q)}, \ \forall F \in L^2(Q) \Leftrightarrow |Z_a g|^2 \leq B, \ a.e. \qquad \square$$

Lemma 9.7.2 and Proposition 9.7.3 put restrictions on the functions g for which $\{E_{mb}T_{na}g\}_{m,n\in\mathbb{Z}}$ can be a Riesz basis for $ab = 1$:

Corollary 9.7.4 *A continuous function $g \in L^2(\mathbb{R})$ with compact support cannot generate a Gabor Riesz basis $\{E_{mb}T_{na}g\}_{m,n\in\mathbb{Z}}$ for any $a, b > 0$.*

We ask the reader to provide the proof in Exercise 9.18. Another consequence of Proposition 9.7.3 is that the Gaussian $g(x) = e^{-x^2/2}$ does not generate a Riesz basis $\{E_{mb}T_{na}g\}_{m,n\in\mathbb{Z}}$ for any $a, b > 0$; in fact, its Zak transform is continuous and has a zero.

For the rest of this section, we consider a *rationally oversampled* Gabor system $\{E_{mb}T_{na}g\}_{m,n\in\mathbb{Z}}$, i.e., we assume that

$$ab \in \mathbb{Q}, \ ab = \frac{p}{q} \ \text{with} \ 1 \leq p \leq q.$$

We always choose p, q such that $\gcd(p, q) = 1$. We state results by Zibulski and Zeevi, resp. Janssen. The references for further information and proofs are [66], [44], [45], and [46].

For a rationally oversampled Gabor system $\{E_{mb}T_{na}g\}_{m,n\in\mathbb{Z}}$, the *Zibulski–Zeevi matrix* is defined by

$$\Phi^g(t,\nu) = p^{-\frac{1}{2}} \left((Z_{\frac{1}{b}}g)(t - \ell\frac{p}{q}, \nu + \frac{k}{p}) \right)_{k=0,\ldots,p-1;\ell=0,\ldots,q-1}, \ a.e. \ t, \nu \in \mathbb{R}.$$

Note that $\Phi^g(t,\nu)$ is a $p \times q$ matrix, whose entries are well defined for a.e. $t, \nu \in \mathbb{R}$. The importance of the rationally oversampled Gabor case is that the frame properties of $\{E_{mb}T_{na}g\}_{m,n\in\mathbb{Z}}$ can be formulated in terms of this finite matrix. For example, one can prove that $\{E_{mb}T_{na}g\}_{m,n\in\mathbb{Z}}$ is a Bessel sequence with bound B if and only if $\Phi^g(t,\nu)$ for a.e. $t,\nu\in[0,1[$ defines a bounded linear mapping of $\mathbb{C}^q$ into $\mathbb{C}^p$, with norm at most $B^{\frac{1}{2}}$. If we do not need the information about a specific Bessel bound, this result has a nice formulation:

Theorem 9.7.5 *Let the setup be as above. A rationally oversampled Gabor system $\{E_{mb}T_{na}g\}_{m,n\in\mathbb{Z}}$ is a Bessel sequence if and only if there exists a constant $C > 0$ such that*

$$\left|Z_{\frac{1}{b}}g(t,\nu)\right| \le C, \ a.e. \ t,\nu\in[0,1[.$$

Note that Theorem 9.7.5 generalizes Proposition 9.7.3(ii) to the case of rational oversampling.

Some of the results proved in [46] and [66] are (still assuming rationally oversampling):

- When $A,B > 0$ are given, $\{E_{mb}T_{na}g\}_{m,n\in\mathbb{Z}}$ is a frame with frame bounds A,B if and only if

$$AI \le \Phi^g(t,\nu)\left(\Phi^g(t,\nu)\right)^* \le BI, \ a.e. \ t,\nu\in[0,1].$$

 Here I denotes the identity operator on $\mathbb{C}^p$.

- Two Bessel systems $\{E_{mb}T_{na}g\}_{m,n\in\mathbb{Z}}$ and $\{E_{mb}T_{na}h\}_{m,n\in\mathbb{Z}}$ are dual if and only if for a.e. $t,\nu\in[0,1[$ and all $k = 0,\ldots,p-1$,

$$\frac{1}{p}\sum_{\ell=0}^{q-1}(Z_{\frac{1}{b}}g)\,(t-\ell p/q,\nu+k/p)\,\overline{(Z_{\frac{1}{b}}h)\,(t-\ell p/q,\nu)} = \delta_{k,0}.$$

- If $\{E_{mb}T_{na}g\}_{m,n\in\mathbb{Z}}$ is a Bessel sequence with frame operator S, then for all $f\in L^2(\mathbb{R})$,

$$\Phi^{Sf}(t,\nu) = \Phi^g(t,\nu)\left(\Phi^g(t,\nu)\right)^*\Phi^f(t,\nu), \ a.e. \ t,\nu\in[0,1].$$

9.8 Time–frequency localization of Gabor expansions

It is well-known that no function $g \ne 0$ can have compact support simultaneously in the time-domain and the frequency-domain. However, most signals appearing in practice are *essentially localized* in the time–frequency plane, meaning that the interesting part of the signal takes place on a finite time-interval, with frequencies belonging to a certain finite interval. We will now analyze how this affects the Gabor frame expansion of such signals.

Given a number $T > 0$, define the operator

$$Q_T : L^2(\mathbb{R}) \to L^2(\mathbb{R}), \ (Q_Tf)(x) = \chi_{[-T,T]}(x)f(x).$$

We will use $||(I - Q_T)f||$ as a measure for the content of the function f outside the interval $[-T, T]$. So, intuitively, to say that a function $f \in L^2(\mathbb{R})$ essentially is localized on the interval $[-T, T]$ means that $||(I - Q_T)f||$ is small compared with $||f||$. Similarly, for $\Omega > 0$ we introduce an operator P_Ω (the expression below defines the operator in the Fourier domain) by

$$P_\Omega : L^2(\mathbb{R}) \rightarrow L^2(\mathbb{R}), \quad \widehat{P_\Omega f}(\nu) = \chi_{[-\Omega,\Omega]}(\nu)\hat{f}(\nu);$$

the function $\hat{f}$ being essentially localized on $[-\Omega, \Omega]$ means that $||(I - P_\Omega)f||$ is small compared with $||f||$.

Now assume that the function f is essentially localized in both domains, i.e., on $[-T, T] \times [-\Omega, \Omega]$ for some $T, \Omega > 0$. Let

$$B(T, \Omega) := \{(m, n) \in \mathbb{Z}^2 : \ mb \in [-\Omega, \Omega], \ na \in [-T, T]\}.$$

A natural question is how well frame decompositions capture the localization of the signal f. That is, considering the frame expansion of f in terms of dual Gabor frames $\{E_{mb}T_{na}g\}_{m,n\in\mathbb{Z}}$ and $\{E_{mb}T_{na}h\}_{m,n\in\mathbb{Z}}$,

$$f = \sum_{m,n\in\mathbb{Z}} \langle f, E_{mb}T_{na}h\rangle E_{mb}T_{na}g, \tag{9.53}$$

do we obtain a reasonable approximation of f if we replace the sum over $(m, n) \in \mathbb{Z}^2$ with a sum over $(m, n) \in B(T, \Omega)$?

Since the expansion (9.53) involve inner products between the function f and the functions in the Gabor system $\{E_{mb}T_{na}h\}_{m,n\in\mathbb{Z}}$, it is natural to assume that the function h has good localization properties. We will prove a result by Daubechies [25]. It shows that under this assumption, the question has an affirmative answer, at least for a certain enlargement

$$B(T + \Lambda, \Omega + \Gamma)$$

of $B(T, \Omega)$. Daubechies formulated the result for the classical frame decomposition in terms of a Gabor frame and its canonical dual frame, but the same argument holds for dual Gabor frame pairs.

Theorem 9.8.1 *Assume that the Gabor systems* $\{E_{mb}T_{na}g\}_{m,n\in\mathbb{Z}}$ *and* $\{E_{mb}T_{na}h\}_{m,n\in\mathbb{Z}}$ *form a pair of dual frames for* $L^2(\mathbb{R})$ *with upper frame bounds B and D respectively, and that for some constants $C > 0, \alpha > 1/2$, the decay conditions*

$$|h(x)| \leq C(1 + x^2)^{-\alpha}, \ x \in \mathbb{R}, \quad |\hat{h}(\nu)| \leq C(1 + \nu^2)^{-\alpha}, \ \nu \in \mathbb{R}, \tag{9.54}$$

hold. Then, for any $\epsilon > 0$, there exist numbers $\Lambda, \Gamma > 0$ such that for all $T, \Omega > 0$,

$$\left|\left| f - \sum_{\{(m,n)\in B(T+\Lambda,\Omega+\Gamma)\}} \langle f, E_{mb}T_{na}h\rangle E_{mb}T_{na}g \right|\right|$$
$$\leq \sqrt{BD} \left(||(I - Q_T)f|| + ||(I - P_\Omega)f|| + \epsilon ||f||\right)$$

for all $f \in L^2(\mathbb{R})$.

Proof. Let $f \in L^2(\mathbb{R})$, and consider some fixed numbers $T, \Omega > 0$. Then, for any given $\Lambda, \Gamma > 0$, the assumption of $\{E_{mb}T_{na}g\}_{m,n\in\mathbb{Z}}$ and $\{E_{mb}T_{na}h\}_{m,n\in\mathbb{Z}}$ being dual frames implies that

$$\left\| f - \sum_{\{(m,n)\in B(T+\Lambda,\Omega+\Gamma)\}} \langle f, E_{mb}T_{na}h\rangle E_{mb}T_{na}g \right\|$$

$$= \left\| \sum_{\{(m,n)\notin B(T+\Lambda,\Omega+\Gamma)\}} \langle f, E_{mb}T_{na}h\rangle E_{mb}T_{na}g \right\|.$$

Via Lemma 2.3.4, it follows that

$$\left\| f - \sum_{\{(m,n)\in B(T+\Lambda,\Omega+\Gamma)\}} \langle f, E_{mb}T_{na}h\rangle E_{mb}T_{na}g \right\|$$

$$= \sup_{||\varphi||=1} \left| \left\langle \sum_{\{(m,n)\notin B(T+\Lambda,\Omega+\Gamma)\}} \langle f, E_{mb}T_{na}h\rangle E_{mb}T_{na}g, \varphi \right\rangle \right|$$

$$\leq \sup_{||\varphi||=1} \sum_{\{(m,n)\notin B(T+\Lambda,\Omega+\Gamma)\}} |\langle f, E_{mb}T_{na}h\rangle| \, |\langle E_{mb}T_{na}g, \varphi\rangle|.$$

Observe that
$$B(T+\Lambda, \Omega+\Gamma) \subseteq \{(m,n) : |na| > T+\Lambda\} \cup \{(m,n) : |mb| > \Omega+\Gamma\};$$

thus, we arrive at

$$\left\| f - \sum_{\{(m,n)\in B(T+\Lambda,\Omega+\Gamma)\}} \langle f, E_{mb}T_{na}h\rangle E_{mb}T_{na}g \right\| \tag{9.55}$$

$$\leq \sup_{||\varphi||=1} \sum_{\{(m,n):\, |na|>T+\Lambda\}} |\langle f, E_{mb}T_{na}h\rangle| \, |\langle E_{mb}T_{na}g, \varphi\rangle| \tag{9.56}$$

$$+ \sup_{||\varphi||=1} \sum_{\{(m,n):\, |mb|>\Omega+\Gamma\}} |\langle f, E_{mb}T_{na}h\rangle| \, |\langle E_{mb}T_{na}g, \varphi\rangle|. \tag{9.57}$$

We will now estimate the terms in (9.56) and (9.57) separately. For the term in (9.56), we use that $f = Q_T + (I - Q_T)f$; via the triangle inequality this implies that

$$
\sum_{\{(m,n):\ |na|>T+\Lambda\}} |\langle f, E_{mb}T_{na}h\rangle|\,|\langle E_{mb}T_{na}g, \varphi\rangle|
$$

$$
= \sum_{\{(m,n):\ |na|>T+\Lambda\}} |\langle Q_T + (I - Q_T)f, E_{mb}T_{na}h\rangle|\,|\langle E_{mb}T_{na}g, \varphi\rangle|
$$

$$
\leq \sum_{\{(m,n):\ |na|>T+\Lambda\}} |\langle Q_T f, E_{mb}T_{na}h\rangle|\,|\langle E_{mb}T_{na}g, \varphi\rangle|
$$

$$
+ \sum_{\{(m,n):\ |na|>T+\Lambda\}} |\langle (I - Q_T)f, E_{mb}T_{na}h\rangle|\,|\langle E_{mb}T_{na}g, \varphi\rangle|
$$

$$
\leq \left(\sum_{\{(m,n):\ |na|>T+\Lambda\}} |\langle Q_T f, E_{mb}T_{na}h\rangle|^2 \right)^{1/2} \times
$$

$$
\left(\sum_{\{(m,n):\ |na|>T+\Lambda\}} |\langle E_{mb}T_{na}g, \varphi\rangle|^2 \right)^{1/2}
$$

$$
+ \left(\sum_{\{(m,n):\ |na|>T+\Lambda\}} |\langle (I - Q_T)f, E_{mb}T_{na}h\rangle|^2 \right)^{1/2} \times
$$

$$
\left(\sum_{\{(m,n):\ |na|>T+\Lambda\}} |\langle E_{mb}T_{na}g, \varphi\rangle|^2 \right)^{1/2}.
$$

Using that $\{E_{mb}T_{na}g\}_{m,n\in\mathbb{Z}}$ has the upper frame bound B and that the dual frame $\{E_{mb}T_{na}h\}_{m,n\in\mathbb{Z}}$ has the upper frame bound D, this implies that

$$
\sup_{||\varphi||=1} \sum_{\{(m,n):\ |na|>T+\Lambda\}} |\langle f, E_{mb}T_{na}h\rangle|\,|\langle E_{mb}T_{na}g, \varphi\rangle| \tag{9.58}
$$

$$
\leq \sqrt{B} \left(\sum_{\{(m,n):\ |na|>T+\Lambda\}} |\langle Q_T f, E_{mb}T_{na}h\rangle|^2 \right)^{1/2} + \sqrt{BD}\,||(I - Q_T)f||.
$$

In order to estimate the expression further, we will use the calculation in (9.17). For this reason, we now assume that f is bounded and has compact support. Then,

$$\sum_{\{(m,n):\ |na|>T+\Lambda\}} |\langle Q_T f, E_{mb} T_{na} h\rangle|^2$$

$$= \frac{1}{b} \left| \sum_{\{n:\ |na|>T+\Lambda\}} \int_{-\infty}^{\infty} \overline{Q_T f(x)} h(x-na) \times \right.$$

$$\left. \sum_{k\in\mathbb{Z}} Q_T f(x-k/b) \overline{h(x-na-k/b)}\, dx \right|$$

$$\le \frac{1}{b} \sum_{\{n:\ |na|>T+\Lambda\}} \sum_{k\in\mathbb{Z}} \int_{-\infty}^{\infty} |Q_T f(x)|\, |Q_T f(x-k/b)| \times$$

$$|h(x-na)|\, |h(x-na-k/b)|\, dx.$$

Using that $Q_T f$ has support on $[-T, T]$ and the decay condition on h leads to

$$\sum_{\{(m,n):\ |na|>T+\Lambda\}} |\langle Q_T f, E_{mb} T_{na} h\rangle|^2 \tag{9.59}$$

$$\le \frac{1}{b} \sum_{\{n:\ |na|>T+\Lambda\}} \sum_{k\in\mathbb{Z}} \sup_{|x|\le T, |x-k/b|\le T} |h(x-na)|\, |h(x-na-k/b)| \times$$

$$\int_{-\infty}^{\infty} |Q_T f(x)|\, |Q_T f(x-k/b)|\, dx$$

$$\le \frac{\|Q_T f\|^2}{b} \sum_{\{n:\ |na|>T+\Lambda\}} \sum_{k\in\mathbb{Z}} \sup_{|x|\le T, |x-k/b|\le T} |h(x-na)|\, |h(x-na-k/b)|$$

$$\le \frac{1}{b} \|f\|^2 \times$$

$$\sum_{\{n:\ |na|>T+\Lambda\}} \sum_{k\in\mathbb{Z}} \sup_{|x|\le T, |x-k/b|\le T} \frac{1}{(1+(x-na)^2)^\alpha} \frac{1}{(1+(x-na-k/b)^2)^\alpha}.$$

Now, a careful examination performed in [25] (we will skip it) shows that for some constant κ that is independent of T and Λ,

$$\sum_{\{n:\ |na|>T+\Lambda\}} \sum_{k\in\mathbb{Z}} \sup_{|x|\le T, |x-k/b|\le T} \frac{1}{(1+(x-na)^2)^\alpha} \frac{1}{(1+(x-na-k/b)^2)^\alpha}$$
$$\le \kappa(1+\Lambda^2)^{-2\alpha+1}.$$

Together with the calculation in (9.59) and (9.58), this leads to the following estimate of the term (9.56):

$$\sup_{||\varphi||=1} \sum_{\{(m,n): \ |na|>T+\Lambda\}} |\langle f, E_{mb}T_{na}h\rangle| \, |\langle E_{mb}T_{na}g, \varphi\rangle|$$

$$\leq \sqrt{\frac{B}{b}\kappa(1+\Lambda^2)^{-2\alpha+1}} \, ||f|| + \sqrt{BD} \, ||(I-Q_T)f||. \qquad (9.60)$$

We will now estimate the term (9.57). First,

$$\langle f, E_{mb}T_{na}h\rangle = \langle \hat{f}, \mathcal{F}E_{mb}T_{na}h\rangle$$
$$= e^{-2\pi imbna}\langle \hat{f}, E_{-na}T_{mb}\hat{h}\rangle.$$

Using that $\hat{f} = \widehat{P_\Omega f} + \mathcal{F}(I - P_\Omega)f$, calculations like before lead to

$$\sup_{||\varphi||=1} \sum_{\{(m,n): \ |mb|>\Omega+\Gamma\}} |\langle f, E_{mb}T_{na}h\rangle| \, |\langle E_{mb}T_{na}g, \varphi\rangle|$$

$$= \sup_{||\varphi||=1} \sum_{\{(m,n): \ |mb|>\Omega+\Gamma\}} |\langle \hat{f}, E_{na}T_{mb}\hat{h}\rangle| \, |\langle E_{mb}T_{na}g, \varphi\rangle|$$

$$\leq \sup_{||\varphi||=1} \left(\sum_{\{(m,n): \ |mb|>\Omega+\Gamma\}} |\langle \widehat{P_\Omega f}, E_{na}T_{mb}\hat{h}\rangle|^2 \right)^{1/2} \times$$

$$\left(\sum_{\{(m,n): \ |mb|>\Omega+\Gamma\}} |\langle E_{mb}T_{na}g, \varphi\rangle|^2 \right)^{1/2}$$

$$+ \sup_{||\varphi||=1} \left(\sum_{\{(m,n): \ |mb|>\Omega+\Gamma\}} |\langle \mathcal{F}(I-P_\Omega)f, E_{na}T_{mb}\hat{h}\rangle|^2 \right)^{1/2}$$

$$\left(\sum_{\{(m,n): \ |mb|>\Omega+\Gamma\}} |\langle E_{mb}T_{na}g, \varphi\rangle|^2 \right)^{1/2}$$

$$\leq \sqrt{B} \left(\sum_{\{(m,n): \ |mb|>\Omega+\Gamma\}} |\langle \widehat{P_\Omega f}, E_{na}T_{mb}\hat{h}\rangle|^2 \right)^{1/2} + \sqrt{BD} \, ||(I-P_\Omega)f||.$$

The assumptions that f is bounded and has compact support imply that $\widehat{P_\Omega f}$ is bounded as well. Thus, exactly as before we can use (9.17) to prove that

$$\sum_{\{(m,n): \ |mb|>\Omega+\Gamma\}} |\langle \widehat{P_\Omega f}, E_{na}T_{mb}\hat{h}\rangle|^2 \leq$$

$$\leq \frac{1}{a} ||\widehat{P_\Omega f}||^2 \sum_{\{m: \ |mb|>\Omega+\Gamma\}} \sum_{k\in\mathbb{Z}} \sup_{|\nu|\leq\Omega, |\nu-k/a|\leq\Omega} |\hat{h}(\nu-nb)| \, |\hat{h}(\nu-nb-k/a)|.$$

The decay condition on $\hat{h}$ together with the above calculations now leads to an estimate on (9.57), with some constant $\eta > 0$ that is independent of Ω and Γ:

$$\sup_{||\varphi||=1} \sum_{\{(m,n):|mb|>\Omega+\Gamma\}} |\langle f, E_{mb}T_{na}h\rangle| \, |\langle E_{mb}T_{na}g, \varphi\rangle|$$

$$\leq \sqrt{\frac{B}{a}} \eta (1+\Gamma^2)^{-2\alpha+1} ||f|| + \sqrt{BD} \, ||(I - P_\Omega)f||. \qquad (9.61)$$

Finally, inserting (9.61) and (9.60) in the calculation in (9.55) leads to

$$\left\| f - \sum_{\{(m,n)\in B(T+\Lambda,\Omega+\Gamma)\}} \langle f, E_{mb}T_{na}h\rangle E_{mb}T_{na}g \right\|$$

$$\leq \sqrt{\frac{B}{b}} \kappa (1+\Lambda^2)^{-2\alpha+1} ||f|| + \sqrt{BD} \, ||(I - Q_T)f||$$

$$+ \sqrt{\frac{B}{a}} \eta (1+\Gamma^2)^{-2\alpha+1} ||f|| + \sqrt{BD} \, ||(I - P_\Omega)f||,$$

valid for all bounded functions $f \in L^2(\mathbb{R})$ with compact support. For a given $\epsilon > 0$, one can now find $\Lambda, \Gamma > 0$ such that the conclusion in the theorem holds for all such functions f; because the set of bounded functions with compact support are dense in $L^2(\mathbb{R})$, the result actually holds for all $f \in L^2(\mathbb{R})$ by Lemma 3.1.6. $\qquad\square$

Theorem 9.8.1 shows that if the generator h for the dual frame $\{E_{mb}T_{na}h\}_{m,n\in\mathbb{Z}}$ is well localized in time and frequency, then the frame expansion indeed captures the time–frequency localization of a given signal $f \in L^2(\mathbb{R})$. That is, if we for a given "allowed deviation" $\epsilon > 0$ choose the constants Λ, Γ as in Theorem 9.8.1, and the function f essentially is localized on $[-T, T] \times [-\Omega, \Omega]$, then

$$\sum_{\{(m,n)\in B(T+\Lambda,\Omega+\Gamma)\}} \langle f, E_{mb}T_{na}h\rangle E_{mb}T_{na}g$$

yields a reasonable approximation of f.

9.9   Continuous representations

So far, we have only dealt with discrete Gabor expansions, i.e., representations of functions in $L^2(\mathbb{R})$ in terms of infinite series. We will now give a short presentation of continuous representations, i.e., representations of functions in $L^2(\mathbb{R})$ in terms of integrals. In order to introduce the continuous Gabor representations, we will first motivate the definition of the short-time Fourier transform in Section 9.5. For a signal $f(x)$, the variable

x is often interpreted as time, and the Fourier transform $\hat{f}(\gamma)$ gives information about the content of oscillations with frequency γ. In practice, it is a problem that the time-information is lost in the Fourier transform, i.e., there is no information about which frequencies appear at which time. A way to try to overcome this problem is to "look at the signal at a small time-interval and take the Fourier transform here." Mathematically, this loose formulation means that we multiply the signal f with a *window function* g, which is constant on a small interval, and decays fast and smooth to zero outside the interval; by taking the Fourier transform of this product, we get an idea about the frequency content of f in the small time-interval. In order to obtain information about f on the entire time axis, we repeat the process with translated versions of the window function. This corresponds exactly to our definition of the short-time Fourier transform $\Psi_g(f)$, which we introduced in Section 9.5.

The short-time Fourier transform turns out to be the key to obtain representations of the type (9.2). This point will be clear after the proof of the following statement.

Proposition 9.9.1 *Let* $f_1, f_2, g_1, g_2 \subset L^2(\mathbb{R})$. *Then*

$$\int_{-\infty}^{\infty} \int_{-\infty}^{\infty} \Psi_{g_1}(f_1)(a, b)\overline{\Psi_{g_2}(f_2)(a, b)}\, db\, da = \langle f_1, f_2\rangle\langle g_2, g_1\rangle.$$

Proof. By definition,

$$
\begin{aligned}
\Psi_{g_1}(f_1)(a, b) &= \langle f_1, E_b T_a g_1\rangle \\
&= \int_{-\infty}^{\infty} f_1(x)e^{-2\pi i b x}\overline{g_1(x - a)}\, dx.
\end{aligned}
$$

Consider for a moment a fixed value for a. Then the above expression for $\Psi_{g_1}(f_1)(a, b)$ is the Fourier transform of the function

$$F_1(x) = f_1(x)\overline{g_1(x - a)},$$

evaluated at the point b. Letting

$$F_2(x) = f_2(x)\overline{g_2(x - a)}$$

and using Plancherel's Theorem, it follows that

$$
\begin{aligned}
\int_{-\infty}^{\infty} \int_{-\infty}^{\infty} \Psi_{g_1}(f_1)(a, b)\overline{\Psi_{g_2}(f_2)(a, b)}\, db\, da &= \int_{-\infty}^{\infty} \int_{-\infty}^{\infty} \widehat{F_1}(b)\overline{\widehat{F_2}(b)}\, db\, da \\
&= \int_{-\infty}^{\infty} \int_{-\infty}^{\infty} F_1(b)\overline{F_2(b)}\, db\, da.
\end{aligned}
$$

Finally, inserting the expressions for the functions F_1 and F_2, an application of Fubini's Theorem yields that

$$\int_{-\infty}^{\infty}\int_{-\infty}^{\infty}\Psi_{g_1}(f_1)(a,b)\overline{\Psi_{g_2}(f_2)(a,b)}\,db\,da$$

$$=\int_{-\infty}^{\infty}\int_{-\infty}^{\infty}f_1(b)\overline{g_1(b-a)}\overline{f_2(b)}g_2(b-a)\,db\,da$$

$$=\int_{-\infty}^{\infty}f_1(b)\overline{f_2(b)}\left(\int_{-\infty}^{\infty}\overline{g_1(b-a)}g_2(b-a)\,da\right)db$$

$$=\langle f_1,f_2\rangle\langle g_2,g_1\rangle.$$

This completes the proof. $\qquad\qquad\qquad\qquad\qquad\qquad\qquad\qquad$ □

Formulated in terms of the operators E_b, T_a, Proposition 9.9.1 says that

$$\int_{-\infty}^{\infty}\int_{-\infty}^{\infty}\langle f_1,E_bT_ag_1\rangle\langle E_bT_ag_2,f_2\rangle\,db\,da=\langle f_1,f_2\rangle\langle g_2,g_1\rangle.\qquad(9.62)$$

We now show how one can obtain integral representations like (9.2). Fix $f\in L^2(\mathbb{R})$; then Proposition 9.9.1 shows that the mapping

$$f_2\mapsto\int_{-\infty}^{\infty}\int_{-\infty}^{\infty}\langle f,E_bT_ag_1\rangle\langle E_bT_ag_2,f_2\rangle\,db\,da$$

is a conjugated linear functional on $L^2(\mathbb{R})$. By Theorem 2.3.2, there exists a unique element in $L^2(\mathbb{R})$ – we call it

$$\int_{-\infty}^{\infty}\int_{-\infty}^{\infty}\langle f,E_bT_ag_1\rangle E_bT_ag_2\,db\,da$$

– such that for all $f_2\in L^2(\mathbb{R})$,

$$\left\langle\int_{-\infty}^{\infty}\int_{-\infty}^{\infty}\langle f,E_bT_ag_1\rangle E_bT_ag_2\,db\,da,f_2\right\rangle$$

$$=\int_{-\infty}^{\infty}\int_{-\infty}^{\infty}\langle f,E_bT_ag_1\rangle\langle E_bT_ag_2,f_2\rangle\,db\,da$$

$$=\langle f,f_2\rangle\langle g_2,g_1\rangle$$

$$=\langle\langle g_2,g_1\rangle f,f_2\rangle.$$

Since this holds for all $f_2\in L^2(\mathbb{R})$, we obtain the desired integral representation:

Corollary 9.9.2 *Choose $g_1,g_2\in L^2(\mathbb{R})$ such that $\langle g_2,g_1\rangle\neq0$. Then every $f\in L^2(\mathbb{R})$ has the representation*

$$f=\frac{1}{\langle g_2,g_1\rangle}\int_{-\infty}^{\infty}\int_{-\infty}^{\infty}\langle f,E_bT_ag_1\rangle E_bT_ag_2\,db\,da,\qquad(9.63)$$

where the integral is interpreted in the weak sense.

Thus we have obtained representations like (9.2) and explained how they have to be interpreted. Note that the function $f \in L^2(\mathbb{R})$ is represented as a superposition of time–frequency shifts of *one* function $g_2 \in L^2(\mathbb{R})$, with coefficients given by the short-time Fourier transformation of possibly *another* function g_1.

Proposition 9.9.1 shows that the set of all time–frequency shifts of an arbitrary function $g \neq 0$ forms a continuous frame:

Corollary 9.9.3 *Let $g \in L^2(\mathbb{R}) \setminus \{0\}$. Then $\{E_b T_a g\}_{a,b \in \mathbb{R}}$ is a continuous frame for $L^2(\mathbb{R})$ with respect to $M = \mathbb{R}^2$ equipped with the Lebesgue measure.*

With the theory for discrete frames in mind, it is natural to think of the set $\{E_b T_a g_1\}_{a,b \in \mathbb{R}}$ in (9.63) as a *dual frame* of $\{E_b T_a g_2\}_{a,b \in \mathbb{R}}$. With this interpretation, we see that there is a large freedom in the choice of generators for the frame and the dual frame in the continuous setting: *any* pair of functions $g_1, g_2 \in L^2(\mathbb{R})$ for which $\langle g_2, g_1 \rangle \neq 0$ generate dual frames $\{E_b T_a g_1\}_{a,b \in \mathbb{R}}$ and $\{E_b T_a g_2\}_{a,b \in \mathbb{R}}$.

.

9.10   Exercises

9.1 Show by an example (maybe with $a = b = 1$) that the necessary condition in Proposition 9.1.2 does not suffice for $\{E_{mb} T_{na} g\}_{m,n \in \mathbb{Z}}$ being a Gabor frame.

9.2 Prove (9.9) under the assumptions in Lemma 9.1.4 and justify all the following interchanges of summation and integration in the proof.

9.3 Prove that $\{E_m T_{na} \chi_{[0,1]}\}_{m,n \in \mathbb{Z}}$ is a frame for $L^2(\mathbb{R})$ for all $a \in]0, 1]$.

9.4 Show that condition (CC) is satisfied for all $a, b > 0$ if $g \in W$.

9.5 Prove Corollary 9.1.6.

9.6 Show that for the B-spline B_2, the system $\{E_{mb} T_{2n} B_2\}_{m,n \in \mathbb{Z}}$ cannot be a frame for any $b > 0$.

9.7 Prove Corollary 9.1.8.

9.8 Derive Theorem 9.2.1 via Theorem 8.1.7.

9.9 Prove Corollary 9.2.2.

9.10 Assume that $\{E_{mb}T_{na}g\}_{m,n\in\mathbb{Z}}$ is an overcomplete frame. Show that there exist dual frames not having the Gabor structure. (Hint: check the proof of Lemma 5.2.3.)

9.11 Let $g \in L^2(\mathbb{R})$ and $a, b > 0$, and assume that $\{E_{mb}T_{na}g\}_{m,n\in\mathbb{Z}}$ is a frame for $L^2(\mathbb{R})$. Show that the dual Gabor frames having Gabor structure are exactly the families $\{E_{mb}T_{na}h\}_{m,n\in\mathbb{Z}}$, where

$$h = S^{-1}g + f - \sum_{m,n\in\mathbb{Z}} \langle S^{-1}g, E_{mb}T_{na}g\rangle E_{mb}T_{na}f$$

for some function $f \in L^2(\mathbb{R})$ that generates a Bessel sequence $\{E_{mb}T_{na}f\}_{m,n\in\mathbb{Z}}$. (Hint: apply Theorem 5.7.4 to a Bessel sequence $\{E_{mb}T_{na}f\}_{m,n\in\mathbb{Z}}$ and use Lemma 9.3.1.)

9.12 Prove Theorem 9.3.5 via Theorem 8.1.7.

9.13 Let W denote the Wiener space.

(i) Prove (9.40).

(ii) Prove that $W \subset L^1(\mathbb{R}) \cap L^2(\mathbb{R})$.

(iii) Prove that every bounded measurable function with compact support belongs to W, and that

$$\|g\|_{W,1} \le (\,|\mathrm{supp}(g)| + 1)\|g\|_\infty.$$

9.14 Consider the function $g(x) = \frac{1}{1+x^2}$.

(i) Show that $g \in W$ and find an estimate for $\|g\|_{W,1}$.

(ii) Find a constant C such that

$$\sum_{n\in\mathbb{Z}} |g(x - n)|^2 \ge C, \ \forall x \in \mathbb{R}.$$

(iii) Show that for all $N \in \mathbb{N}$,

$$\sum_{|n|\ge N} \|g\chi_{[n,n+1]}\|_\infty \le \pi - 2\arctan(N - 1) + \frac{1}{1 + (N - 1)^2}.$$

9.15 This exercise concerns condition (R) and its relationship to Lebesgue points.

(i) Assume that $g \in L^2(\mathbb{R})$ satisfies condition (R). Show that all integers are Lebesgue points for g.

(ii) Assume that g is a bounded compactly supported function for which every integer is a Lebesgue point. Show that g satisfies condition (R).

(iii) Prove via (ii) that condition (R) is satisfied on a dense subset of $L^2(\mathbb{R})$.

(iv) Prove that the Gaussian $g(x) = e^{-\frac{1}{2}x^2}$ satisfies condition (R).

9.16 Prove (i) and (ii) in Lemma 9.7.2.

9.17 Complete the proof of Proposition 9.7.3 by proving the equivalence in the last line of the proof and the similar statement for the lower bound.

9.18 Prove Corollary 9.7.4.

9.19 Let $Q = [0, 1[\times[0, 1[$. Prove that $L^2(Q) \subset L^1(Q)$, and find a function $f \in L^1(Q)$ that does not belong to $L^2(Q)$.

9.20 Describe how a sequence $\{e_{m,n}\}_{m,n\in\mathbb{Z}}$ can be reindexed as $\{e_k\}_{k=1}^\infty$.

10
Gabor Frames in $\ell^2(\mathbb{Z})$

Every numerical calculation with functions in $L^2(\mathbb{R})$ will involve a discrete model, where all calculations are done with (finite) sequences in $\ell^2(\mathbb{Z})$. Therefore, it is important to know that certain conditions on a Gabor frame $\{E_{mb}T_{na}g\}_{m,n\in\mathbb{Z}}$ for $L^2(\mathbb{R})$ imply that we can construct a frame for $\ell^2(\mathbb{Z})$ having a similar structure. The relevant conditions were discovered by Janssen, and the main part of this chapter will deal with his results.

One can also consider frames in $\ell^2(\mathbb{Z})$ with a Gabor-like structure without referring to Gabor frames in $L^2(\mathbb{R})$. The theory for these frames is very similar to the Gabor theory in $L^2(\mathbb{R})$ and will not be discussed in detail.

10.1   Translation and modulation on $\ell^2(\mathbb{Z})$

In this chapter, we will change the notation slightly. In fact, for a sequence $g \in \ell^2(\mathbb{Z})$, we will (with a few exceptions) denote the j-th coordinate by $g(j)$ rather than g_j. Thus

$$g = (\ldots, g(-1), g(0), g(1), \ldots).$$

This change in the notation will make the similarity between Gabor theory in $L^2(\mathbb{R})$ and $\ell^2(\mathbb{Z})$ more transparent.

In order to introduce the Gabor systems on $\ell^2(\mathbb{Z})$, we have to define suitable modulation operators and translation operators. We first define the *modulation operator* $E_b, b \in \mathbb{R}$, on $\ell^2(\mathbb{Z})$. Given a sequence $g \in \ell^2(\mathbb{Z})$ and $b \in \mathbb{R}$, we define $E_b g$ to be the sequence in $\ell^2(\mathbb{Z})$ whose j-th coordinate is

O. Christensen, *Frames and Bases*. DOI: 10.1007/978-0-8176-4678-3_10,
© Springer Science+Business Media, LLC 2008

$$E_b g(j) := e^{2\pi i b j} g(j). \tag{10.1}$$

Even though the definition of E_b makes sense for all $b \in \mathbb{R}$, we will only use modulations of the form $E_{m/M}$, where $M \in \mathbb{N}$ is fixed and $m \in \mathbb{Z}$. In the terminology used for Gabor systems in $L^2(\mathbb{R})$, this means that the modulation parameter is $1/M$. There is, however, one important difference between the two settings: in the $L^2(\mathbb{R})$-setting, modulation operators with different parameters are necessarily different, but this is not the case in the discrete setting discussed here. In fact, with the definition (10.1),

$$E_{\frac{m}{M}} = E_{\frac{m+kM}{M}} \text{ for all } k \in \mathbb{Z}.$$

Therefore, $\{E_{m/M}g\}_{m \in \mathbb{Z}}$ cannot be a Bessel sequence in $\ell^2(\mathbb{Z})$, except for the case $g = 0$. For this reason, we will only consider modulations $E_{m/M}$ with $m = 0, \dots, M - 1$.

We now introduce the *translation operator* on $\ell^2(\mathbb{Z})$. Given $n \in \mathbb{Z}$ and $g \in \ell^2(\mathbb{Z})$, we let $T_n g$ be the sequence in $\ell^2(\mathbb{Z})$ whose j-th coordinate is

$$T_n g(j) = g(j - n). \tag{10.2}$$

The *discrete Gabor system* generated by a sequence $g \in \ell^2(\mathbb{Z})$ and with modulation parameter $1/M$ and translation parameter N, $(M, N \in \mathbb{N})$ is now defined as the family of sequences $\{E_{m/M}T_{nN}g\}_{n \in \mathbb{Z}, m=0,\dots,M-1}$; specifically, $E_{m/M}T_{nN}g$ is the sequence in $\ell^2(\mathbb{Z})$ whose j-th coordinate is

$$E_{m/M}T_{nN}g(j) = e^{2\pi i j m/M}g(j - nN).$$

Many results for Gabor systems in $L^2(\mathbb{R})$ have analog counterparts for Gabor systems in $\ell^2(\mathbb{Z})$, with similar proofs. For example, a necessary condition for $\{E_{m/M}T_{nN}g\}_{n \in \mathbb{Z}, m=0,\dots,M-1}$ to be a frame for $\ell^2(\mathbb{Z})$ is that $\frac{N}{M} \leq 1$; and if $\{E_{m/M}T_{nN}g\}_{n \in \mathbb{Z}, m=0,\dots,M-1}$ is a frame, it is a Riesz basis if and only if $M = N$. We will not go into the general theory (see, e.g., [23] or [63]) but concentrate on a method for constructing a Gabor frame for $\ell^2(\mathbb{Z})$ based on a Gabor frame for $L^2(\mathbb{R})$.

10.2   Gabor systems in $\ell^2(\mathbb{Z})$ through sampling

Janssen proved in [48] that there is a natural way to obtain Gabor frames in $\ell^2(\mathbb{Z})$ via Gabor frames for $L^2(\mathbb{R})$ through sampling. We present some of his results here. We mention that the original results by Janssen are more general than the results presented here: in order to avoid technical complications, we will assume that the generator for the considered Gabor system is continuous, but it is actually sufficient that the function contains all integers among its Lebesgue points.

The starting point is a Gabor system for $L^2(\mathbb{R})$; we assume it to have the form $\{E_{m/M}T_{nN}g\}_{m,n\in\mathbb{Z}}$, where $g \in L^2(\mathbb{R})$ and $M, N \in \mathbb{N}$. The first question is how one can construct sequences in $\ell^2(\mathbb{Z})$ based on the Gabor system in $L^2(\mathbb{R})$. If we assume that the function g is continuous, then we can easily obtain sequences indexed by $\mathbb{Z}$ by *sampling*. That is, for each $m, n \in \mathbb{Z}$ we consider the sequence

$$\{E_{m/M}T_{nN}g(j)\}_{j\in\mathbb{Z}} = \{e^{2\pi ijm/M}g(j - nN)\}_{j\in\mathbb{Z}}. \tag{10.3}$$

The basic idea by Janssen is to ask for conditions such that the family of all the sequences in (10.3), where $m = 0, \ldots, M - 1, n \in \mathbb{Z}$, constitutes a frame for $\ell^2(\mathbb{Z})$. The first point is to ensure that the sequences in (10.3) do in fact belong to $\ell^2(\mathbb{Z})$. The condition given in Lemma 10.2.1 will guarantee this.

Let us introduce some notation. Given a continuous function $f \in L^2(\mathbb{R})$, we denote the discrete sequence obtained by sampling f at the integers by

$$f^D := \{f(j)\}_{j\in\mathbb{Z}}.$$

With this notation, we can consider the sequence $E_{m/M}T_{nN}(f^D)$, obtained by letting the discrete Gabor system act on the sequence f^D; or we can consider the discrete sequence $(E_{m/M}T_{nN}f)^D$, obtained by sampling of the function $E_{m/M}T_{nN}f \in L^2(\mathbb{R})$. The two procedures lead to the same outcome, and we simply write

$$E_{m/M}T_{nN}f^D = \{e^{2\pi ijm/M}f(j - nN)\}_{j\in\mathbb{Z}}. \tag{10.4}$$

The first result gives a condition on the Gabor system $\{E_{m/M}T_{nN}g\}_{m,n\in\mathbb{Z}}$ in $L^2(\mathbb{R})$, which implies that the discrete time–frequency shifts of g^D belong to $\ell^2(\mathbb{Z})$.

Lemma 10.2.1 *Let $M, N \in \mathbb{N}$ and $g \in L^2(\mathbb{R})$ be a continuous function. Assume that $\{E_{m/M}T_{nN}g\}_{m,n\in\mathbb{Z}}$ is a Bessel sequence in $L^2(\mathbb{R})$. Then*

$$\sum_{j\in\mathbb{Z}} |g(j)|^2 \leq \frac{BN}{M}.$$

In particular, $E_{m/M}T_{nN}g^D \in \ell^2(\mathbb{Z})$ for all $m, n \in \mathbb{Z}$.

Proof. Letting B denote an upper frame bound for $\{E_{m/M}T_{nN}g\}_{m,n\in\mathbb{Z}}$, we know from Proposition 9.1.2 that

$$\sum_{n\in\mathbb{Z}} |g(x + nN)|^2 \leq \frac{B}{M}, \text{ a.e. } x \in \mathbb{R}. \tag{10.5}$$

Because of the continuity of the function g, the inequality (10.5) actually holds for all $x \in \mathbb{R}$. Thus,

$$\sum_{j \in \mathbb{Z}} |g(j)|^2 = \sum_{k=0}^{N-1} \sum_{n \in \mathbb{Z}} |g(k+nN)|^2$$

$$\leq \frac{BN}{M}.$$

This proves that $g^D \in \ell^2(\mathbb{Z})$. Now, for any $m, n \in \mathbb{Z}$,

$$\sum_{j \in \mathbb{Z}} |E_{m/M} T_{nN} g^D(j)|^2 = \sum_{j \in \mathbb{Z}} |g(j-nN)|^2 = \sum_{j \in \mathbb{Z}} |g(j)|^2;$$

this completes the proof. $\qquad\square$

The next lemma is an important step from Gabor systems in $L^2(\mathbb{R})$ to Gabor systems in $\ell^2(\mathbb{Z})$. It contains an identity involving functions in $L^2(\mathbb{R})$, which "approaches discrete sequences" for small values of ϵ:

Lemma 10.2.2 *Let $g \in L^2(\mathbb{R})$ and $M, N \in \mathbb{N}$ be given, and assume that $\{E_{m/M} T_{nN} g\}_{m,n \in \mathbb{Z}}$ is a Bessel sequence in $L^2(\mathbb{R})$. Given $\epsilon \in]0, \frac{1}{2}[$, let*

$$\delta^\epsilon = \frac{1}{\epsilon} \chi_{]-\frac{1}{2}\epsilon, \frac{1}{2}\epsilon[}.$$

Consider a finite linear combination of translates of δ^ϵ,

$$f^\epsilon = \sum_j c_j T_j \delta^\epsilon. \tag{10.6}$$

Then

$$\sum_{m,n \in \mathbb{Z}} |\langle f^\epsilon, E_{m/M} T_{nN} g \rangle|^2$$

$$= \sum_{n \in \mathbb{Z}} \sum_{m=0}^{M-1} \sum_{j,k} c_j \overline{c_k} \frac{1}{\epsilon^2} \int_{-\frac{1}{2}\epsilon}^{\frac{1}{2}\epsilon} \overline{E_{m/M} T_{nN} g(x+j)} E_{m/M} T_{nN} g(x+k) \, dx.$$

Proof. First, we use the definition of f^ϵ to write

$$\sum_{m,n \in \mathbb{Z}} |\langle f^\epsilon, E_{m/M} T_{nN} g \rangle|^2$$

$$= \sum_{m,n \in \mathbb{Z}} \sum_{j,k} c_j \overline{c_k} \langle T_j \delta^\epsilon, E_{m/M} T_{nN} g \rangle \langle E_{m/M} T_{nN} g, T_k \delta^\epsilon \rangle$$

$$= \sum_{n \in \mathbb{Z}} \sum_{m=0}^{M-1} \sum_{\ell \in \mathbb{Z}} \sum_{j,k} c_j \overline{c_k} \langle T_j \delta^\epsilon, E_{\ell+m/M} T_{nN} g \rangle \langle E_{\ell+m/M} T_{nN} g, T_k \delta^\epsilon \rangle.$$

Now, via Lemma 9.1.3,

$$
\langle T_j\delta^\epsilon, E_{\ell+m/M}T_{nN}g\rangle
$$
$$
= \langle E_{-m/M}T_j\delta^\epsilon, E_\ell T_{nN}g\rangle
$$
$$
= \int_0^1 \left(\sum_{r\in\mathbb{Z}} T_j\delta^\epsilon(x-r)\overline{E_{m/M}T_{nN}g(x-r)}\right)e^{-2\pi i\ell x}\,dx,
$$

which is the ℓ-th Fourier coefficient of the 1-periodic function

$$
\alpha_j(x) = \sum_{r\in\mathbb{Z}} T_j\delta^\epsilon(x-r)\overline{E_{m/M}T_{nN}g(x-r)}.
$$

Note that for $x \in [-1/2, 1/2]$,

$$
\alpha_j(x) = \delta^\epsilon(x)\overline{E_{m/M}T_{nN}g(x+j)}
$$
$$
= \frac{1}{\epsilon}\chi_{]-\frac{1}{2}\epsilon,\frac{1}{2}\epsilon[}(x)\overline{E_{m/M}T_{nN}g(x+j)}.
$$

Via Lemma 3.5.2,

$$
\sum_{l\in\mathbb{Z}}\langle T_j\delta^\epsilon, E_{l+m/M}T_{nN}g\rangle\langle E_{l+m/M}T_{nN}g, T_k\delta^\epsilon\rangle
$$
$$
= \langle\alpha_j,\alpha_k\rangle
$$
$$
= \int_{-\frac{1}{2}}^{\frac{1}{2}} \alpha_j(x)\overline{\alpha_k(x)}\,dx
$$
$$
= \frac{1}{\epsilon^2}\int_{-\frac{1}{2}\epsilon}^{\frac{1}{2}\epsilon} \overline{E_{m/M}T_{nN}g(x+j)}E_{m/M}T_{nN}g(x+k)\,dx,
$$

and the result follows. $\qquad\square$

We are now ready to show how one can obtain a Gabor frame for $\ell^2(\mathbb{Z})$ by sampling of a Gabor frame $\{E_{mb}T_{na}g\}_{m,n\in\mathbb{Z}}$ for $L^2(\mathbb{R})$. We will assume that the function g satisfies condition (R), introduced on page 223.

Theorem 10.2.3 *Let $M, N \in \mathbb{N}$. Assume that $g \in L^2(\mathbb{R})$ is a continuous function satisfying condition (R) and that $\{E_{m/M}T_{nN}g\}_{m,n\in\mathbb{Z}}$ is a frame for $L^2(\mathbb{R})$ with frame bounds A, B. Then the discrete Gabor system $\{E_{m/M}T_{nN}g^D\}_{n\in\mathbb{Z},m=0,\ldots,M-1}$ is a frame for $\ell^2(\mathbb{Z})$ with frame bounds A, B.*

Proof. In order to prove that $\{E_{m/M}T_{nN}g^D\}_{n\in\mathbb{Z},m=0,\ldots,M-1}$ is a frame for $\ell^2(\mathbb{Z})$ with bounds A, B, we have to prove that for all finite sequences $\{c_k\}_{k\in\mathbb{Z}}$,

$$
A\sum_{j\in\mathbb{Z}}|c_j|^2 \le \sum_{n\in\mathbb{Z}}\sum_{m=0}^{M-1}\left|\sum_j c_j\overline{E_{m/M}T_{nN}g(j)}\right|^2 \le B\sum_{j\in\mathbb{Z}}|c_j|^2. \qquad (10.7)
$$

In fact, if the frame condition (10.7) holds for all finite sequences, then Lemma 5.1.2 shows that it holds for all $\{c_k\}_{k\in\mathbb{Z}} \in \ell^2(\mathbb{Z})$. Now, consider a finite sequence $\{c_k\}_{k\in\mathbb{Z}}$. Then, for any $\epsilon \in]0, \frac{1}{2}[$, the square of the $L^2(\mathbb{R})$-norm of the function f^ϵ in (10.6) is

$$\|f^\epsilon\|^2 = \frac{1}{\epsilon} \sum_{j\in\mathbb{Z}} |c_j|^2.$$

Applying the frame condition for $\{E_{m/M}T_{nN}g\}_{m,n\in\mathbb{Z}}$ on f^ϵ gives that for all $\epsilon \in]0, \frac{1}{2}[$,

$$A \sum_{j\in\mathbb{Z}} |c_j|^2 \le \epsilon \sum_{m,n\in\mathbb{Z}} |\langle f^\epsilon, E_{m/M}T_{nN}g\rangle|^2 \le B \sum_{j\in\mathbb{Z}} |c_j|^2.$$

For the proof of Theorem 10.2.3, it is therefore enough to show that

$$\liminf_{\epsilon\to 0} \epsilon \sum_{m,n\in\mathbb{Z}} |\langle f^\epsilon, E_{m/M}T_{nN}g\rangle|^2 = \sum_{n\in\mathbb{Z}} \sum_{m=0}^{M-1} \left| \sum_j c_j \overline{E_{m/M}T_{nN}g(j)} \right|^2. \quad (10.8)$$

In order to prove (10.8), we first note that

$$\sum_{n\in\mathbb{Z}} \sum_{m=0}^{M-1} \left| \sum_j c_j \overline{E_{m/M}T_{nN}g(j)} \right|^2$$

$$= \sum_{n\in\mathbb{Z}} \sum_{m=0}^{M-1} \sum_{j,k} c_j \overline{c_k} \overline{E_{m/M}T_{nN}g(j)} E_{m/M}T_{nN}g(k),$$

while by Lemma 10.2.2

$$\epsilon \sum_{m,n\in\mathbb{Z}} |\langle f^\epsilon, E_{m/M}T_{nN}g\rangle|^2$$

$$= \sum_{n\in\mathbb{Z}} \sum_{m=0}^{M-1} \sum_{j,k} c_j \overline{c_k} \frac{1}{\epsilon} \int_{-\frac{1}{2}\epsilon}^{\frac{1}{2}\epsilon} \overline{E_{m/M}T_{nN}g(x+j)} E_{m/M}T_{nN}g(x+k)\, dx.$$

Comparing the two expressions, we see that (10.8) follows if we can prove that

$$\sum_{n\in\mathbb{Z}} \frac{1}{\epsilon} \int_{-\frac{1}{2}\epsilon}^{\frac{1}{2}\epsilon} \overline{E_{m/M}T_{nN}g(x+j)} E_{m/M}T_{nN}g(x+k)\, dx$$

$$\to \sum_{n\in\mathbb{Z}} \overline{E_{m/M}T_{nN}g(j)} E_{m/M}T_{nN}g(k) \text{ as } \epsilon \to 0$$

for all $m = 0, \ldots, M - 1$ and $j, k \in \mathbb{Z}$ (recall that the sums over j, k are finite). Now,

$$\left| \frac{1}{\epsilon} \int_{-\frac{1}{2}\epsilon}^{\frac{1}{2}\epsilon} \overline{E_{m/M}T_{nN}g(x+j)}E_{m/M}T_{nN}g(x+k)\, dx \right.$$

$$\left. - \overline{E_{m/M}T_{nN}g(j)}E_{m/M}T_{nN}g(k) \right|$$

$$\leq \frac{1}{\epsilon} \int_{-\frac{1}{2}\epsilon}^{\frac{1}{2}\epsilon} \left| \overline{E_{m/M}T_{nN}g(x+j)}E_{m/M}T_{nN}g(x+k) \right.$$

$$\left. - \overline{E_{m/M}T_{nN}g(j)}E_{m/M}T_{nN}g(k) \right|\, dx$$

$$= \frac{1}{\epsilon} \int_{-\frac{1}{2}\epsilon}^{\frac{1}{2}\epsilon} \left| \overline{g(x+j-nN)}g(x+k-nN) - \overline{g(j-nN)}g(k-nN) \right|\, dx$$

$$\leq \frac{1}{\epsilon} \int_{-\frac{1}{2}\epsilon}^{\frac{1}{2}\epsilon} \left| \overline{g(x+j-nN)} - \overline{g(j-nN)} \right|\, |g(x+k-nN)|\, dx$$

$$+ \frac{1}{\epsilon} \int_{-\frac{1}{2}\epsilon}^{\frac{1}{2}\epsilon} |\overline{g(j-nN)}|\, |g(x+k-nN) - g(k-nN)|\, dx.$$

It follows that

$$\left| \sum_{n\in\mathbb{Z}} \frac{1}{\epsilon} \int_{-\frac{1}{2}\epsilon}^{\frac{1}{2}\epsilon} \overline{E_{m/M}T_{nN}g(x+j)}E_{m/M}T_{nN}g(x+k)\, dx \right.$$

$$\left. - \sum_{n\in\mathbb{Z}} \overline{E_{m/M}T_{nN}g(j)}E_{m/M}T_{nN}g(k) \right|$$

$$\leq \frac{1}{\epsilon} \sum_{n\in\mathbb{Z}} \int_{-\frac{1}{2}\epsilon}^{\frac{1}{2}\epsilon} \left| \overline{g(x+j-nN)} - \overline{g(j-nN)} \right|\, |g(x+k-nN)|\, dx \quad (10.9)$$

$$+ \frac{1}{\epsilon} \sum_{n\in\mathbb{Z}} \int_{-\frac{1}{2}\epsilon}^{\frac{1}{2}\epsilon} |\overline{g(j-nN)}|\, |g(x+k-nN) - g(k-nN)|\, dx. \quad (10.10)$$

Both (10.9) and (10.10) converge to zero as $\epsilon \to 0$; we give the argument for (10.9). Applying Cauchy–Schwarz' inequality twice, first on the integral and then on the sum,

$$\frac{1}{\epsilon} \sum_{n \in \mathbb{Z}} \int_{-\frac{1}{2}\epsilon}^{\frac{1}{2}\epsilon} \overline{|g(x+j-nN) - g(j-nN)|} \; |g(x+k-nN)| \, dx$$

$$\leq \frac{1}{\epsilon} \sum_{n \in \mathbb{Z}} \left(\int_{-\frac{1}{2}\epsilon}^{\frac{1}{2}\epsilon} |g(x+j-nN) - g(j-nN)|^2 dx \right)^{1/2}$$

$$\times \left(\int_{-\frac{1}{2}\epsilon}^{\frac{1}{2}\epsilon} |g(x+k-nN)|^2 \, dx \right)^{1/2}$$

$$\leq \frac{1}{\epsilon} \left(\sum_{n \in \mathbb{Z}} \int_{-\frac{1}{2}\epsilon}^{\frac{1}{2}\epsilon} |g(x+j-nN) - g(j-nN)|^2 dx \right)^{1/2}$$

$$\times \left(\sum_{n \in \mathbb{Z}} \int_{-\frac{1}{2}\epsilon}^{\frac{1}{2}\epsilon} |g(x+k-nN)|^2 \, dx \right)^{1/2} = (*).$$

Via Lemma 10.2.1, the second term in $(*)$ can be estimated by

$$\left(\sum_{n \in \mathbb{Z}} \int_{-\frac{1}{2}\epsilon}^{\frac{1}{2}\epsilon} |g(x+k-nN)|^2 \, dx \right)^{1/2} \leq \sqrt{\frac{BN}{M}} \, \epsilon;$$

thus

$$(*) \quad \leq \quad \sqrt{\frac{BN}{M}} \left(\frac{1}{\epsilon} \sum_{n \in \mathbb{Z}} \int_{-\frac{1}{2}\epsilon}^{\frac{1}{2}\epsilon} |g(x+j-nN) - g(j-nN)|^2 dx \right)^{1/2},$$

which converges to zero for $\epsilon \to 0$ because of condition (R); the proof is completed. $\quad\square$

As conclusion of this section, we now state a few results without proofs. Both are due to Janssen [47]. The first result concerns condition (R):

Lemma 10.2.4 *Suppose that $g \in L^2(\mathbb{R})$ satisfies condition (R) and that $\{E_{m/M}T_{nN}g\}_{m,n \in \mathbb{Z}}$ is a Bessel sequence in $L^2(\mathbb{R})$ for some $M, N \in \mathbb{N}$. Then any function of the form*

$$\phi = \sum_{m,n \in \mathbb{Z}} c_{mn} E_{m/N} T_{nM} g, \text{ where } \{c_{mn}\} \in \ell^1(\mathbb{Z}^2) \qquad (10.11)$$

also satisfies condition (R).

Note that the functions ϕ in (10.11) are linear combinations of the Gabor system with respect to the dual lattice $\{(nM, m/N)\}_{m,n \in \mathbb{Z}}$. Since the Gaussian $g(x) = e^{-\frac{1}{2}x^2}$ satisfies condition (R), see Exercise 9.15, and $\{E_m T_n g\}_{m,n \in \mathbb{Z}}$ is complete in $L^2(\mathbb{R})$, Lemma 10.2.4 implies that condition (R) is satisfied on a dense set of functions in $L^2(\mathbb{R})$.

Finally, we state a result, showing how one can sample the frame operator:

Proposition 10.2.5 *Let* $g \in L^2(\mathbb{R})$, $M, N \in \mathbb{N}$, *and assume that* $\{E_{m/M}T_{nN}g\}_{m,n\in\mathbb{Z}}$ *is a Bessel sequence in* $L^2(\mathbb{R})$ *and satisfies condition (A). Denote the frame operator by* S. *Then, for any* $f \in L^2(\mathbb{R})$ *that satisfies condition (R) and for which* $\{E_{m/M}T_{nN}f\}_{m,n\in\mathbb{Z}}$ *is a Bessel sequence,*

$$Sf(j) = \frac{M}{N} \sum_{m,n\in\mathbb{Z}} \langle g, E_{m/N}T_{nM}g \rangle E_{m/N}T_{nM}f(j), \ j \in \mathbb{Z}. \qquad (10.12)$$

If furthermore g *satisfies condition (R) and we denote the frame operator for* $\{E_{m/M}T_{nN}g^D\}_{n\in\mathbb{Z},m=0,\dots,M-1}$ *by* $S^D : \ell^2(\mathbb{Z}) \to \ell^2(\mathbb{Z})$, *then*

$$(Sf)^D = S^D f^D; \qquad (10.13)$$

if we also add the assumption that $\{E_{m/M}T_{nN}g\}_{m,n\in\mathbb{Z}}$ *is a frame, then*

$$(S^{-1}g)^D = (S^D)^{-1}g^D. \qquad (10.14)$$

One can prove that the canonical dual frame associated with a frame $\{E_{m/M}T_{nN}g^D\}_{n\in\mathbb{Z},m=0,\dots,M-1}$ is $\{E_{m/M}T_{nN}(S^D)^{-1}g^D\}_{n\in\mathbb{Z},m=0,\dots,M-1}$; that is, as for Gabor frames in $L^2(\mathbb{R})$, it consists of time–frequency shifts of a single function.

10.3 Shift-invariant systems

As in the $L^2(\mathbb{R})$-case, the discrete Gabor systems in $\ell^2(\mathbb{Z})$ are special cases of general shift-invariant systems. In the discrete case, these systems consist of sequences of the form

$$\{g_m(j - nN)\}_{j\in\mathbb{Z}},$$

where $n \in \mathbb{Z}, m = 0, \dots, M - 1$, and each g_m is a sequence in $\ell^2(\mathbb{Z})$. We will always let m, n run through the index set given above, so we will skip the index and simply write $\{g_{nm}\}$ for the shift-invariant system.

The results for continuous shift-invariant systems in Section 8.1 have discrete counterparts, which are stated in [44]. In order to formulate the results, define the *Fourier transform* of a sequence $h \in \ell^2(\mathbb{Z})$ by

$$\hat{h}(\nu) = \sum_{j\in\mathbb{Z}} h(j)e^{-2\pi ij\nu}, \ a.e. \ \nu \in \mathbb{R}.$$

Given a shift-invariant system $\{g_{nm}\}$ we define, analogous to (8.13), the matrix-valued function

$$H(\nu) = (\widehat{g_m}(\nu - k/N))_{k=0,\dots,N-1,m=0,\dots M-1}, \ a.e. \ \nu \in \mathbb{R}.$$

Observe that this is an $N \times M$ matrix.

Theorem 10.3.1 *In the setting above, the following holds:*

(i) $\{g_{nm}\}$ *is a Bessel sequence in* $\ell^2(\mathbb{Z})$ *with upper bound* B *if and only if* $H(\nu)$ *for a.e.* $\nu \in \mathbb{R}$ *defines a bounded linear mapping from* $\mathbb{C}^M$ *into* $\mathbb{C}^N$ *of norm at most* $\sqrt{NB}$.

(ii) *A Bessel sequence* $\{g_{nm}\}$ *is a frame for* $\ell^2(\mathbb{Z})$ *with lower frame bound* A *if and only if*

$$NAI \leq H(\nu)H(\nu)^*, \ a.e. \ \nu \in \mathbb{R}.$$

(ii) $\{g_{nm}\}$ *is a tight frame for* $\ell^2(\mathbb{Z})$ *if and only if there is a constant* $c > 0$ *such that*

$$\sum_{m=0}^{M-1} \widehat{g_m}(\nu - k/N)\overline{\widehat{g_m}(\nu)} = c\delta_{k,0}, \ k \in \mathbb{Z}, \ a.e. \ \nu \in \mathbb{R}.$$

(iv) *Two shift-invariant systems* $\{g_{nm}\}$ *and* $\{h_{nm}\}$*, which form Bessel sequences in* $\ell^2(\mathbb{Z})$*, are dual frames if and only if*

$$\sum_{m=0}^{M-1} \widehat{g_m}(\nu - k/N)\overline{\widehat{h_m}(\nu)} = N\delta_{k,0}, \ k \in \mathbb{Z}, \ a.e. \ \nu \in \mathbb{R}.$$

Most proofs follow by repeating the arguments from the continuous setting. Again, the statements have direct consequences for discrete Gabor frames (Exercise 10.1).

10.4   Exercises

10.1 Derive characterizations of frames, tight frames, and dual frame pairs for discrete Gabor systems via Theorem 10.3.1.

10.2 Here we ask the reader to prove an extension of Proposition 1.2.3. In fact, show that for a bi-infinite matrix $\Lambda = \{\lambda_{m,n}\}_{m,n \in \mathbb{Z}}$, the following are equivalent:

(i) There exist constants $A, B > 0$ such that

$$A\sum |c_k|^2 \leq ||\Lambda\{c_k\}||^2 \leq B\sum |c_k|^2 \text{ for all finite sequences } \{c_k\}.$$

(ii) The columns in Λ constitute a Riesz basis for their closed span in $\ell^2(\mathbb{Z})$.

(iii) The rows in Λ constitute a frame for $\ell^2(\mathbb{Z})$.

11

Wavelet Frames in $L^2(\mathbb{R})$

Wavelet theory is based on two classes of operators on $L^2(\mathbb{R})$, namely,

Translation by $b \in \mathbb{R}$, $T_b : L^2(\mathbb{R}) \to L^2(\mathbb{R})$, $(T_b f)(x) = f(x - b)$;

Dilation by $a \neq 0$, $D_a : L^2(\mathbb{R}) \to L^2(\mathbb{R})$, $(D_a f)(x) = \dfrac{1}{\sqrt{|a|}} f(\dfrac{x}{a})$.

The fundamental question in wavelet analysis is what conditions we have to impose on a function ψ such that a given signal $f \in L^2(\mathbb{R})$ can be expanded via translated and scaled versions of ψ, i.e., via functions

$$\psi^{a,b}(x) := (T_b D_a \psi)(x) = \frac{1}{|a|^{1/2}} \psi(\frac{x - b}{a}), \ a \neq 0, \ b \in \mathbb{R}. \qquad (11.1)$$

Thus, there is a basic similarity between wavelet analysis and Gabor analysis: both concern sequences of functions defined by letting a special class of operators act on a fixed function, i.e., in both cases we are dealing with coherent systems. As in Gabor analysis, there are two ways in which one can think about expansions of a signal f in terms of the functions $\psi^{a,b}$. One way is to ask for representations of f as integrals involving $\psi^{a,b}$ over $\mathbb{R}^2$. Alternatively, one can restrict the parameters a, b to a discrete subset Λ of $\mathbb{R}^2$ and ask for series expansions of f in terms of the corresponding functions $\psi^{a,b}$. For applications, the latter is the most convenient choice; connecting with the main theme of this book, the natural question is how we can choose the discrete subset Λ and the function ψ such that $\{\psi^{a,b}\}_{(a,b) \in \Lambda}$ is a frame for $L^2(\mathbb{R})$.

This chapter will deal with different aspects related to overcompleteness of collections of functions of the form (11.1). As discussed in Section 4.3,

O. Christensen, *Frames and Bases*. DOI: 10.1007/978-0-8176-4678-3_11,
© Springer Science+Business Media, LLC 2008

overcompleteness is introduced in order to obtain more flexibility and be able to make constructions with properties that cannot be obtained with orthonormal bases or Riesz bases.

The central part of the chapter is formed by the sections dealing with the unitary extension principle and variants hereof. In Section 11.1, we introduce the dyadic wavelet frames and discuss some of their properties. The section is mainly meant as an introduction to the constructions in Sections 11.2–11.3. In Section 11.2, we prove the unitary extension principle, which in its original version goes back to the fundamental papers [55] and [56] by Ron and Shen. It describes how one can choose functions $\psi_1, \ldots, \psi_n$ such that the multiwavelet system $\{D_2^j T_k \psi_\ell\}_{\ell=1,\ldots,n,j,k\in\mathbb{Z}}$ forms a tight frame for $L^2(\mathbb{R})$. The structure of a multiresolution analysis is maintained in the construction. A reformulation of the unitary extension, known as the oblique extension principle, is derived in Section 11.3. It provides more freedom in the construction and is useful in order to improve the approximation theoretic properties, a topic that is discussed in Section 11.4. In Section 11.5, the oblique extension principle is generalized to a construction of dual wavelet pairs. Section 11.6 describes the unitary extension principle in signal processing terms and relates it to the perfect reconstruction property for certain filter banks. Section 11.7 gives a short overview of the theory for general wavelet frames. Finally, Section 11.8 presents the continuous wavelet transform, which delivers integral representations of each $f \in L^2(\mathbb{R})$ of the type

$$f = \int_{-\infty}^{\infty} \int_{-\infty}^{\infty} c_f(a, b) \psi^{a,b} da\, db, \tag{11.2}$$

provided that ψ satisfies some admissibility conditions and that the integral is interpreted in the right sense.

A few words about terminology are needed. The word *wavelet* is usually reserved for a function ψ for which the functions

$$\{2^{j/2}\psi(2^j x - k)\}_{j,k\in\mathbb{Z}} = \{\psi^{2^{-j}, 2^{-j}k}\}_{j,k\in\mathbb{Z}} \tag{11.3}$$

form an orthonormal basis for $L^2(\mathbb{R})$. We will follow this tradition, but the word "wavelet" will appear in several constellations. Since we are interested in more general ways of choosing the translates and dilates than in (11.3), we will call *any* discrete family of the type $\{\psi^{a,b}\}_{(a,b)\in\Lambda}$, $\Lambda \subset \mathbb{R}^2$, a *wavelet system*. A family of functions that consists of translated and dilated versions of a single function is said to have *wavelet structure*.

11.1   Dyadic wavelet frames

Already in Section 3.6, we considered orthonormal bases for $L^2(\mathbb{R})$ having wavelet structure. In the current section, we will concentrate on *dyadic wavelet systems*, i.e., we will only consider scalings in terms of powers of two and translates by integers. Letting

$$(Df)(x) = 2^{1/2} f(2x),$$

this means that we will consider wavelet systems of the form $\{D^j T_k \psi\}_{j,k \in \mathbb{Z}}$ for some function $\psi \in L^2(\mathbb{R})$.

Definition 11.1.1 *Let* $\psi \in L^2(\mathbb{R})$. *A frame for* $L^2(\mathbb{R})$ *of the form* $\{D^j T_k \psi\}_{j,k \in \mathbb{Z}}$ *is called a dyadic wavelet frame.*

Let us investigate some of the basic properties of dyadic wavelet frames $\{D^j T_k \psi\}_{j,k \in \mathbb{Z}}$. First, for such a frame, the associated frame operator is given by

$$S : L^2(\mathbb{R}) \to L^2(\mathbb{R}), \ Sf = \sum_{j,k \in \mathbb{Z}} \langle f, D^j T_k \psi \rangle D^j T_k \psi.$$

The frame decomposition, see Theorem 5.1.7, takes the form

$$f = \sum_{j,k \in \mathbb{Z}} \langle f, S^{-1} D^j T_k \psi \rangle D^j T_k \psi, \ f \in L^2(\mathbb{R}).$$

As stated here, the frame decomposition is rather inconvenient: in order to find the coefficients $\langle f, S^{-1} D^j T_k \psi \rangle$, we need to calculate the action of the inverse frame operator on all the functions $D^j T_k \psi$, $j, k \in \mathbb{Z}$. A slight improvement can be obtained via a calculation showing that (Exercise 11.1)

$$S^{-1} D^j T_k \psi = D^j S^{-1} T_k \psi;$$

thus, it is enough to find the action of S^{-1} on the functions $T_k \psi$ for all $k \in \mathbb{Z}$ – then we obtain the rest of the functions by applying the operators D^j. In case one could prove that S^{-1} commutes with T_k for all $k \in \mathbb{Z}$, a further simplification would be obtained; but unfortunately, in general

$$D^j S^{-1} T_k \psi \neq D^j T_k S^{-1} \psi.$$

For this reason, one cannot expect the canonical dual frame to have wavelet structure. We will illustrate this with a concrete example, appearing in [26] and [17]. It describes a Riesz basis, for which the canonical dual frame can be calculated explicitly; in particular, we show that it does not have wavelet structure. As in Section 3.6, we use the notation

$$\psi_{j,k} = D^j T_k \psi.$$

Example 11.1.2 Let $\{\psi_{j,k}\}_{j,k \in \mathbb{Z}}$ be a wavelet orthonormal basis for $L^2(\mathbb{R})$. Given $\epsilon \in]0, 1[$, we define a function θ by

$$\theta = \psi + \epsilon D\psi.$$

We want to prove that $\{\theta_{j,k}\}_{j,k \in \mathbb{Z}}$ is a Riesz basis and find the dual Riesz basis. The idea is to consider θ as a small perturbation of ψ and use a stability result for frames to conclude that $\{\theta_{j,k}\}_{j,k\in\mathbb{Z}}$ is a Riesz basis. First, the commutator relation (2.22) shows that

$$\psi_{j,k} - \theta_{j,k} = -\epsilon D^j T_k D\psi = -\epsilon D^{j+1} T_{2k}\psi. \tag{11.4}$$

Using that $\{D^{j+1} T_{2k}\psi\}_{j,k\in\mathbb{Z}}$ is a subfamily of the orthonormal basis $\{\psi_{j,k}\}_{j,k\in\mathbb{Z}}$, it follows that for any finite scalar sequence $\{c_{j,k}\}$,

$$\left\| \sum_{j,k} c_{j,k}(\psi_{j,k} - \theta_{j,k}) \right\|^2 = \epsilon^2 \left\| \sum_{j,k} c_{j,k} D^{j+1} T_{2k}\psi \right\|^2 = \epsilon^2 \sum_{j,k} |c_{j,k}|^2.$$

Via the perturbation result stated in Theorem 5.6.1, we see that $\{\theta_{j,k}\}_{j,k\in\mathbb{Z}}$ is a Riesz basis for $L^2(\mathbb{R})$. By the definition of a Riesz basis, we can define a bounded invertible operator

$$U : L^2(\mathbb{R}) \to L^2(\mathbb{R}), \quad U\psi_{j,k} := \theta_{j,k}, \; j,k \in \mathbb{Z}.$$

Via Exercise 5.20, the frame operator for $\{\theta_{j,k}\}_{j,k\in\mathbb{Z}}$ is $S = UU^*$, so the canonical dual frame associated with $\{\theta_{j,k}\}_{j,k\in\mathbb{Z}}$ is

$$\{S^{-1}\theta_{j,k}\}_{j,k\in\mathbb{Z}} = \{(U^*)^{-1} U^{-1}\theta_{j,k}\}_{j,k\in\mathbb{Z}} = \{(U^*)^{-1}\psi_{j,k}\}_{j,k\in\mathbb{Z}}. \tag{11.5}$$

We will now calculate the functions in (11.5) explicitly. The idea is first to calculate the operator $I - U$; then we can find the adjoint operator $I - U^*$, and finally use Neumann's theorem to find an expression for $(U^*)^{-1}$.

In terms of the operator U, (11.4) means that

$$(I - U)\psi_{j,k} = -\epsilon D^{j+1} T_{2k}\psi = -\epsilon\psi_{j+1,2k};$$

expanding an arbitrary $f \in L^2(\mathbb{R})$ in the orthonormal basis $\{\psi_{j,k}\}_{j,k\in\mathbb{Z}}$, it follows that

$$(I - U)f = (I - U) \sum_{j,k\in\mathbb{Z}} \langle f, \psi_{j,k}\rangle\psi_{j,k} = -\epsilon \sum_{j,k\in\mathbb{Z}} \langle f, \psi_{j,k}\rangle\psi_{j+1,2k}.$$

Thus, for $f, g \in L^2(\mathbb{R})$,

$$
\begin{aligned}
\langle f, (I - U)^* g\rangle &= \langle (I - U)f, g\rangle \\
&= -\epsilon \sum_{j,k\in\mathbb{Z}} \langle f, \psi_{j,k}\rangle\langle\psi_{j+1,2k}, g\rangle \\
&= \langle f, -\epsilon \sum_{j,k\in\mathbb{Z}} \langle g, \psi_{j+1,2k}\rangle\psi_{j,k}\rangle.
\end{aligned}
$$

It follows that

$$(I - U^*)g = (I - U)^* g = -\epsilon \sum_{j,k\in\mathbb{Z}} \langle g, \psi_{j+1,2k}\rangle\psi_{j,k}. \tag{11.6}$$

In particular, $||I - U^*|| = \epsilon < 1$, which implies that $(U^*)^{-1}$ can be expanded in a Neumann series, see Theorem 2.2.3:

$$(U^*)^{-1} = \sum_{n=0}^{\infty} (I - U^*)^n .$$

Now (11.5) implies that the dual Riesz basis of $\{\theta_{j,k}\}_{j,k\in\mathbb{Z}}$ is

$$\{S^{-1}\theta_{j,k}\}_{j,k\in\mathbb{Z}} = \left\{ \sum_{n=0}^{\infty} (I - U^*)^n \psi_{j,k} \right\}_{j,k\in\mathbb{Z}}. \tag{11.7}$$

We can go one step further. In fact, the action of $I - U^*$ on the functions $\psi_{j,k}, j, k \in \mathbb{Z}$ can be found via (11.6) using that $\{\psi_{j,k}\}_{j,k\in\mathbb{Z}}$ is an orthonormal basis. The outcome depends on k being even or odd:

$$(I - U^*)\psi_{j,2k} = -\epsilon\psi_{j-1,k}, \quad \text{while} \quad (I - U^*)\psi_{j,2k+1} = 0, \quad \forall j, k \in \mathbb{Z}. \tag{11.8}$$

In particular, via (11.7),

$$S^{-1}\theta_{j,2k+1} = \psi_{j,2k+1} \text{ for all } j, k \in \mathbb{Z}.$$

Also, for any $k \neq 0$, the equations in (11.8) show that there exists a value of $n \in \mathbb{N}$ for which

$$(I - U^*)^n \psi_{j,2k} = 0.$$

Thus, $S^{-1}\theta_{j,2k}$ is a finite linear combination of functions $\{\psi_{j,k}\}_{j,k\in\mathbb{Z}}$,

$$\begin{aligned} S^{-1}\theta_{j,2k} &= \psi_{j,2k} + (I - U^*)\psi_{j,2k} + \cdots + (I - U^*)^n\psi_{j,2k} \\ &= \psi_{j,2k} - \epsilon\psi_{j-1,k} + \cdots + 0, \quad j \in \mathbb{Z}, k \neq 0. \end{aligned}$$

For $k = 0$, (11.7) and (11.8) imply that

$$S^{-1}\theta_{j,0} = \sum_{n=0}^{\infty} (I - U^*)^n \psi_{j,0} = \sum_{n=0}^{\infty}(-\epsilon)^n\psi_{j-n,0}, \quad j \in \mathbb{Z}. \tag{11.9}$$

In particular, the canonical dual frame of $\{\theta_{j,k}\}_{j,k\in\mathbb{Z}}$ does *not* have the wavelet structure; the functions $\{S^{-1}\theta_{j,k}\}_{j,k\in\mathbb{Z}}$ do not even have the same norm. This is in contrast with the situation for Gabor frames and frames of translates, where we saw that the canonical dual frame has the same structure as the frame itself.

The above calculations show that there are other properties that are not inherited by the canonical dual frame. For example, if we assume that the function ψ has compact support, then θ also has compact support, and all the functions $\{\theta_{j,k}\}_{j,k\in\mathbb{Z}}$ have compact support. If we look at the canonical dual frame $\{S^{-1}\theta_{j,k}\}_{j,k\in\mathbb{Z}}$, then we obtain functions with compact support when $k \neq 0$ because the functions $S^{-1}\theta_{j,k}$ are finite linear combinations of the functions in $\{\psi_{j,k}\}_{j,k\in\mathbb{Z}}$ in this case. However, for $k = 0$, the expression (11.9) shows that the functions $S^{-1}\theta_{j,0}$ do not have compact support. $\square$

The calculations in Example 11.1.2 are quite tedious, so it is clear that calculation of the canonical dual frame for a general dyadic wavelet frame will be very complicated. From the general frame theory discussed in Chapter 5, we know two ways of avoiding inconvenient frame decompositions: we can restrict our attention to tight frames or we can look at overcomplete frames and search for dual frames that are easier to calculate than the canonical dual frame. In the concrete setting of wavelet frames, some other aspects arise. In fact, the popular wavelet bases considered in Section 3.6 were based on multiresolution analysis, which leads to a very convenient algorithmic structure. As we saw in Section 3.6, this implies a special form for the function ψ generating the wavelet basis: it has the form

$$\psi = \sum_{k \in \mathbb{Z}} c_k DT_k \phi \tag{11.10}$$

for a certain function ϕ satisfying a scaling equation, i.e., an equation of the form

$$\hat{\phi}(2\gamma) = H_0(\gamma)\hat{\phi}(\gamma)$$

for some 1-periodic function H_0. The algorithmic structure offered by a multiresolution analysis is a great advantage compared with the use of general wavelet orthonormal bases. Thus, while constructing wavelet frames, it is very natural to ask the constructions to maintain the important aspects of the multiresolution analysis. Therefore, it is also natural to require the generator ψ for a tight frame $\{D^j T_k \psi\}_{j,k \in \mathbb{Z}}$ to have the form (11.10) for some function ϕ; and, if we want to construct two wavelet systems $\{D^j T_k \psi\}_{j,k \in \mathbb{Z}}$ and $\{D^j T_k \tilde{\psi}\}_{j,k \in \mathbb{Z}}$ such that they form a pair of dual frames, it is natural to require both the functions ψ and $\tilde{\psi}$ to have the form (11.10). In order to facilitate processing, we even want the coefficients in these formulas to be finite sequences. The B-splines B_m defined in (6.11) are obvious candidates for the function ϕ. However, as shown in [19] and [28], we cannot obtain all of these properties simultaneously:

Theorem 11.1.3 *Let B_m denote the m-th order B-spline for some $m > 1$. Then there does not exist pairs of dual wavelet frames $\{D^j T_k \psi\}_{j,k \in \mathbb{Z}}$ and $\{D^j T_k \tilde{\psi}\}_{j,k \in \mathbb{Z}}$ for which ψ and $\tilde{\psi}$ are finite linear combinations of functions $DT_k B_m$, $j, k \in \mathbb{Z}$.*

Thus, neither the approach of looking at tight frames, nor the idea of considering wavelet frame pairs, work if we want the generator (respectively, generators) to have the form (11.10) with ϕ being a B-spline.

It turns out that there is a solution to this problem: we will gain extra freedom by considering systems of the wavelet-type, but generated by more than one function.

Definition 11.1.4 *Consider two sequences of functions*

$$\psi_1, \ldots, \psi_n \in L^2(\mathbb{R}) \text{ and } \tilde{\psi}_1, \ldots, \tilde{\psi}_n \in L^2(\mathbb{R}).$$

We say that $\{D^j T_k \psi_\ell\}_{j,k\in\mathbb{Z}, \ell=1,\ldots,n}$ *and* $\{D^j T_k \tilde{\psi}_\ell\}_{j,k\in\mathbb{Z}, \ell=1,\ldots,n}$ *are a pair of dual multiwavelet frames if both are Bessel sequences and*

$$f = \sum_{\ell=1}^{n} \sum_{j,k\in\mathbb{Z}} \langle f, D^j T_k \psi_\ell \rangle D^j T_k \tilde{\psi}_\ell, \ \forall f \in L^2(\mathbb{R}). \tag{11.11}$$

That Bessel sequences $\{D^j T_k \psi_\ell\}_{j,k\in\mathbb{Z}, \ell=1,\ldots,n}$ and $\{D^j T_k \tilde{\psi}_\ell\}_{j,k\in\mathbb{Z}, \ell=1,\ldots,n}$ are frames if they satisfy (11.11) follows from Lemma 5.7.1. A pair of dual multiwavelet frames is called *sibling frames* in [19] and *bi-frames* in [28]. The frame $\{D^j T_k \psi_\ell\}_{j,k\in\mathbb{Z}, \ell=1,\ldots,n}$ itself is called a *multiwavelet frame*.

We note that a characterization of all dual multiwavelet frame pairs was obtained by Frazier et al. [34]. We will not need the result in our constructions in Section 11.2, so we state it without proof.

Theorem 11.1.5 *Let* $\psi_1, \ldots, \psi_n, \tilde{\psi}_1, \ldots, \tilde{\psi}_n \in L^2(\mathbb{R})$ *and assume that* $\{D^j T_k \psi_\ell\}_{j,k\in\mathbb{Z}, \ell=1,\ldots,n}$ *and* $\{D^j T_k \tilde{\psi}_\ell\}_{j,k\in\mathbb{Z}, \ell=1,\ldots,n}$ *are Bessel sequences. Then* $\{D^j T_k \psi_\ell\}_{j,k\in\mathbb{Z}, \ell=1,\ldots,n}$ *and* $\{D^j T_k \tilde{\psi}_\ell\}_{j,k\in\mathbb{Z}, \ell=1,\ldots,n}$ *are a pair of dual multiwavelet frames if and only if the two equations*

$$\begin{cases} \displaystyle\sum_{\ell=1}^{n} \sum_{j\in\mathbb{Z}} \widehat{\psi_\ell}(2^j \gamma) \overline{\widehat{\tilde{\psi}_\ell}(2^j \gamma)} = 1, \\ \displaystyle\sum_{\ell=1}^{n} \sum_{j=0}^{\infty} \widehat{\psi_\ell}(2^j \gamma) \overline{\widehat{\tilde{\psi}_\ell}(2^j(\gamma+q))} = 0 \text{ for all odd integers } q \end{cases}$$

hold for a.e. $\gamma \in \mathbb{R}$.

The functions $\psi_1, \ldots, \psi_n$ for which $\{D^j T_k \psi_\ell\}_{j,k\in\mathbb{Z}, \ell=1,\ldots,n}$ forms a tight multiwavelet frame can be characterized using Theorem 11.1.5. We formulate the result in the case of one generator ψ, and leave the proof to the reader (Exercise 11.2).

Theorem 11.1.6 *A function* $\psi \in L^2(\mathbb{R})$ *generates a tight wavelet frame* $\{\psi_{j,k}\}_{j,k\in\mathbb{Z}}$ *with frame bound A if and only if the equations*

$$\begin{cases} \displaystyle\sum_{j\in\mathbb{Z}} |\hat{\psi}(2^j \gamma)|^2 = A, \\ \displaystyle\sum_{j=0}^{\infty} \hat{\psi}(2^j \gamma) \overline{\hat{\psi}(2^j(\gamma+q))} = 0 \ \text{ for all odd integers } q \end{cases} \tag{11.12}$$

hold for a.e. $\gamma \in \mathbb{R}$.

Note that the only difference between the conditions for $\{\psi_{j,k}\}_{j,k\in\mathbb{Z}}$ being a tight frame with frame bound equal to 1, and the characterization of wavelets on page 77, is that the condition $||\psi|| = 1$ does not appear in Theorem 11.1.6.

11.2   The unitary extension principle

The purpose of this section is to prove the unitary extension principle of Ron and Shen [55], which enables us to construct tight frames for $L^2(\mathbb{R})$ of the form $\{D^jT_k\psi_\ell\}_{j,k\in\mathbb{Z},\ell=1,\dots,n}$. We state the main result in Theorem 11.2.7, but we need some preparation first. We follow the approach by Benedetto and Treiber [3].

The following proofs are based on standard Fourier analysis for 1-periodic functions. It will be convenient to write the integrals appearing, e.g., in the expression for the Fourier coefficients and in Parseval's equation, as integrals over the interval $] - \frac{1}{2}, \frac{1}{2}[$ rather than $]0, 1[$. The interval $] - \frac{1}{2}, \frac{1}{2}[$ is identified with the torus $\mathbb{T}$, and the class of 1-periodic functions on $\mathbb{R}$ whose restriction to $] - \frac{1}{2}, \frac{1}{2}[$ belongs to $L^p(-\frac{1}{2}, \frac{1}{2})$, $p = 1, 2$, is denoted by $L^p(\mathbb{T})$. Similarly, $L^\infty(\mathbb{T})$ consists of the bounded measurable 1-periodic functions on $\mathbb{R}$. With this notation, $L^\infty(\mathbb{T}) \subset L^2(\mathbb{T})$. We note that the spaces $L^p(\mathbb{T})$ actually consist of equivalence classes of functions that are identical almost everywhere, so when we speak about pointwise relationships between functions, it is understood that they can only be expected to hold almost everywhere.

The functions $\psi_1, \dots, \psi_n$ will be constructed on the basis of a function satisfying a refinement equation. Because we will work with all these functions simultaneously, it is convenient to change the notation used in Section 3.6 slightly and denote the refinable function by ψ_0 instead of ϕ.

We now list the standing assumptions and conventions for this section.

General setup: Let $\psi_0 \in L^2(\mathbb{R})$ and assume that

(i) There exists a function $H_0 \in L^\infty(\mathbb{T})$ such that

$$\widehat{\psi_0}(2\gamma) = H_0(\gamma)\widehat{\psi_0}(\gamma). \qquad (11.13)$$

(ii) $\lim_{\gamma\to 0} \widehat{\psi_0}(\gamma) = 1$.

Further, let $H_1, \dots, H_n \in L^\infty(\mathbb{T})$, and define $\psi_1, \dots, \psi_n \in L^2(\mathbb{R})$ by

$$\widehat{\psi_\ell}(2\gamma) = H_\ell(\gamma)\widehat{\psi_0}(\gamma), \quad \ell = 1, \dots, n. \qquad (11.14)$$

Finally, let H denote the $(n+1) \times 2$ matrix-valued function defined by

$$H(\gamma) = \begin{pmatrix} H_0(\gamma) & T_{1/2}H_0(\gamma) \\ H_1(\gamma) & T_{1/2}H_1(\gamma) \\ \cdot & \cdot \\ \cdot & \cdot \\ H_n(\gamma) & T_{1/2}H_n(\gamma) \end{pmatrix}, \quad \gamma \in \mathbb{R}. \tag{11.15}$$

With this setup, our purpose is to find conditions on the functions $H_1, \ldots, H_n$ such that $\psi_1, \ldots, \psi_n$ defined by (11.14) generate a multiwavelet frame for $L^2(\mathbb{R})$. It turns out to be convenient to formulate the results in terms of the matrices $H(\gamma)$, $\gamma \in \mathbb{R}$. Note that if we know the functions H_ℓ, then we can find an explicit expression for the functions ψ_ℓ: in fact, expanding H_ℓ in a Fourier series, $H_\ell(\gamma) = \sum_{k \in \mathbb{Z}} c_{k,\ell} e^{2\pi i k \gamma}$, Lemma 3.6.3 shows that

$$\psi_\ell(x) = \sqrt{2} \sum_{k \in \mathbb{Z}} c_{k,\ell} DT_{-k} \psi_0(x) = 2 \sum_{k \in \mathbb{Z}} c_{k,\ell} \psi_0(2x+k). \tag{11.16}$$

Recall that we prefer the functions H_ℓ to be trigonometric polynomials: this implies that the sums in (11.16) are finite and therefore that the functions ψ_ℓ have compact support if ψ_0 has compact support.

We note that the general setup presented here preserves the algorithmic structure of a multiresolution analysis: by Theorem 3.6.6, the spaces

$$V_j := \overline{\text{span}}\{D^j T_k \psi_0\}_{k \in \mathbb{Z}}, \ j \in \mathbb{Z},$$

satisfy the conditions for a multiresolution analysis in Definition 3.6.2, except (v). Also, by (11.16) we have that $\psi_1, \ldots, \psi_n \in V_1$.

One of the main tools will be to consider the *periodization* of a function $f : \mathbb{R} \to \mathbb{C}$, which formally is defined by

$$\mathcal{P}f(\gamma) = \sum_{n \in \mathbb{Z}} f(\gamma + n), \ \gamma \in \mathbb{R}.$$

We first show that the periodization is well defined if $f \in L^1(\mathbb{R})$.

Lemma 11.2.1 *If $f \in L^1(\mathbb{R})$, then $\sum_{n \in \mathbb{Z}} f(\gamma + n)$ converges absolutely for a.e. $\gamma \in \mathbb{R}$, and $\mathcal{P}f \in L^1(\mathbb{T})$. Furthermore,*

$$\int_{-\infty}^{\infty} f(\gamma) \, d\gamma = \int_{-\frac{1}{2}}^{\frac{1}{2}} \mathcal{P}f(\gamma) \, d\gamma. \tag{11.17}$$

Proof. If $f \in L^1(\mathbb{R})$, then

$$\int_{-\frac{1}{2}}^{\frac{1}{2}} \sum_{n \in \mathbb{Z}} |f(\gamma + n)| \, d\gamma = \sum_{n \in \mathbb{Z}} \int_{-\frac{1}{2}}^{\frac{1}{2}} |f(\gamma + n)| \, d\gamma$$

$$= \int_{-\infty}^{\infty} |f(\gamma)| \, d\gamma < \infty. \tag{11.18}$$

Thus, $\sum_{n \in \mathbb{Z}} f(\gamma + n)$ is absolutely convergent for almost all $\gamma \in \mathbb{R}$. This proves that $\mathcal{P}f$ is a well-defined 1-periodic function. Since

$$|\mathcal{P}f(\gamma)| \le \sum_{n \in \mathbb{Z}} |f(\gamma + n)|, \text{ a.e. } \gamma \in \mathbb{R},$$

it follows from (11.18) that $\mathcal{P}f \in L^1(\mathbb{T})$. Repeating the above argument, an application of Lebesgue's dominated convergence theorem finally shows that (11.17) holds.   □

Remember that we use the notation E_k, $k \in \mathbb{Z}$, to denote the function

$$E_k(\gamma) = e^{2\pi i k \gamma}, \ \gamma \in \mathbb{R}.$$

Lemma 11.2.2 *Let $g, \psi_0 \in L^2(\mathbb{R})$ and assume that $\mathcal{P}(g\overline{\widehat{\psi_0}}) \in L^2(\mathbb{T})$. Then*

$$\mathcal{P}(g\overline{\widehat{\psi_0}}) = \sum_{k \in \mathbb{Z}} \langle g, \widehat{\psi_0} E_k \rangle E_k \qquad (11.19)$$

and

$$\int_{-\frac{1}{2}}^{\frac{1}{2}} \left| \mathcal{P}(g\overline{\widehat{\psi_0}})(\gamma) \right|^2 d\gamma = \sum_{k \in \mathbb{Z}} |\langle g, \widehat{\psi_0} E_k \rangle|^2. \qquad (11.20)$$

Proof. Since $g, \psi_0 \in L^2(\mathbb{R})$, we know that $g\overline{\widehat{\psi_0}} \in L^1(\mathbb{R})$; that is, by Lemma 11.2.1 the function

$$\mathcal{P}(g\overline{\widehat{\psi_0}})(\gamma) = \sum_{n \in \mathbb{Z}} g(\gamma + n)\overline{\widehat{\psi_0}(\gamma + n)}$$

is well defined. Now, using (11.17),

$$
\begin{aligned}
\langle g, \widehat{\psi_0} E_k \rangle &= \int_{-\infty}^{\infty} g(\gamma)\overline{\widehat{\psi_0}(\gamma)}e^{-2\pi i k \gamma} d\gamma \\
&= \int_{-\frac{1}{2}}^{\frac{1}{2}} \sum_{n \in \mathbb{Z}} \left(g(\gamma + n)\overline{\widehat{\psi_0}(\gamma + n)}e^{-2\pi i k(\gamma + n)} \right) d\gamma \\
&= \int_{-\frac{1}{2}}^{\frac{1}{2}} \left(\sum_{n \in \mathbb{Z}} g(\gamma + n)\overline{\widehat{\psi_0}(\gamma + n)} \right) e^{-2\pi i k \gamma} d\gamma,
\end{aligned}
$$

which is the k-th Fourier coefficient for the 1-periodic function $\mathcal{P}(g\overline{\widehat{\psi_0}})$. Because this function belongs to $L^2(\mathbb{T})$ by assumption, the lemma follows: (11.19) is just the expansion of $\mathcal{P}(g\overline{\widehat{\psi_0}})$ in a Fourier series, and (11.20) is Parseval's equation.   □

The first main result, proved in Theorem 11.2.7, will show that a condition on the matrices $H(\gamma)$ in (11.15) implies that the multiwavelet system $\{D^j T_k \psi_\ell\}_{j,k \in \mathbb{Z}, \ell = 1, \ldots, n}$ is a tight frame for $L^2(\mathbb{R})$. In the proof of this, it is enough to show that the frame condition is satisfied on a dense subset of $L^2(\mathbb{R})$, see Lemma 5.1.2. In the following lemmas, we will work with the dense subspace consisting of functions f for which the Fourier transform $\hat{f}$ is continuous and has compact support:

$$\mathcal{D} := \{f \in L^2(\mathbb{R})|\ \hat{f} \in C_c(\mathbb{R})\}. \tag{11.21}$$

Lemma 11.2.3 *Let $\psi_0 \in L^2(\mathbb{R})$ and assume that $\lim_{\gamma \to 0} \widehat{\psi_0}(\gamma) = 1$. Let $f \in \mathcal{D}$. Then, for any $\epsilon > 0$ there exists $J \in \mathbb{Z}$ such that*

$$(1-\epsilon)||f||^2 \leq \sum_{k \in \mathbb{Z}} |\langle f, D^j T_k \psi_0 \rangle|^2 \leq (1+\epsilon)||f||^2 \text{ for all } j \geq J.$$

Proof.   Let $j \in \mathbb{Z}$ and $f \in \mathcal{D}$. As a product of $L^2(\mathbb{R})$-functions, the function $(D^j \hat{f})\overline{\widehat{\psi_0}}$ belongs to $L^1(\mathbb{R})$; thus $\mathcal{P}((D^j \hat{f})\overline{\widehat{\psi_0}})$ is well defined by Lemma 11.16. We will now prove that actually $\mathcal{P}((D^j \hat{f})\overline{\widehat{\psi_0}}) \in L^2(\mathbb{T})$. In order to do so, we use that $D^j \hat{f}$ has compact support, say, in the interval $[-N, N]$. This implies that for $\gamma \in \mathbb{T}$,

$$\left| \mathcal{P}((D^j \hat{f})\overline{\widehat{\psi_0}}) \right| = \left| \sum_{n \in \mathbb{Z}} (D^j \hat{f})(\gamma + n)\overline{\widehat{\psi_0}(\gamma + n)} \right|$$

$$= \left| \sum_{n=-N}^{N} (D^j \hat{f})(\gamma + n)\overline{\widehat{\psi_0}(\gamma + n)} \right|$$

$$\leq ||D^j \hat{f}||_\infty \sum_{n=-N}^{N} |\overline{\widehat{\psi_0}(\gamma + n)}|;$$

as a finite linear combination of translates of a function in $L^2(\mathbb{R})$, the function in the last expression clearly belongs to $L^2(\mathbb{T})$, which implies that $\mathcal{P}((D^j \hat{f})\overline{\widehat{\psi_0}}) \in L^2(\mathbb{T})$. Via the Fourier transform and the commutator relations (2.23),

$$\langle f, D^j T_k \psi_0 \rangle = \langle \mathcal{F}f, \mathcal{F}D^j T_k \psi_0 \rangle = \langle D^j \hat{f}, E_{-k}\widehat{\psi_0} \rangle; \tag{11.22}$$

therefore Lemma 11.2.2 shows that

$$\sum_{k \in \mathbb{Z}} |\langle f, D^j T_k \psi_0 \rangle|^2 = \sum_{k \in \mathbb{Z}} |\langle D^j \hat{f}, E_{-k}\widehat{\psi_0} \rangle|^2$$

$$= \int_{-\frac{1}{2}}^{\frac{1}{2}} \left| \mathcal{P}((D^j \hat{f})\overline{\widehat{\psi_0}})(\gamma) \right|^2 d\gamma$$

$$= \int_{-\frac{1}{2}}^{\frac{1}{2}} \left| \sum_{n \in \mathbb{Z}} (D^j \hat{f})(\gamma + n)\overline{\widehat{\psi_0}(\gamma + n)} \right|^2 d\gamma.$$

Now let $\epsilon > 0$ be given. By the assumption that $\lim_{\gamma \to 0} \widehat{\psi_0}(\gamma) = 1$, we can choose $b \in]0, 1/2[$ such that $1 - \epsilon \le |\widehat{\psi_0}(\gamma)|^2 \le 1 + \epsilon$ whenever $|\gamma| \le b$. By taking $J \in \mathbb{Z}$ such that $D^j \hat{f}$ has support in $[-b, b]$ for $j > J$, we obtain that for all $j > J$,

$$\int_{-\frac{1}{2}}^{\frac{1}{2}} \left| \sum_{n \in \mathbb{Z}} (D^j \hat{f})(\gamma + n) \overline{\widehat{\psi_0}(\gamma + n)} \right|^2 d\gamma = \int_{-b}^{b} |(D^j \hat{f})(\gamma) \widehat{\psi_0}(\gamma)|^2 d\gamma,$$

and therefore

$$(1 - \epsilon) \|D^j \hat{f}\|^2 \le \sum_{k \in \mathbb{Z}} |\langle f, D^j T_k \psi_0 \rangle|^2 \le (1 + \epsilon) \|D^j \hat{f}\|^2.$$

Since D^j and the Fourier transform are unitary operators, the lemma follows. $\qquad\square$

In the rest of this section, we assume that $\{\psi_\ell, H_\ell\}_{\ell=0}^{n}$ is as in the general setup on page 260. For a function $f \in \mathcal{D}$, an argument as in the proof of Lemma 11.2.3 shows that (Exercise 11.3)

$$\{\langle f, D^j T_k \psi_\ell \rangle\}_{k \in \mathbb{Z}} \in \ell^2(\mathbb{Z}) \text{ for all } j \in \mathbb{Z} \text{ and all } \ell = 1, \ldots, n. \quad (11.23)$$

We can therefore define a family of functions $F_{j,\ell} \in L^2(\mathbb{T})$ by the Fourier series

$$F_{j,\ell} := \sum_{k \in \mathbb{Z}} \langle f, D^j T_k \psi_\ell \rangle E_{-k}, \ j \in \mathbb{Z}, \ \ell = 0, 1, \ldots, n. \quad (11.24)$$

Because $F_{j,\ell}$ is defined in terms of ψ_ℓ, which is defined via ψ_0 and H_ℓ, it is natural to search for an expression for $F_{j,\ell}$ in terms of $F_{j,0}$ and H_ℓ. For convenience, we work with $F_{j-1,\ell}$:

Lemma 11.2.4 Let $\{\psi_\ell, H_\ell\}_{\ell=0}^{n}$ be as in the general setup on page 260. Then, for all $j \in \mathbb{Z}$, $\ell = 0, 1, \ldots, n$,

$$F_{j-1,\ell}(\gamma) = 2^{-1/2} (\overline{H_\ell(\gamma/2)} F_{j,0}(\gamma/2) + \overline{T_{1/2} H_\ell(\gamma/2)} \ T_{1/2} F_{j,0}(\gamma/2))$$

for a.e. $\gamma \in \mathbb{R}$.

Proof. First, we use the properties of $D^j, \mathcal{F}$ and their commutator relations in Section 2.9 to see that

$$\begin{aligned}
\langle f, D^{j-1} T_k \psi_\ell \rangle &= \langle D^{-j} f, D^{-1} T_k \psi_\ell \rangle \\
&= \langle D^{-j} f, T_{2k} D^{-1} \psi_\ell \rangle \\
&= \langle \mathcal{F} D^{-j} f, \mathcal{F} T_{2k} D^{-1} \psi_\ell \rangle \\
&= \langle D^j \hat{f}, E_{-2k} D\widehat{\psi_\ell} \rangle.
\end{aligned}$$

By (11.14), we can continue with

$$\langle f, D^{j-1}T_k\psi_\ell\rangle = \langle D^j\hat{f}, E_{-2k}2^{1/2}H_\ell\widehat{\psi_0}\rangle$$

$$= 2^{1/2}\int_{-\infty}^{\infty}(D^j\hat{f})(\gamma)\overline{H_\ell(\gamma)\widehat{\psi_0}(\gamma)}E_{2k}(\gamma)\,d\gamma. \quad (11.25)$$

By Lemma 11.2.1, and using the periodicity of E_{2k},

$$\int_{-\infty}^{\infty}(D^j\hat{f})(\gamma)\overline{H_\ell(\gamma)\widehat{\psi_0}(\gamma)}E_{2k}(\gamma)\,d\gamma = \int_{-\frac{1}{2}}^{\frac{1}{2}}\mathcal{P}((D^j\hat{f})\overline{H_\ell\widehat{\psi_0}}E_{2k})(\gamma)\,d\gamma$$

$$= \int_{-\frac{1}{2}}^{\frac{1}{2}}\mathcal{P}((D^j\hat{f})\overline{H_\ell\widehat{\psi_0}})(\gamma)E_{2k}(\gamma)\,d\gamma;$$

thus, (11.25) implies that

$$\langle f, D^{j-1}T_k\psi_\ell\rangle$$

$$= 2^{1/2}\int_{-\frac{1}{2}}^{\frac{1}{2}}\mathcal{P}((D^j\hat{f})\overline{H_\ell\widehat{\psi_0}})(\gamma)E_{2k}(\gamma)\,d\gamma$$

$$= 2^{1/2}\int_{0}^{\frac{1}{2}}\mathcal{P}((D^j\hat{f})\overline{H_\ell\widehat{\psi_0}})(\gamma)E_{2k}(\gamma)\,d\gamma$$

$$+2^{1/2}\int_{0}^{\frac{1}{2}}T_{1/2}\mathcal{P}((D^j\hat{f})\overline{H_\ell\widehat{\psi_0}})(\gamma)\,T_{1/2}E_{2k}(\gamma)\,d\gamma$$

$$= 2^{1/2}\int_{0}^{\frac{1}{2}}\left(\mathcal{P}((D^j\hat{f})\overline{H_\ell\widehat{\psi_0}})(\gamma) + T_{1/2}\mathcal{P}((D^j\hat{f})\overline{H_\ell\widehat{\psi_0}})(\gamma)\right)E_{2k}(\gamma)\,d\gamma.$$

This calculation shows that $\langle f, D^{j-1}T_k\psi_\ell\rangle$ is the $-k$-th coefficient in the Fourier expansion for the $\frac{1}{2}$-periodic function

$$\mathcal{P}((D^j\hat{f})\overline{H_\ell\widehat{\psi_0}}) + T_{1/2}\mathcal{P}((D^j\hat{f})\overline{H_\ell\widehat{\psi_0}})$$

with respect to the orthonormal basis $\{2^{1/2}E_{2k}\}_{k\in\mathbb{Z}}$ for $L^2(0, 1/2)$. Using the definition of $F_{j-1,\ell}$ and that

$$E_{-k}(\gamma) = 2^{-1/2}2^{1/2}E_{-2k}(\gamma/2),$$

it follows that for a.e. $\gamma\in\mathbb{R}$,

$$F_{j-1,\ell}(\gamma) = \sum_{k\in\mathbb{Z}}\langle f, D^{j-1}T_k\psi_\ell\rangle E_{-k}(\gamma)$$

$$= 2^{-1/2}\sum_{k\in\mathbb{Z}}\langle f, D^{j-1}T_k\psi_\ell\rangle 2^{1/2}E_{-2k}(\gamma/2)$$

$$= 2^{-1/2}\left(\mathcal{P}((D^j\hat{f})\overline{H_\ell\widehat{\psi_0}}) + T_{1/2}\mathcal{P}((D^j\hat{f})\overline{H_\ell\widehat{\psi_0}})\right)(\gamma/2). \quad (11.26)$$

The function H_ℓ is 1-periodic, so

$$\mathcal{P}((D^j\hat{f})\overline{H_\ell\widehat{\psi_0}})(\gamma) = \overline{H_\ell(\gamma)}\mathcal{P}((D^j\hat{f})\overline{\widehat{\psi_0}})(\gamma), \quad a.e.\ \gamma\in\mathbb{R}. \quad (11.27)$$

Also, by the calculation in (11.22) we have

$$\langle f, D^j T_k \psi_0 \rangle = \langle D^j \hat{f}, E_{-k} \widehat{\psi_0} \rangle;$$

via Lemma 11.2.2 (check the assumptions),

$$F_{j,0}(\gamma) = \sum_{k \in \mathbb{Z}} \langle f, D^j T_k \psi_0 \rangle E_{-k}(\gamma) \quad = \quad \sum_{k \in \mathbb{Z}} \langle D^j \hat{f}, E_{-k} \widehat{\psi_0} \rangle E_{-k}(\gamma)$$

$$= \quad \mathcal{P}((D^j \hat{f}) \overline{\widehat{\psi_0}})(\gamma). \qquad (11.28)$$

Inserting (11.27) and (11.28) in the expression (11.26) for $F_{j-1,\ell}$ finally gives the result. □

In terms of the matrix H defined in (11.15), the result in Lemma 11.2.4 shows that for a.e. $\gamma \in \mathbb{R}$,

$$\begin{pmatrix} F_{j-1,0}(\gamma) \\ F_{j-1,1}(\gamma) \\ \cdot \\ \cdot \\ \cdot \\ F_{j-1,n}(\gamma) \end{pmatrix}$$

$$= \quad 2^{-1/2} \begin{pmatrix} \overline{(H_0(\gamma/2)}F_{j,0}(\gamma/2) + \overline{T_{1/2}H_0(\gamma/2)} \, T_{1/2}F_{j,0}(\gamma/2) \\ \overline{(H_1(\gamma/2)}F_{j,0}(\gamma/2) + \overline{T_{1/2}H_1(\gamma/2)} \, T_{1/2}F_{j,0}(\gamma/2) \\ \cdot \\ \overline{(H_n(\gamma/2)}F_{j,0}(\gamma/2) + \overline{T_{1/2}H_n(\gamma/2)} \, T_{1/2}F_{j,0}(\gamma/2) \end{pmatrix}$$

$$= \quad 2^{-1/2} \overline{\begin{pmatrix} H_0(\gamma/2) & T_{1/2}H_0(\gamma/2) \\ H_1(\gamma/2) & T_{1/2}H_1(\gamma/2) \\ \cdot & \cdot \\ \cdot & \cdot \\ H_n(\gamma/2) & T_{1/2}H_n(\gamma/2) \end{pmatrix}} \begin{pmatrix} F_{j,0}(\gamma/2) \\ T_{1/2}F_{j,0}(\gamma/2) \end{pmatrix}$$

$$= \quad 2^{-1/2} \overline{H(\gamma/2)} \begin{pmatrix} F_{j,0}(\gamma/2) \\ T_{1/2}F_{j,0}(\gamma/2) \end{pmatrix}. \qquad (11.29)$$

The following lemmas will be based on the assumption that the matrix $H(\gamma)$ satisfies that

$$H(\gamma)^* H(\gamma) = I, \quad a.e. \ \gamma \in \mathbb{T}. \qquad (11.30)$$

Note that the matrix $H(\gamma)^* H(\gamma)$ in (11.30) is a 2×2 matrix. Later, it turns out that (11.30) is the essential assumption in the unitary extension principle: in fact, given the general setup, it is the only condition we need.

Lemma 11.2.5 *Let $\{\psi_\ell, H_\ell\}_{\ell=0}^n$ be as in the general setup on page 260, and assume that $H(\gamma)^* H(\gamma) = I$ for a.e. $\gamma \in \mathbb{T}$. Then, for all $j \in \mathbb{Z}$ and all $f \in \mathcal{D}$,*

$$\sum_{k \in \mathbb{Z}} |\langle f, D^j T_k \psi_0 \rangle|^2 = \sum_{\ell=0}^n \sum_{k \in \mathbb{Z}} |\langle f, D^{j-1} T_k \psi_\ell \rangle|^2.$$

Proof. The definition of $F_{j-1,\ell}$ and Parseval's equation show that

$$\sum_{\ell=0}^n \sum_{k \in \mathbb{Z}} |\langle f, D^{j-1} T_k \psi_\ell \rangle|^2 = \sum_{\ell=0}^n \int_{-\frac{1}{2}}^{\frac{1}{2}} |F_{j-1,\ell}(\gamma)|^2 \, d\gamma. \tag{11.31}$$

The assumption on the matrix $H(\gamma)$ means that we can consider $H(\gamma)$ as an isometry from $\mathbb{C}^2$ into $\mathbb{C}^{n+1}$ for a.e. $\gamma \in \mathbb{T}$. Using this together with (11.29), it follows from (11.31) that

$$
\begin{aligned}
\sum_{\ell=0}^n \sum_{k \in \mathbb{Z}} |\langle f, D^{j-1} T_k \psi_\ell \rangle|^2 &= 2^{-1} \int_{-\frac{1}{2}}^{\frac{1}{2}} \left\| \overline{H(\gamma/2)} \begin{pmatrix} F_{j,0}(\gamma/2) \\ T_{1/2} F_{j,0}(\gamma/2) \end{pmatrix} \right\|_{\mathbb{C}^{n+1}}^2 d\gamma \\
&= 2^{-1} \int_{-\frac{1}{2}}^{\frac{1}{2}} \left\| \begin{pmatrix} F_{j,0}(\gamma/2) \\ T_{1/2} F_{j,0}(\gamma/2) \end{pmatrix} \right\|_{\mathbb{C}^2}^2 d\gamma \\
&= 2^{-1} \int_{-\frac{1}{2}}^{\frac{1}{2}} \left(|F_{j,0}(\gamma/2)|^2 + |T_{1/2} F_{j,0}(\gamma/2)|^2 \right) d\gamma \\
&= \int_{-\frac{1}{4}}^{\frac{1}{4}} |F_{j,0}(\gamma)|^2 d\gamma + \int_{-\frac{3}{4}}^{-\frac{1}{4}} |F_{j,0}(\gamma)|^2 d\gamma.
\end{aligned}
$$

Using the 1-periodicity of the function $F_{j,0}$, we conclude that

$$
\begin{aligned}
\sum_{\ell=0}^n \sum_{k \in \mathbb{Z}} |\langle f, D^{j-1} T_k \psi_\ell \rangle|^2 &= \int_{-\frac{1}{2}}^{\frac{1}{2}} |F_{j,0}(\gamma)|^2 d\gamma \\
&= \sum_{k \in \mathbb{Z}} |\langle f, D^j T_k \psi_0 \rangle|^2.
\end{aligned}
$$

$\square$

Lemma 11.2.6 *Let $\{\psi_\ell, H_\ell\}_{\ell=0}^n$ be as in the general setup on page 260, and assume that $H(\gamma)^* H(\gamma) = I$ for a.e. $\gamma \in \mathbb{T}$. Then the following hold:*

(i) *$\{T_k \psi_0\}_{k \in \mathbb{Z}}$ is a Bessel sequence with bound 1.*

(ii) *If $f \in L^2(\mathbb{R})$, then*

$$\lim_{j \to -\infty} \sum_{k \in \mathbb{Z}} |\langle f, D^j T_k \psi_0 \rangle|^2 = 0.$$

Proof. Consider a function $f \in \mathcal{D}$. Lemma 11.2.5 shows that for any $j \in \mathbb{Z}$,

$$\sum_{k \in \mathbb{Z}} |\langle f, D^{j-1} T_k \psi_0 \rangle|^2 \leq \sum_{k \in \mathbb{Z}} |\langle f, D^j T_k \psi_0 \rangle|^2. \tag{11.32}$$

Let $\epsilon > 0$ be given. Via Lemma 11.2.3, we can find $j > 0$ such that

$$\sum_{k \in \mathbb{Z}} |\langle f, D^j T_k \psi_0 \rangle|^2 \leq (1 + \epsilon) \|f\|^2.$$

Applying (11.32) j times leads to

$$\sum_{k \in \mathbb{Z}} |\langle f, T_k \psi_0 \rangle|^2 \leq \sum_{k \in \mathbb{Z}} |\langle f, D^j T_k \psi_0 \rangle|^2 \leq (1 + \epsilon) \|f\|^2.$$

Because $\epsilon > 0$ was arbitrary, it follows that

$$\sum_{k \in \mathbb{Z}} |\langle f, T_k \psi_0 \rangle|^2 \leq \|f\|^2.$$

Because this inequality holds on a dense subset of $L^2(\mathbb{R})$, it holds on $L^2(\mathbb{R})$ by Lemma 3.1.6. Thus, $\{T_k \psi_0\}_{k \in \mathbb{Z}}$ is a Bessel sequence with bound 1.

For the proof of (ii), let $f \in L^2(\mathbb{R})$. By (i) and the fact that D^j is unitary, we know that $\{D^j T_k \psi_0\}_{k \in \mathbb{Z}}$ is a Bessel sequence with bound 1 for all $j \in \mathbb{Z}$ (same argument as in Exercise 5.15). Letting $I \subset \mathbb{R}$ be any bounded interval, we can write

$$f = f \chi_I + f(1 - \chi_I).$$

Using the inequality $|a + b|^2 \leq 2 (|a|^2 + |b|^2)$, $a, b \in \mathbb{C}$, we obtain that

$$
\begin{aligned}
|\langle f, D^j T_k \psi_0 \rangle|^2 &= |\langle f \chi_I + f(1 - \chi_I), D^j T_k \psi_0 \rangle|^2 \\
&= |\langle f \chi_I, D^j T_k \psi_0 \rangle + \langle f(1 - \chi_I), D^j T_k \psi_0 \rangle|^2 \\
&\leq 2 \left(|\langle f \chi_I, D^j T_k \psi_0 \rangle|^2 + |\langle f(1 - \chi_I), D^j T_k \psi_0 \rangle|^2 \right).
\end{aligned}
$$

This implies that

$$
\begin{aligned}
\sum_{k \in \mathbb{Z}} |\langle f, D^j T_k \psi_0 \rangle|^2 &\leq 2 \sum_{k \in \mathbb{Z}} |\langle f \chi_I, D^j T_k \psi_0 \rangle|^2 \\
&\quad + 2 \sum_{k \in \mathbb{Z}} |\langle f(1 - \chi_I), D^j T_k \psi_0 \rangle|^2 \\
&\leq 2 \sum_{k \in \mathbb{Z}} |\langle f \chi_I, D^j T_k \psi_0 \rangle|^2 + 2 \|f(1 - \chi_I)\|^2.
\end{aligned}
$$

By choosing I sufficiently large, we can make $\|f(1 - \chi_I)\|^2$ arbitrarily small. Thus, it is enough to show that

$$\sum_{k \in \mathbb{Z}} |\langle f \chi_I, D^j T_k \psi_0 \rangle|^2 \to 0 \text{ as } j \to -\infty.$$

Now,

$$
\begin{aligned}
\sum_{k\in\mathbb{Z}}|\langle f\chi_I, D^j T_k\psi_0\rangle|^2 &= 2^j\sum_{k\in\mathbb{Z}}\left|\int_I f(x)\overline{\psi_0(2^j x-k)}\,dx\right|^2\\
&\le \|f\|^2 2^j\sum_{k\in\mathbb{Z}}\int_I |\psi_0(2^j x-k)|^2\,dx\\
&= \|f\|^2\sum_{k\in\mathbb{Z}}\int_{2^j I-k}|\psi_0(x)|^2\,dx.
\end{aligned}
$$

An application of Lebesgue's dominated convergence theorem yields that the final expression goes to zero as $j\to-\infty$, which concludes the proof. $\square$

We are now ready to formulate and prove the *unitary extension principle*.

Theorem 11.2.7 *Let $\{\psi_\ell, H_\ell\}_{\ell=0}^n$ be as in the general setup on page 260, and assume that $H(\gamma)^* H(\gamma) = I$ for a.e. $\gamma\in\mathbb{T}$. Then the multiwavelet system $\{D^j T_k\psi_\ell\}_{j,k\in\mathbb{Z},\ell=1,\dots,n}$ constitutes a tight frame for $L^2(\mathbb{R})$ with frame bound equal to 1, and*

$$
f = \sum_{\ell=1}^n\sum_{j\in\mathbb{Z}}\sum_{k\in\mathbb{Z}}\langle f, D^j T_k\psi_\ell\rangle D^j T_k\psi_\ell, \quad \forall f\in L^2(\mathbb{R}). \tag{11.33}
$$

Proof. Let $\epsilon > 0$ be given, and consider a function $f\in\mathcal{D}$. By Lemma 11.2.3, we can choose $J > 0$ such that for all $j > J$,

$$
(1-\epsilon)\|f\|^2 \le \sum_{k\in\mathbb{Z}}|\langle f, D^j T_k\psi_0\rangle|^2 \le (1+\epsilon)\|f\|^2. \tag{11.34}
$$

For *any* $j\in\mathbb{Z}$, Lemma 11.2.5 shows that

$$
\begin{aligned}
\sum_{k\in\mathbb{Z}}|\langle f, D^j T_k\psi_0\rangle|^2 &= \sum_{\ell=0}^n\sum_{k\in\mathbb{Z}}|\langle f, D^{j-1}T_k\psi_\ell\rangle|^2\\
&= \sum_{k\in\mathbb{Z}}|\langle f, D^{j-1}T_k\psi_0\rangle|^2 + \sum_{\ell=1}^n\sum_{k\in\mathbb{Z}}|\langle f, D^{j-1}T_k\psi_\ell\rangle|^2;
\end{aligned}
$$

iterating the argument on $\sum_{k\in\mathbb{Z}}|\langle f, D^{j-1}T_k\psi_0\rangle|^2$, it follows that for all $m < j$,

$$
\sum_{k\in\mathbb{Z}}|\langle f, D^j T_k\psi_0\rangle|^2 = \sum_{k\in\mathbb{Z}}|\langle f, D^m T_k\psi_0\rangle|^2 + \sum_{\ell=1}^n\sum_{p=m}^{j-1}\sum_{k\in\mathbb{Z}}|\langle f, D^p T_k\psi_\ell\rangle|^2.
$$

Via (11.34), we deduce that for all $j > J$ and $m < j$,

$$
\begin{aligned}
(1 - \epsilon)\|f\|^2 &\leq \sum_{k \in \mathbb{Z}} |\langle f, D^m T_k \psi_0 \rangle|^2 + \sum_{\ell=1}^n \sum_{p=m}^{j-1} \sum_{k \in \mathbb{Z}} |\langle f, D^p T_k \psi_\ell \rangle|^2 \\
&\leq (1 + \epsilon)\|f\|^2.
\end{aligned}
\tag{11.35}
$$

By Lemma 11.2.6(ii),

$$
\lim_{m \to -\infty} \sum_{k \in \mathbb{Z}} |\langle f, D^m T_k \psi_0 \rangle|^2 = 0.
$$

Therefore, letting $m \to -\infty$ in (11.35) yields that for all $j > J$,

$$
(1 - \epsilon)\|f\|^2 \leq \sum_{\ell=1}^n \sum_{p=-\infty}^{j-1} \sum_{k \in \mathbb{Z}} |\langle f, D^p T_k \psi_\ell \rangle|^2 \leq (1 + \epsilon)\|f\|^2.
$$

Letting $j \to \infty$,

$$
(1 - \epsilon)\|f\|^2 \leq \sum_{\ell=1}^n \sum_{p=-\infty}^{\infty} \sum_{k \in \mathbb{Z}} |\langle f, D^p T_k \psi_\ell \rangle|^2 \leq (1 + \epsilon)\|f\|^2.
$$

Because $\epsilon > 0$ was arbitrary, we conclude that

$$
\sum_{\ell=1}^n \sum_{p \in \mathbb{Z}} \sum_{k \in \mathbb{Z}} |\langle f, D^p T_k \psi_\ell \rangle|^2 = \|f\|^2
$$

for all $f \in \mathcal{D}$; therefore the equality holds for all $f \in L^2(\mathbb{R})$ by Lemma 5.1.2. The expansion property (11.33) follows from Corollary 5.1.8. $\quad\square$

The matrix $H(\gamma)^* H(\gamma)$ has four entries, so at a first glance it seems that we have to solve four scalar equations in order to apply Theorem 11.2.7. However, it turns out that it is enough to verify two sets of equations (Exercise 11.4):

Corollary 11.2.8 *Let $\{\psi_\ell, H_\ell\}_{\ell=0}^n$ be as in the general setup on page 260, and assume that*

$$
\begin{cases}
\displaystyle\sum_{\ell=0}^n |H_\ell(\gamma)|^2 = 1, \\[2mm]
\displaystyle\sum_{\ell=0}^n \overline{H_\ell(\gamma)} T_{1/2} H_\ell(\gamma) = 0,
\end{cases}
\tag{11.36}
$$

for a.e. $\gamma \in \mathbb{T}$. Then the multiwavelet system $\{D^j T_k \psi_\ell\}_{j,k \in \mathbb{Z}, \ell=1,\dots,n}$ constitutes a tight frame for $L^2(\mathbb{R})$ with frame bound equal to 1.

As an application of Corollary 11.2.8, we show how one can construct compactly supported tight multiwavelet frames based on B-splines. In contrast with the Battle–Lemarié wavelets discussed on page 76, the generators will be *finite* linear combinations of splines $B_m(2x - k), k \in \mathbb{Z}$, and thus have compact support. As we have seen in Theorem 11.1.3, such a construction cannot be based on a single generator: the price to pay is that we need multiple generators.

Example 11.2.9 For any $m = 1, 2, \ldots$, we consider the B-spline

$$\psi_0 := B_{2m}$$

of order $2m$ as defined in (6.11). By Corollary 6.2.1,

$$\widehat{\psi_0}(\gamma) = \left(\frac{\sin(\pi\gamma)}{\pi\gamma}\right)^{2m}.$$

It is clear that $\lim_{\gamma \to 0} \widehat{\psi_0}(\gamma) = 1$. Furthermore, the result in Exercise 6.7 shows that

$$\widehat{\psi_0}(2\gamma) = \cos^{2m}(\pi\gamma)\widehat{\psi_0}(\gamma).$$

Thus ψ_0 satisfies a refinement equation with two-scale symbol

$$H_0(\gamma) = \cos^{2m}(\pi\gamma). \tag{11.37}$$

Note that Exercise 6.7 also explains why we are restricting to the case of even-order B-splines. Now, consider the binomial coefficient

$$\binom{2m}{\ell} := \frac{(2m)!}{(2m - \ell)!\ell!},$$

and define the functions $H_1, \ldots, H_{2m} \in L^\infty(\mathbb{T})$ by

$$H_\ell(\gamma) = \sqrt{\binom{2m}{\ell}} \sin^\ell(\pi\gamma) \cos^{2m-\ell}(\pi\gamma), \ \ell = 1, \ldots, 2m. \tag{11.38}$$

Using that $\cos(\pi(\gamma - 1/2)) = \sin(\pi\gamma)$ and $\sin(\pi(\gamma - 1/2)) = -\cos(\pi\gamma)$, it follows that

$$T_{1/2}H_\ell(\gamma) = \sqrt{\binom{2m}{\ell}}(-1)^\ell \cos^\ell(\pi\gamma) \sin^{2m-\ell}(\pi\gamma), \ \ell = 1, \ldots, 2m. \tag{11.39}$$

Thus, the matrix H in (11.15) is given by

$$H(\gamma) = \begin{pmatrix} H_0(\gamma) & T_{1/2}H_0(\gamma) \\ H_1(\gamma) & T_{1/2}H_1(\gamma) \\ \cdot & \cdot \\ \cdot & \cdot \\ H_{2m}(\gamma) & T_{1/2}H_{2m}(\gamma) \end{pmatrix} =$$

$$
\begin{pmatrix}
\cos^{2m}(\pi\gamma) & \sin^{2m}(\pi\gamma) \\
\sqrt{\begin{pmatrix} 2m \\ 1 \end{pmatrix}} \sin(\pi\gamma)\cos^{2m-1}(\pi\gamma) & -\sqrt{\begin{pmatrix} 2m \\ 1 \end{pmatrix}} \cos(\pi\gamma)\sin^{2m-1}(\pi\gamma) \\
\sqrt{\begin{pmatrix} 2m \\ 2 \end{pmatrix}} \sin^2(\pi\gamma)\cos^{2m-2}(\pi\gamma) & \sqrt{\begin{pmatrix} 2m \\ 2 \end{pmatrix}} \cos^2(\pi\gamma)\sin^{2m-2}(\pi\gamma) \\
\vdots & \vdots \\
\sqrt{\begin{pmatrix} 2m \\ 2m \end{pmatrix}} \sin^{2m}(\pi\gamma) & \sqrt{\begin{pmatrix} 2m \\ 2m \end{pmatrix}} \cos^{2m}(\pi\gamma)
\end{pmatrix}.
$$

We now verify the conditions in Corollary 11.2.8. Using the binomial formula

$$
(x+y)^{2m} = \sum_{\ell=0}^{2m} \begin{pmatrix} 2m \\ \ell \end{pmatrix} x^\ell y^{2m-\ell}, \tag{11.40}
$$

we see via (11.38) that

$$
\begin{aligned}
\sum_{\ell=0}^{2m} |H_\ell(\gamma)|^2 &= \sum_{\ell=0}^{2m} \begin{pmatrix} 2m \\ \ell \end{pmatrix} \sin^{2\ell}(\pi\gamma)\cos^{2(2m-\ell)}(\pi\gamma) \\
&= \left(\sin^2(\pi\gamma) + \cos^2(\pi\gamma)\right)^{2m} \\
&= 1, \ \gamma \in \mathbb{T}.
\end{aligned}
$$

Using the binomial formula with $x = -1, y = 1$, the expressions in (11.38) and (11.39) yield that

$$
\begin{aligned}
\sum_{\ell=0}^{2m} \overline{H_\ell(\gamma)} T_{1/2} H_\ell(\gamma) &= \sin^{2m}(\pi\gamma)\cos^{2m}(\pi\gamma) \sum_{\ell=0}^{2m} (-1)^\ell \begin{pmatrix} 2m \\ \ell \end{pmatrix} \\
&= \sin^{2m}(\pi\gamma)\cos^{2m}(\pi\gamma)(1-1)^{2m} \\
&= 0.
\end{aligned}
$$

Now Corollary 11.2.8 implies that the $2m$ functions $\psi_1, \ldots, \psi_{2m}$ defined by

$$
\begin{aligned}
\widehat{\psi_\ell}(\gamma) &= H_\ell(\gamma/2)\widehat{\psi_0}(\gamma/2) \\
&= \sqrt{\begin{pmatrix} 2m \\ \ell \end{pmatrix}} \frac{\sin^{2m+\ell}(\pi\gamma/2)\cos^{2m-\ell}(\pi\gamma/2)}{(\pi\gamma/2)^{2m}}
\end{aligned}
$$

generate a tight multiwavelet frame $\{D^j T_k \psi_\ell\}_{j,k\in\mathbb{Z},\ell=1,\ldots,2m}$ for $L^2(\mathbb{R})$. $\square$

We want to study the properties of the frame constructed in Example 11.2.9, but we first change the definition slightly by multiplying each of the functions H_ℓ in (11.38) with a complex number of absolute value 1. This modification will not change the frame properties for the generated wavelet system.

Example 11.2.10 We continue Example 11.2.9, but now we define

$$H_\ell(\gamma) \;=\; i^\ell \sqrt{\binom{2m}{\ell}} \sin^\ell(\pi\gamma) \cos^{2m-\ell}(\pi\gamma), \;\; \ell = 1, \ldots, 2m. \quad (11.41)$$

H_ℓ only differs from the choice in (11.38) by a constant of absolute value 1, so the functions $\psi_1, \ldots, \psi_{2m}$ given by

$$\widehat{\psi_\ell}(2\gamma) = H_\ell(\gamma)\widehat{\psi_0}(\gamma), \;\; \ell = 1, \ldots, 2m, \quad (11.42)$$

also generate a tight multiwavelet frame. Instead of inserting the expression for $\widehat{\psi_0}$ in (11.42), we now rewrite $H_\ell(\gamma)$ using Euler's formula:

$$
\begin{aligned}
H_\ell(\gamma) \;&=\; i^\ell \sqrt{\binom{2m}{\ell}} \left(\frac{e^{\pi i\gamma} - e^{-\pi i\gamma}}{2i}\right)^\ell \left(\frac{e^{\pi i\gamma} + e^{-\pi i\gamma}}{2}\right)^{2m-\ell} \\
&=\; 2^{-2m} \sqrt{\binom{2m}{\ell}} \left(e^{\pi i\gamma} - e^{-\pi i\gamma}\right)^\ell \left(e^{\pi i\gamma} + e^{-\pi i\gamma}\right)^{2m-\ell}.
\end{aligned}
$$
$$(11.43)$$

Via the binomial formula we see that $H_\ell(\gamma)$ is a finite linear combination of terms

$$e^{-2\pi i m\gamma}, e^{-2\pi i(m-1)\gamma}, \ldots, e^{2\pi i(m-1)\gamma}, e^{2\pi i m\gamma}.$$

All coefficients in the linear combination are real. Writing

$$H_\ell(\gamma) = \sum_{k=-m}^{m} c_{k,\ell} e^{2\pi i k\gamma},$$

Lemma 3.6.3 shows that

$$\psi_\ell = \sqrt{2} \sum_{k=-m}^{m} c_{k,\ell} DT_{-k}\psi_0. \quad (11.44)$$

That is, ψ_ℓ is a real-valued spline. Since $DT_m\psi_0$ has support in $[0, m]$ and $DT_{-m}\psi_0$ has support in $[-m, 0]$, the spline ψ_ℓ has support in $[-m, m]$. Our arguments also show that the splines ψ_ℓ inherit other properties from ψ_0: they have degree $2m - 1$, belong to $C^{2m-2}(\mathbb{R})$, and have knots at $\mathbb{Z}/2$. $\square$

Let us find an explicit expression for the generators in Example 11.2.10 in the case $m = 1$:

Example 11.2.11 In the case $m = 1$, the construction in Example 11.2.10 leads to two generators ψ_1 and ψ_2. Via the expression (11.43) for H_1,

$$
\begin{aligned}
H_1(\gamma) &= \frac{1}{4}\sqrt{\binom{2}{1}}(e^{\pi i\gamma} - e^{-\pi i\gamma})(e^{\pi i\gamma} + e^{-\pi i\gamma}) \\
&= \frac{1}{2\sqrt{2}}(e^{2\pi i\gamma} - e^{-2\pi i\gamma}).
\end{aligned}
$$

By Lemma 3.6.3, we conclude that

$$
\psi_1(x) = \frac{1}{\sqrt{2}}(B_2(2x + 1) - B_2(2x - 1)). \tag{11.45}
$$

See Figure 11.1. Similarly, one proves (Exercise 11.6) that

$$
\psi_2(x) = \frac{1}{2}\left(B_2(2x + 1) - 2B_2(2x) + B_2(2x - 1)\right), \tag{11.46}
$$

which is shown in Figure 11.2.   □

We note that the computational effort in Example 11.2.10 increases with the order of the B-spline B_{2m} we start with: the number of generators $\psi_1, \dots, \psi_{2m}$ increases with the order of the spline B_{2m}, and (11.44) shows that computation of ψ_ℓ involves calculation of a large number of coefficients for high-order B-splines. In practice, it is annoying that the number of generators is forced to increase if we want to obtain generators with higher smoothness. The results in Section 11.3 will allow us to construct multiwavelet frames with two generators based on any B-spline B_{2m}, i.e., with any prescribed regularity.

Example 11.2.12 In continuation of Example 11.2.10, we can also construct spline frames with support on $[0, 2m]$. We ask the reader to provide the details in Exercise 11.7. Letting $\psi_0 := N_{2m}$ be the B-spline of order $2m$ defined in (6.2), one can prove that

$$
\widehat{\psi_0}(2\gamma) = H_0(\gamma)\widehat{\psi_0}(\gamma) \tag{11.47}
$$

with

$$
H_0(\gamma) = \left(\frac{1 + e^{-2\pi i\gamma}}{2}\right)^{2m} = e^{-2\pi im\gamma}\cos^{2m}(\pi\gamma). \tag{11.48}
$$

Because H_0 appears from the corresponding function in (11.37) simply by multiplication with $e^{-2\pi im\gamma}$, the functions

$$
H_\ell(\gamma) = e^{-2\pi im\gamma}\sqrt{\binom{2m}{\ell}}\sin^\ell(\pi\gamma)\cos^{2m-\ell}(\pi\gamma), \ \ell = 1, \dots, 2m
$$

satisfy the conditions in the unitary extension principle. We prefer to multiply the functions with a complex number, i.e., to consider

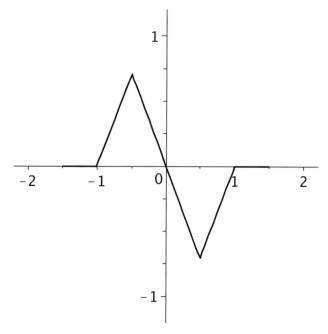

Figure 11.1. The function ψ_1 given by (11.45).

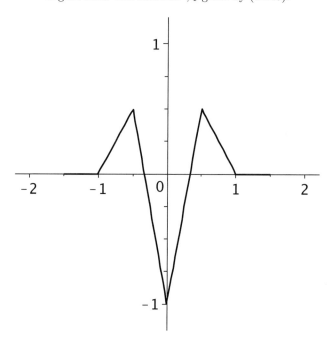

Figure 11.2. The function ψ_2 given by (11.46).

$$H_\ell(\gamma) = i^\ell e^{-2\pi i m \gamma} \sqrt{\binom{2m}{\ell}} \sin^\ell(\pi\gamma) \cos^{2m-\ell}(\pi\gamma), \ \ell = 1, \ldots, 2m;$$

with this choice, we conclude that the functions $\psi_1, \ldots, \psi_{2m}$ defined by

$$
\begin{aligned}
\widehat{\psi_\ell}(\gamma) &= H_\ell(\gamma/2)\widehat{\psi_0}(\gamma/2) \\
&= i^\ell e^{-2\pi i m \gamma} \sqrt{\binom{2m}{\ell}} \frac{\sin^{2m+\ell}(\pi\gamma/2)\cos^{2m-\ell}(\pi\gamma/2)}{(\pi\gamma/2)^{2m}}
\end{aligned}
$$

generate a tight multiwavelet frame for $L^2(\mathbb{R})$. The spline functions $\psi_1, \ldots, \psi_{2m}$ now have support on $[0, 2m]$. We return to the case $m = 1$ in Example 11.3.7. □

11.3 The oblique extension principle

In this section, we keep the assumptions in the general setup on page 260. Our purpose is to prove a more flexible version of the unitary extension principle; let us first give some reasons why we want to do so.

   In the context of approximation theory, it is desirable that a multiwavelet frame is generated by functions $\{\psi_\ell\}_{\ell=1}^n$ having a large number of vanishing moments; see, e.g., Theorem 4.3.1, which shows that a large number of vanishing moments for a smooth compactly supported function ψ implies that the expansion coefficients $\langle f, \psi_{j,k} \rangle$ decay fast for $j \to \infty$. If $\{\psi_\ell\}_{\ell=1}^n$ is constructed via the unitary extension principle, we know that $\widehat{\psi_\ell}(\gamma) = H_\ell(\gamma/2)\widehat{\psi_0}(\gamma/2)$ and that $\widehat{\psi_0}(0) = 1$; it follows from here that the number of vanishing moments for the function ψ_ℓ is equal to the order of zero for H_ℓ at $\gamma = 0$. This actually puts a restriction on the number of vanishing moments one can obtain for generators constructed via the unitary extension principle:

Example 11.3.1 We return to the B-spline B_{2m} of order $2m$ considered in Example 11.2.9; it satisfies a refinement equation with two-scale symbol

$$H_0(\gamma) = \cos^{2m}(\pi\gamma).$$

If we want to construct a frame via the unitary extension principle, the conditions in Corollary 11.2.8 in particular state that

$$1 = \sum_{\ell=0}^n |H_\ell(\gamma)|^2,$$

i.e., that

$$\sum_{\ell=1}^{n} |H_\ell(\gamma)|^2 = 1 - \cos^{4m}(\pi\gamma). \tag{11.49}$$

The order of the zero at $\gamma = 0$ for the function $1 - \cos^{4m}(\pi\gamma)$ is 2, so also on the left-hand side of (11.49) we can only factor γ^2 out; this implies that at least one of the functions $|H_\ell|^2$ at most can have a zero at $\gamma = 0$ of order 2, and therefore at least one of the functions ψ_ℓ can at most have one vanishing moment. $\qquad\square$

An important reformulation of Theorem 11.2.7 was simultaneously obtained by Daubechies, Han, Ron, and Shen in [28] and Chui, He, and Stöckler in [19]. It gives a more flexible recipe for construction of frames than Theorem 11.2.7, and can, e.g., be used to construct tight frames with a higher number of vanishing moments. The result is called the *oblique extension principle*:

Theorem 11.3.2 *Let $\{\psi_\ell, H_\ell\}_{\ell=0}^{n}$ be as in the general setup on page 260. Assume that there exists a strictly positive function $\theta \in L^\infty(\mathbb{T})$ for which*

$$\lim_{\gamma\to 0} \theta(\gamma) = 1$$

and such that for a.e. $\gamma \in \mathbb{T}$,

$$H_0(\gamma)\overline{H_0(\gamma+\nu)}\theta(2\gamma) \; + \; \sum_{\ell=1}^{n} H_\ell(\gamma)\overline{H_\ell(\gamma+\nu)}$$

$$= \begin{cases} \theta(\gamma) & \text{if } \nu = 0, \\ 0 & \text{if } \nu = \frac{1}{2}. \end{cases} \tag{11.50}$$

Then the functions $\{D^j T_k \psi_\ell\}_{j,k\in\mathbb{Z},\ell=1,\ldots,n}$ constitute a tight frame for $L^2(\mathbb{R})$ with frame bound equal to 1.

Proof. Assume that the conditions in Theorem 11.3.2 are satisfied, and define the function $\widetilde{\psi}_0 \in L^2(\mathbb{R})$ by

$$\widehat{\widetilde{\psi}_0}(\gamma) = \sqrt{\theta(\gamma)}\widehat{\psi}_0(\gamma). \tag{11.51}$$

Define the 1-periodic functions $\widetilde{H}_0, \ldots, \widetilde{H}_n$ by

$$\widetilde{H}_0(\gamma) = \sqrt{\frac{\theta(2\gamma)}{\theta(\gamma)}} H_0(\gamma), \quad \widetilde{H}_\ell(\gamma) = \sqrt{\frac{1}{\theta(\gamma)}} H_\ell(\gamma), \ \ell = 1, \ldots, n. \tag{11.52}$$

The idea in the proof is to apply the unitary extension principle to $\widetilde{\psi}_0, \widetilde{H}_0, \ldots, \widetilde{H}_n$ and thereby obtain a tight frame $\{D^j T_k \widetilde{\psi}_\ell\}_{j,k\in\mathbb{Z},\ell=1,\ldots,n}$; finally, it turns out that $\widetilde{\psi}_\ell = \psi_\ell, \ell = 1, \ldots, n$.

We now prove that $\widetilde{\psi}_0, \widetilde{H}_0, \ldots, \widetilde{H}_n$ satisfy the conditions in the general setup. First,

$$\widehat{\widetilde{\psi}_0}(2\gamma) = \sqrt{\theta(2\gamma)}\widehat{\psi}_0(2\gamma) \quad = \quad \sqrt{\theta(2\gamma)}H_0(\gamma)\widehat{\psi}_0(\gamma)$$

$$= \quad \sqrt{\frac{\theta(2\gamma)}{\theta(\gamma)}}\,H_0(\gamma)\widehat{\widetilde{\psi}_0}(\gamma)$$

$$= \quad \widetilde{H}_0(\gamma)\widehat{\widetilde{\psi}_0}(\gamma).$$

Also,

$$\lim_{\gamma \to 0} \widehat{\widetilde{\psi}_0}(\gamma) = \lim_{\gamma \to 0} \left(\sqrt{\theta(\gamma)}\widehat{\psi}_0(\gamma) \right) = 1.$$

Via the definition (11.52) and (11.50) with $\nu = 0$,

$$\sum_{\ell=0}^{n} |\widetilde{H}_\ell(\gamma)|^2 \quad = \quad \frac{\theta(2\gamma)}{\theta(\gamma)}|H_0(\gamma)|^2 + \sum_{\ell=1}^{n} \frac{|H_\ell(\gamma)|^2}{\theta(\gamma)}$$

$$= \quad 1, \ a.e. \ \gamma \in \mathbb{T}.$$

Thus, $\widetilde{H}_0, \ldots, \widetilde{H}_n \in L^\infty(\mathbb{T})$. Because $\theta(2(\gamma + \frac{1}{2})) = \theta(2\gamma)$, we also see that

$$\sum_{\ell=0}^{n} \widetilde{H}_\ell(\gamma)\overline{\widetilde{H}_\ell(\gamma + \tfrac{1}{2})} \quad = \quad \frac{\theta(2\gamma)}{\sqrt{\theta(\gamma)\theta(\gamma + \frac{1}{2})}}H_0(\gamma)\overline{H_0(\gamma + \tfrac{1}{2})}$$

$$+ \frac{1}{\sqrt{\theta(\gamma)\theta(\gamma + \frac{1}{2})}} \sum_{\ell=1}^{n} \overline{H_\ell(\gamma)H_\ell(\gamma + \tfrac{1}{2})}$$

$$= \quad 0, \ a.e. \ \gamma \in \mathbb{T}.$$

Defining the functions $\widetilde{\psi}_1, \ldots, \widetilde{\psi}_n$ by

$$\widehat{\widetilde{\psi}_\ell}(2\gamma) = \widetilde{H}_\ell(\gamma)\widehat{\widetilde{\psi}_0}(\gamma), \ \ell = 1, \ldots, n, \tag{11.53}$$

it follows from Theorem 11.2.7 that the functions $\{D^j T_k \widetilde{\psi}_\ell\}_{j,k \in \mathbb{Z}, \ell=1,\ldots,n}$ constitute a tight frame for $L^2(\mathbb{R})$ with frame bound equal to 1. The proof is now completed by the observation that for $\ell = 1, \ldots, n$,

$$\widehat{\psi}_\ell(2\gamma) = H_\ell(\gamma)\widehat{\psi}_0(\gamma) = \sqrt{\theta(\gamma)}\widetilde{H}_\ell(\gamma)\frac{1}{\sqrt{\theta(\gamma)}}\widehat{\widetilde{\psi}_0}(\gamma) = \widehat{\widetilde{\psi}_\ell}(2\gamma),$$

which shows that $\psi_\ell = \widetilde{\psi}_\ell$. □

By taking $\theta = 1$ in Theorem 11.3.2, we obtain Theorem 11.2.7. From the extra freedom in Theorem 11.3.2 concerning the choice of θ, one could expect it to be a more general result than Theorem 11.2.7, but the proof shows that the class of frames that can be constructed is the same for the two theorems. However, in practice Theorem 11.3.2 gives more flexibility because it naturally leads to some constructions one would not expect from Theorem 11.2.7. Let us explain this in more detail.

Suppose that ψ_0 is a compactly supported function satisfying (11.13) for some function $H_0 \in L^\infty(\mathbb{T})$, and that θ and $H_\ell, \ell = 1, \ldots, n$ are trigonometric polynomials satisfying the conditions in Theorem 11.3.2. Lemma 3.6.3 shows that the generators ψ_ℓ for the frame $\{D^j T_k \psi_\ell\}_{j,k \in \mathbb{Z}, \ell = 1, \ldots, n}$ have compact support. Now, the proof of Theorem 11.3.2 shows that the same frame can be constructed via Theorem 11.2.7: if we define $\widetilde{\psi_0}$ by (11.51), then the functions $\widetilde{\psi_\ell}$ defined via (11.53) and (11.52) will satisfy the conditions in the unitary extension principle, and $\psi_\ell = \widetilde{\psi_\ell}$. However, in general $\widetilde{\psi_0}$ is not compactly supported, so the fact that the resulting frame $\{D^j T_k \widetilde{\psi_\ell}\}_{j,k \in \mathbb{Z}, \ell = 1, \ldots, n}$ is generated by compactly supported functions is somewhat miraculous and could certainly not be predicted in advance. In short, this shows that there are constructions that appear naturally via Theorem 11.3.2, but one would not even think about constructing them via Theorem 11.2.7.

In practice, it is desirable that a multiwavelet frame contains as few generators as possible. We will now show how to construct frames with two or three generators based on the oblique extension. In order to apply the oblique extension principle, one needs to choose the functions θ and $H_1, \ldots, H_n$ simultaneously such that (11.50) is satisfied. It is not clear how to do this in general, but we now prove that an extra condition on the choice of θ will make it easy to construct frames.

Corollary 11.3.3 *Let ψ_0 and H_0 be as in the general setup on page 260. Let $\theta \in L^\infty(\mathbb{T})$ be a strictly positive function for which $\lim_{\gamma \to 0} \theta(\gamma) = 1$, chosen such that the function*

$$\eta(\gamma) := \theta(\gamma) - \theta(2\gamma)\left(|H_0(\gamma)|^2 + |H_0(\gamma + \tfrac{1}{2})|^2\right) \tag{11.54}$$

is positive as well. Fix an integer $n \geq 2$ and let $\{G_\ell\}_{\ell=2}^n$ be 1-periodic trigonometric polynomials for which

$$\sum_{\ell=2}^n |G_\ell(\gamma)|^2 = 1, \ \text{ and } \ \sum_{\ell=2}^n G_\ell(\gamma)\overline{G_\ell(\gamma + \tfrac{1}{2})} = 0, \ \gamma \in \mathbb{R}. \tag{11.55}$$

Let ρ, σ be 1-periodic functions such that

$$|\rho(\gamma)|^2 = \theta(\gamma), \ \ |\sigma(\gamma)|^2 = \eta(\gamma), \tag{11.56}$$

and define the 1-periodic functions $\{H_\ell\}_{\ell=1}^n$ by

$$H_1(\gamma) = e^{2\pi i \gamma} \rho(2\gamma)\overline{H_0(\gamma + \tfrac{1}{2})}, \ \ H_\ell(\gamma) = G_\ell(\gamma)\sigma(\gamma), \ \ell = 2, \ldots, n.$$

Then the functions $\{\psi_\ell\}_{\ell=1}^n$ given by (11.14) generate a tight frame $\{D^j T_k \psi_\ell\}_{j,k \in \mathbb{Z}, \ell=1,\ldots,n}$ for $L^2(\mathbb{R})$.

Proof. We check that the functions θ and H_ℓ satisfy (11.50). First, for $\gamma \in \mathbb{T}$,

$$|H_0(\gamma)|^2\theta(2\gamma) + \sum_{\ell=1}^{n} |H_\ell(\gamma)|^2$$

$$= |H_0(\gamma)|^2\theta(2\gamma) + |H_0(\gamma + \tfrac{1}{2})|^2|\rho(2\gamma)|^2 + |\sigma(\gamma)|^2 \sum_{\ell=2}^{n} |G_\ell(\gamma)|^2$$

$$= |H_0(\gamma)|^2\theta(2\gamma) + |H_0(\gamma + \tfrac{1}{2})|^2\theta(2\gamma) + \eta(\gamma)$$

$$= \theta(\gamma).$$

Similarly,

$$\overline{H_0(\gamma)}\,\overline{H_0(\gamma + \tfrac{1}{2})}\theta(2\gamma) + \sum_{\ell=1}^{n} \overline{H_\ell(\gamma)}\,\overline{H_\ell(\gamma + \tfrac{1}{2})}$$

$$= \overline{H_0(\gamma)}\,\overline{H_0(\gamma + \tfrac{1}{2})}\theta(2\gamma)$$

$$\quad + \rho(2\gamma)\overline{\rho(2(\gamma + \tfrac{1}{2}))}e^{2\pi i\gamma}e^{-2\pi i(\gamma + 1/2)}\overline{H_0(\gamma)}\,\overline{H_0(\gamma + \tfrac{1}{2})}$$

$$\quad + \sigma(\gamma)\overline{\sigma(\gamma + \tfrac{1}{2})}\sum_{\ell=2}^{n} \overline{G_\ell(\gamma)}\,\overline{G_\ell(\gamma + \tfrac{1}{2})}$$

$$= \overline{H_0(\gamma)}\,\overline{H_0(\gamma + \tfrac{1}{2})}\theta(2\gamma) - \theta(2\gamma)\overline{H_0(\gamma)}\,\overline{H_0(\gamma + \tfrac{1}{2})}$$

$$= 0.$$

$\square$

If the condition (11.54) is satisfied for an appropriate function θ, Corollary 11.3.3 makes it relatively easy to obtain frames with for example three generators. For example, (11.55) is satisfied with

$$G_2(\gamma) = \frac{1}{\sqrt{2}}, \quad G_3(\gamma) = \frac{1}{\sqrt{2}}e^{2\pi i\gamma}. \tag{11.57}$$

Thus, in order to apply Corollary 11.3.3, the remaining work consists in finding ρ, σ such that (11.56) is satisfied. It turns out that if the functions θ and η are trigonometric polynomials, then the functions ρ and σ can be chosen to be trigonometric polynomials as well. This is the outcome of the so-called *spectral factorization*, which is described in Lemma 3.5.5.

The assumption (11.54) even implies that we can construct a tight multiwavelet frame generated by two functions. The reader is asked to provide the proof in Exercise 11.8:

Corollary 11.3.4 *Let ψ_0 and H_0 be as in the general setup on page 260. Let $\theta \in L^\infty(\mathbb{T})$ be a strictly positive function for which $\lim_{\gamma \to 0} \theta(\gamma) = 1$, chosen such that the function η in (11.54) is positive as well. Define the functions ρ, σ as in (11.56) and let*

$$H_1(\gamma) = e^{2\pi i \gamma} \rho(2\gamma) \overline{H_0(\gamma + \frac{1}{2})}, \quad H_2(\gamma) = H_0(\gamma) \sigma(2\gamma). \qquad (11.58)$$

Then the functions $\{\psi_\ell\}_{\ell=1}^2$ given by (11.14) generate a tight frame $\{D^j T_k \psi_\ell\}_{j,k \in \mathbb{Z}, \ell=1,2}$ for $L^2(\mathbb{R})$.

Note that if θ and H_0 are trigonometric polynomials, then η defined in (11.54) also is a trigonometric polynomial. The assumption that θ and η are positive implies by Lemma 3.5.5 that we can choose ρ, σ in (11.56) to be trigonometric polynomials. In this case, the generators ψ_ℓ in Corollary 11.3.3 and Corollary 11.3.4 are finite linear combinations of functions $DT_k\psi_0$ by Lemma 3.6.3.

The oblique extension principle is very useful in order to construct multiwavelet frames based on B-splines. Even the extra assumptions for reduction to two or three generators can be fulfilled:

Theorem 11.3.5 *Let B_{2m} denote the B-spline of order $2m$ with two-scale symbol $H_0(\gamma) = \cos^{2m}(\pi\gamma)$. Then, for each positive integer $M \leq 2m$, there exists a trigonometric polynomial θ of the form*

$$\theta(\gamma) = 1 + \sum_{j=1}^{M-1} c_j \sin^{2j}(\pi\gamma), \qquad (11.59)$$

for which the following hold:

(i) $c_j \geq 0$ for all $j = 1, \ldots, M-1$, i.e., $\theta(\gamma) > 0$ for all $\gamma \in \mathbb{R}$;

(ii) The function η in (11.54) is positive;

(iii) The generators in the tight wavelet frames constructed via the oblique extension principle and its corollaries have M vanishing moments.

The coefficients c_j, $j = 1, \ldots, M-1$ can be determined via the requirement that

$$\left(1 + \sum_{j=1}^{\infty} \frac{(2j-1)!}{(2j)!(2j+1)} y^j\right)^{4m} = 1 + \sum_{j=1}^{M-1} c_j y^j + O(|y|^M) \text{ as } y \to 0.$$

$$(11.60)$$

Theorem 11.3.5 is proved in [28]. Thus, we can apply the result in Corollary 11.3.4 to construct multiwavelet frames with two generators based on *any* B-spline B_{2m}. As already mentioned, if we choose the functions ρ, σ in (11.56) to be trigonometric polynomials, then the associated frame generators ψ_1 and ψ_2 are finite linear combinations of functions $B_{2m}(2x - k), k \in \mathbb{Z}$. By choosing m large enough, we can thus obtain generators belonging to any prescribed smoothness class $C^N(\mathbb{R})$. In contrast, in the application of the unitary extension principle in Example 11.2.9, the number of generators was forced to grow with the desired smoothness.

Let us demonstrate the calculation of the coefficients c_j in (11.59) in the case $M = 2$:

Example 11.3.6 Let us find the trigonometric polynomial associated with the B-spline B_{2m}, $m \in \mathbb{N}$, and $M = 2$. Note that

$$\left(1 + \sum_{j=1}^{\infty} \frac{(2j-1)!}{(2j)!\,(2j+1)} y^j\right)^{4m} = \left(1 + \frac{1}{6}y + \frac{1}{20}y^2 + \cdots\right)^{4m}$$

$$= 1 + \frac{2m}{3}y + O(|y|^2).$$

This proves that for $M = 2$, (11.60) is satisfied with $c_1 = 2m/3$. Thus, the desired trigonometric polynomial is

$$\theta(\gamma) = 1 + \frac{2m}{3}\sin^2(\pi\gamma) = 1 + \frac{2m}{3}\frac{1 - \cos(2\pi\gamma)}{2} = \frac{3+m}{3} - \frac{m}{3}\cos(2\pi\gamma). \qquad \square$$

We now give an example of frame constructions via Theorem 11.2.7 and Theorem 11.3.2.

Example 11.3.7 We consider the function $\psi_0 = N_2$. As we have seem in Example 11.2.12,

$$\widehat{N_2}(2\gamma) = H_0(\gamma)\widehat{N_2}(\gamma),$$

where

$$H_0(\gamma) = \frac{(1 + e^{-2\pi i\gamma})^2}{4} = e^{-2\pi i\gamma}\cos^2(\pi\gamma).$$

We first revisit Example 11.2.12 and then give constructions via the oblique extension principle and its corollaries.

(i) Defining H_1 and H_2 by

$$H_1(\gamma) = ie^{-2\pi i\gamma}\sqrt{2}\sin(\pi\gamma)\cos(\pi\gamma) = \frac{1}{\sqrt{2}}e^{-2\pi i\gamma}i\sin(2\pi\gamma)$$

$$= \frac{\sqrt{2}}{4}(1 - e^{-4\pi i\gamma}),$$

$$H_2(\gamma) = -e^{-2\pi i\gamma}\sin^2(\pi\gamma) = \frac{(1 - e^{-2\pi i\gamma})^2}{4}, \qquad (11.61)$$

it follows from Example 11.2.12 that the functions $\psi_1^{(i)} := \psi_1$ and ψ_2 defined via (11.14) generate a tight frame. They are given by (Exercise 11.9)

$$\psi_1^{(i)}(x) = \frac{1}{\sqrt{2}}(N_2(2x) - N_2(2x - 2)), \tag{11.62}$$

$$\psi_2(x) = \frac{1}{2}(N_2(2x) - 2N_2(2x - 1) + N_2(2x - 2)). \tag{11.63}$$

See Figures 11.3–11.4.

(ii) An alternative construction can be obtained via the oblique extension principle. Let

$$\theta(\gamma) = \frac{4 - \cos(2\pi\gamma)}{3}. \tag{11.64}$$

Note that this is exactly the function we constructed in Example 11.3.6 corresponding to the piecewise linear B-spline and $M = 1$.

In this example, we keep the choice of H_2 in (11.61). Thus, if we want to use the oblique extension principle, we have to choose H_1 such that the two conditions in (11.50) are satisfied; that is, we require that

$$|H_1(\gamma)|^2 = \theta(\gamma) - |H_0(\gamma)|^2\theta(2\gamma) - |H_2(\gamma)|^2,$$

and

$$\overline{H_1(\gamma)}H_1(\gamma + \frac{1}{2}) = -\overline{H_0(\gamma)}H_0(\gamma + \frac{1}{2})\theta(2\gamma) - \overline{H_2(\gamma)}H_2(\gamma + \frac{1}{2}).$$

Inserting θ, H_0, and H_1 leads to the equations

$$\begin{cases} |H_1(\gamma)|^2 = \frac{1}{6}(\cos(2\pi\gamma) + 2)^2(\cos(2\pi\gamma) - 1)^2, \\ \overline{H_1(\gamma)}H_1(\gamma + \frac{1}{2}) = \frac{1}{6}(\cos(2\pi\gamma) + 2)(\cos(2\pi\gamma) - 2) \\ \qquad\qquad\qquad \times(\cos(2\pi\gamma) - 1)(\cos(2\pi\gamma) + 1) \end{cases}$$

These equations are satisfied if we let

$$H_1(\gamma) = \frac{1}{\sqrt{6}}(\cos(2\pi\gamma) + 2)(\cos(2\pi\gamma) - 1)$$

$$= \frac{1}{\sqrt{6}}(\cos^2(2\pi\gamma) + \cos(2\pi\gamma) - 2)$$

$$= \frac{1}{4\sqrt{6}}(e^{4\pi i\gamma} + e^{-4\pi i\gamma} + 2e^{2\pi i\gamma} + 2e^{-2\pi i\gamma} - 6).$$

Via the choice of ψ_1 in the general setup, this leads to (Exercise 11.9)

$$\psi_1(\gamma) = \frac{1}{2\sqrt{6}}(N_2(2\gamma - 2) + 2N_2(2\gamma - 1) - 6N_2(2\gamma))$$

$$+ \frac{1}{2\sqrt{6}}(2N_2(2\gamma + 1) + N_2(2\gamma + 2)). \tag{11.65}$$

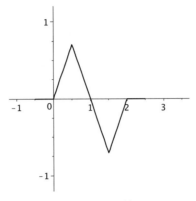

Figure 11.3. The function $\psi_1^{(i)}$ given by (11.62).

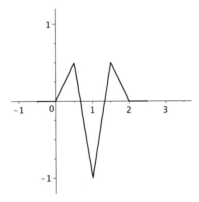

Figure 11.4. The function ψ_2 given by (11.63).

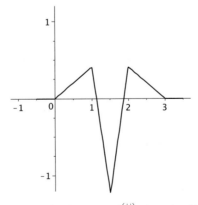

Figure 11.5. The function $\psi_1^{(ii)}$ given by (11.66).

This function has support on $[-1, 2]$. Instead of taking this generator, we take

$$
\begin{aligned}
\psi_1^{(ii)}(\gamma) \; &:= \; \psi_1(\gamma - 1) \tag{11.66} \\
&= \; \frac{1}{2\sqrt{6}} \left(N_2(2\gamma - 4) + 2N_2(2\gamma - 3) - 6N_2(2\gamma - 2) \right) \\
&\quad + \frac{1}{2\sqrt{6}} \left(2N_2(2\gamma - 1) + N_2(2\gamma) \right),
\end{aligned}
$$

which generate the same wavelet system and has support on $[0, 3]$. The function $\psi_1^{(ii)}$ is shown in Figure 11.5. $\qquad \square$

11.4 Approximation orders

In this section, we give some more reasons for constructing frames via the oblique extension principle. More information can be found in [28]. We assume again that $\{H_\ell, \psi_\ell\}_{\ell=0}^n$ is as in the general setup, and that $\{D^j T_k \psi_\ell\}_{j,k\in\mathbb{Z}, \ell=1,\ldots,n}$ is a tight frame constructed via the oblique extension principle. We restrict our discussion to the case of tight wavelet frames based on even-order B-splines, i.e., $\psi_0 = B_{2m}$ for some $m \in \mathbb{N}$. Based on the refinable function ψ_0, we let

$$
V_j = \overline{\text{span}}\{D^j T_k \psi_0\}_{j,k\in\mathbb{Z}}.
$$

For $s > 0$, consider the *Sobolev space*

$$
H_s(\mathbb{R}) = \left\{ f : \mathbb{R} \to \mathbb{C} \mid \int_{-\infty}^{\infty} |\hat{f}(\gamma)|^2 (1 + |\gamma|^2)^s d\gamma < \infty \right\}.
$$

$H_s(\mathbb{R})$ is a Banach space with respect to the natural norm,

$$
||f||_{H_s} = \left(\int_{-\infty}^{\infty} |\hat{f}(\gamma)|^2 (1 + |\gamma|^2)^s d\gamma \right)^{1/2}.
$$

Compared with the unitary extension principle, the oblique extension principle and its corollaries give more freedom in the construction of tight frames, due to the different choices of θ one can start with. However, for practical purposes, the main point is which properties we can expect of the constructed frame, and it turns out that some desirable properties will restrict the class of usable functions θ considerably.

We say that ψ_0 *provides approximation order s* if for all f in the Sobolev space $H^s(\mathbb{R})$,

$$
\text{dist}(f, V_j) = O(2^{-js}),
$$

i.e., if there exists a constant $C > 0$ such that

$$\text{dist}(f, V_j) \leq C2^{-js}, \ \forall j \in \mathbb{Z}.$$

For the tight frame $\{D^j T_k \psi_\ell\}_{j,k\in\mathbb{Z},\ell=1,\ldots,n}$, we know from the frame decomposition (5.9) that for all $f \in L^2(\mathbb{R})$,

$$f = \sum_{\ell=1}^{n}\sum_{j\in\mathbb{Z}}\sum_{k\in\mathbb{Z}}\langle f, D^j T_k \psi_\ell\rangle D^j T_k \psi_\ell.$$

As an approximation of f, we can use

$$Q_J f := \sum_{\ell=1}^{n}\sum_{j<J}\sum_{k\in\mathbb{Z}}\langle f, D^j T_k \psi_\ell\rangle D^j T_k \psi_\ell$$

for a reasonably large value of $J \in \mathbb{Z}$. We say that *the frame* $\{D^j T_k \psi_\ell\}_{j,k\in\mathbb{Z},\ell=1,\ldots,n}$ *provides approximation order s* if for all $f \in H^s(\mathbb{R})$,

$$\|f - Q_J f\| = O(2^{-sJ}).$$

When speaking about "the approximation order," it is in both cases understood that we mean the largest possible order.

We know that $\psi_1, \ldots, \psi_n \in V_1$, so $Q_J f \in V_J$ for all $J \in \mathbb{Z}$; thus, the approximation order of the frame $\{D^j T_k \psi_\ell\}_{j,k\in\mathbb{Z},\ell=1,\ldots,n}$ cannot exceed the approximation order of the underlying refinable function ψ_0. Note that in the case of a classical multiresolution analysis, where a refinable function leads to the construction of an orthonormal basis $\{D^j T_k \psi\}_{j,k\in\mathbb{Z}}$ for $L^2(\mathbb{R})$, the operator Q_J is the orthogonal projection onto V_j and the two types of approximation orders coincide; in general they might be different.

Since every implementation has to be done with a finite collection of vectors, the approximation order of $\{D^j T_k \psi_\ell\}_{j,k\in\mathbb{Z},\ell=1,\ldots,n}$ is clearly important in many applications: we usually want it to be as large as possible. One can prove that the refinable function $\psi_0 = B_{2m}$ provides approximation order $2m$. With the function θ chosen as in the oblique extension principle, one can prove that the approximation order of $\{D^j T_k \psi_\ell\}_{j,k\in\mathbb{Z},\ell=1,\ldots,n}$ is $min(2m, 2M)$. Thus, by choosing M sufficiently large, see Theorem 11.3.5, we can obtain the approximation order $2m$; this is the best possible approximation order we can hope for with the given function $\psi_0 = B_{2m}$.

11.5   Construction of pairs of dual wavelet frames

So far, the constructions via the extension principles have concerned tight frames. However, the technique is more far-reaching, and one can actually extend the results and construct dual multiwavelet pairs. We cite a result from [19] and [28] concerning construction of dual pairs of multiwavelet frames:

Theorem 11.5.1 *Let* $\{H_\ell, \psi_\ell\}_{\ell=0}^n$ *and* $\{K_\ell, \widetilde{\psi}_\ell\}_{\ell=0}^n$ *be two sets of functions, satisfying the conditions in the general setup on page 260, and such that for some* $C > 0$ *and* $\rho > \frac{1}{2}$,

$$|\widehat{\psi_0}(\gamma)|, |\widehat{\widetilde{\psi}_0}(\gamma)| \le \frac{C}{|\gamma|^\rho}, \quad a.e.\ \gamma \in \mathbb{R}. \tag{11.67}$$

Assume that there exists a function $\theta \in L^\infty(\mathbb{T})$ *such that* $\lim_{\gamma \to 0} \theta(\gamma) = 1$ *and*

$$H_0(\gamma)\overline{K_0(\gamma + \nu)}\theta(2\gamma) \ + \ \sum_{\ell=1}^n H_\ell(\gamma)\overline{K_\ell(\gamma + \nu)}$$

$$= \begin{cases} \theta(\gamma) & if\ \nu = 0, \\ 0 & if\ \nu = \frac{1}{2}. \end{cases} \tag{11.68}$$

Then $\{D^j T_k \psi_\ell\}_{j,k \in \mathbb{Z}, \ell=1,\dots,n}$ *and* $\{D^j T_k \widetilde{\psi}_\ell\}_{j,k \in \mathbb{Z}, \ell=1,\dots,n}$ *are a pair of dual multiwavelet frames.*

To find a pair of dual frames via Theorem 11.5.1 is actually much easier than to construct tight frames via the oblique extension principle. One reason is that the function θ is not required to be positive. Another reason is that we have freedom to chose *two* sets of trigonometric polynomials H_ℓ and K_ℓ: in fact, the condition (11.50) in the oblique extension principle corresponds exactly to (11.68) with $H_\ell = K_\ell$, and is more complicated to satisfy.

Similar to what we saw for the oblique extension principle, we can use Theorem 11.5.1 to provide explicit constructions of dual pairs of frames with multiple generators. In the rest of this section, we will use the following:

Setup for construction of pairs of dual wavelet frames:

Let $\{\psi_0, H_0\}, \{K_0, \widetilde{\psi}_0\}$ be as in the general setup on page 260, and assume that (11.67) is satisfied. Let $\theta \in L^\infty(\mathbb{T})$ be a real-valued function for which $\lim_{\gamma \to 0} \theta(\gamma) = 1$, and assume that the function

$$\eta(\gamma) := \theta(\gamma) - \theta(2\gamma)\left(H_0(\gamma)\overline{K_0(\gamma)} + H_0(\gamma + \frac{1}{2})\overline{K_0(\gamma + \frac{1}{2})}\right) \tag{11.69}$$

is real-valued and has a zero of order at least 2 at the origin. Choose real-valued functions $\eta_1, \eta_2 \in L^\infty(\mathbb{T})$ such that

$$\eta(\gamma) = 2\eta_1(\gamma)\eta_2(\gamma), \quad and\ \eta_1(0) = \eta_2(0) = 0, \tag{11.70}$$

and choose two $\frac{1}{2}$-periodic and real-valued functions θ_1, θ_2 such that

$$\theta(2\gamma) = \theta_1(\gamma)\theta_2(\gamma). \tag{11.71}$$

$\square$

Let us comment on these assumptions and choices. First, the choice of $\frac{1}{2}$-periodic functions in (11.71) is possible because $\gamma \mapsto \theta(2\gamma)$ has period $\frac{1}{2}$.

In the construction of tight multiwavelet frames in, e.g., Corollary 11.3.3, we had to perform a spectral factorization of the functions θ and η. The choices of the functions $\eta_1, \eta_2, \theta_1, \theta_2$ in (11.70) and (11.71) will replace the spectral factorization: in fact, we now prove how one can construct a multiwavelet frame based on these functions. We note that in general it is much easier to find functions satisfying (11.70) and (11.71) than to perform a spectral factorization; the price we have to pay is that we in general do not obtain a tight frame.

Corollary 11.5.2 *Assume the setup on page 287 and define* $\{H_\ell\}_{\ell=1}^3$ *and* $\{K_\ell\}_{\ell=1}^3$ *by*

$$H_1(\gamma) = e^{2\pi i \gamma} \theta_1(\gamma) \overline{K_0(\gamma + \frac{1}{2})}, \qquad K_1(\gamma) = e^{2\pi i \gamma} \theta_2(\gamma) \overline{H_0(\gamma + \frac{1}{2})},$$
$$H_2(\gamma) = \eta_1(\gamma), \qquad K_2(\gamma) = \eta_2(\gamma),$$
$$H_3(\gamma) = e^{2\pi i \gamma} \eta_1(\gamma), \qquad K_3(\gamma) = e^{2\pi i \gamma} \eta_2(\gamma).$$

Define the associated functions $\{\psi_\ell\}_{\ell=1}^3$ *and* $\{\tilde{\psi}_\ell\}_{\ell=1}^3$ *as in the general setup on page 260. Then* $\{D^j T_k \psi_\ell\}_{j,k \in \mathbb{Z}, \ell=1,2,3}$ *and* $\{D^j T_k \tilde{\psi}_\ell\}_{j,k \in \mathbb{Z}, \ell=1,2,3}$ *constitute a pair of dual multiwavelet frames.*

Proof. For $\nu = 0$,

$$H_0(\gamma) \overline{K_0(\gamma)} \theta(2\gamma) + \sum_{\ell=1}^3 H_\ell(\gamma) \overline{K_\ell(\gamma)}$$

$$= H_0(\gamma) \overline{K_0(\gamma)} \theta(2\gamma) + \theta_1(\gamma) \theta_2(\gamma) \overline{K_0(\gamma + \frac{1}{2}) H_0(\gamma + \frac{1}{2})} + 2\eta_1(\gamma)\eta_2(\gamma)$$

$$= H_0(\gamma) \overline{K_0(\gamma)} \theta(2\gamma) + \theta(2\gamma) \overline{K_0(\gamma + \frac{1}{2}) H_0(\gamma + \frac{1}{2})}$$

$$\quad + \theta(\gamma) - \theta(2\gamma) \left(H_0(\gamma) \overline{K_0(\gamma)} + H_0(\gamma + \frac{1}{2}) \overline{K_0(\gamma + \frac{1}{2})} \right)$$

$$= \theta(\gamma).$$

The proof that (11.68) holds for $\nu = \frac{1}{2}$ is similar and is left to the reader (Exercise 11.5). $\qquad \square$

Corollary 11.5.3 *Assume the setup on page 287 and let*

$$H_1(\gamma) = e^{2\pi i \gamma} \theta_1(\gamma) \overline{K_0(\gamma + \frac{1}{2})}, \qquad K_1(\gamma) = e^{2\pi i \gamma} \theta_2(\gamma) \overline{H_0(\gamma + \frac{1}{2})},$$
$$H_2(\gamma) = \eta_1(2\gamma) H_0(\gamma), \qquad K_2(\gamma) = \eta_2(2\gamma) K_0(\gamma).$$

Then $\{D^j T_k \psi_\ell\}_{j,k \in \mathbb{Z}, \ell=1,2}$ *and* $\{D^j T_k \tilde{\psi}_\ell\}_{j,k \in \mathbb{Z}, \ell=1,2}$ *constitute a pair of dual multiwavelet frames.*

We have assumed the factorizations of $\theta(2\cdot)$ and η to be real-valued. This is not strictly necessary. However, if θ, H_0, and K_0 are trigonometric polynomials and η_1, η_2 and θ_1, θ_2 are real-valued trigonometric polynomials, then the frame generators $\{\psi_\ell\}_{\ell=1}^3$ and $\{\tilde{\psi}_\ell\}_{\ell=1}^3$ are symmetric if the refinable functions ψ_0 and $\tilde{\psi}_0$ are symmetric real-valued functions. Thus, the above process will lead to symmetric dual wavelet pairs when applied to even-order B-splines.

Example 11.5.4 We give an example of a frame construction with two generators. We will base the choices of H_1, H_2 and K_1, K_2 on the same refinable function, namely a translated B-spline of order 2. That is, we take $\psi_0 = \tilde{\psi}_0 = T_1 B_2$; the associated two-scale symbol is

$$H_0(\gamma) = \frac{(1 + e^{-2\pi i \gamma})^2}{4} = e^{-2\pi i \gamma} \cos^2(\pi \gamma).$$

We again take

$$\theta(\gamma) = \frac{4 - \cos(2\pi\gamma)}{3};$$

as proved in Example 11.3.7 this leads to

$$\eta(\gamma) \;=\; \frac{2}{3}(8\cos^4(\pi\gamma) + 1)(\cos(\pi\gamma) - 1)^2(\cos(\pi\gamma) + 1)^2. \quad (11.72)$$

If we want to apply Corollary 11.5.3, we need to find functions $\eta_1, \eta_2, \theta_1, \theta_2$ satisfying (11.70) and (11.71). This is easy: the expression (11.72) immediately gives several choices for η_1, η_2, for example

$$\eta_1(\gamma) \;=\; \frac{1}{3}(8\cos^4(\pi\gamma) + 1)(\cos(\pi\gamma) - 1)(\cos(\pi\gamma) + 1)^2,$$
$$\eta_2(\gamma) \;=\; (\cos(\pi\gamma) - 1).$$

Concerning θ_1, θ_2 we simply take

$$\theta_1(\gamma) = 1, \quad \theta_2(\gamma) = \theta(2\gamma) = \frac{4 - \cos(4\pi\gamma)}{3}.$$

The functions in Corollary 11.5.3 are now as follows:

$$H_1(\gamma) \;=\; e^{2\pi i \gamma}\theta_1(\gamma)\overline{K_0(\gamma + \tfrac{1}{2})}$$
$$\;=\; e^{2\pi i \gamma}\frac{(1 - e^{2\pi i \gamma})^2}{4};$$

$$
\begin{aligned}
K_1(\gamma) &= e^{2\pi i\gamma}\theta_2(\gamma)\overline{H_0\left(\gamma + \frac{1}{2}\right)} \\
&= e^{2\pi i\gamma}\left(\frac{4}{3} - \frac{e^{4\pi i\gamma} + e^{-4\pi i\gamma}}{6}\right)\frac{(1 - e^{2\pi i\gamma})^2}{4}; \\
H_2(\gamma) &= \eta_1(2\gamma)H_0(\gamma) \\
&= \frac{1}{3}\left(8\left(\frac{e^{2\pi i\gamma} + e^{-2\pi i\gamma}}{2}\right)^4 + 1\right)\left(\frac{e^{2\pi i\gamma} + e^{-2\pi i\gamma}}{2} - 1\right) \\
&\quad \times \left(\frac{e^{2\pi i\gamma} + e^{-2\pi i\gamma}}{2} + 1\right)^2\frac{(1 + e^{-2\pi i\gamma})^2}{4}; \\
K_2(\gamma) &= \eta_2(2\gamma)K_0(\gamma) \\
&= \left(\frac{e^{2\pi i\gamma} + e^{-2\pi i\gamma}}{2} - 1\right)\frac{(1 + e^{-2\pi i\gamma})^2}{4}.
\end{aligned}
$$

With these choices, $\{D^j T_k \psi_\ell\}_{j,k\in\mathbb{Z},\ell=1,2}$ and $\{D^j T_k \tilde{\psi}_\ell\}_{j,k\in\mathbb{Z},\ell=1,2}$ constitute a pair of dual multiwavelet frames.   □

11.6   The signal processing perspective

In Section 11.2, we gave a functional analytic presentation of the unitary extension principle. We will now look at this result once more and formulate it in signal processing terms.

We will first reformulate the equations in Corollary 11.2.8 in terms of the *Z-transform*. Formally, the Z-transform of a sequences $\{h_k\}_{k\in\mathbb{Z}}$ is defined as the infinite series (depending on a variable $z \in \mathbb{C}$)

$$
\tilde{H}(z) := \sum_{k\in\mathbb{Z}} h_k z^{-k}.
$$

We will not worry too much about the exact domain of $z \in \mathbb{Z}$ for which the Z-transform of a given sequence $\{h_k\}_{k\in\mathbb{Z}}$ converges. The reason is that we mainly are interested in finite sequences $\{h_k\}_{k\in\mathbb{Z}}$, for which the Z-transform is defined for all $z \neq 0$. Besides such finite sequences, we will only consider the Z-transform of sequences $\{h_k\}_{k\in\mathbb{Z}}$, which are Fourier coefficients; for such sequences, the Z-transform converges for a.e. $z \in \mathbb{C}$ with $|z| = 1$, and this turns out to be sufficient for our purpose. In engineering language, the sequence $\{h_k\}_{k\in\mathbb{Z}}$ is often called a *filter*.

Consider the 1-periodic functions H_ℓ, $\ell = 0, \ldots, n$, in the general setup on page 260. We can write these functions in terms of their Fourier series, with Fourier coefficients $h_{k,\ell}$, $k \in \mathbb{Z}$:

$$
H_\ell(\gamma) = \sum_{k\in\mathbb{Z}} h_{k,\ell} e^{2\pi i k\gamma}.
$$

Note that in terms of the Z-transform, this means that

$$H_\ell(\gamma) = \sum_{k \in \mathbb{Z}} h_{k,\ell} \left(e^{-2\pi i \gamma}\right)^{-k} = \widetilde{H}_\ell(e^{-2\pi i \gamma}).$$

We can now formulate the main condition in the unitary extension principle in terms of the Z-transform:

Theorem 11.6.1 *Assume that the functions H_ℓ, $\ell = 0, \ldots, n$, have real Fourier coefficients $h_{k,\ell}$, $k \in \mathbb{Z}$. Then the conditions (11.36) hold if and only if the equations*

$$\begin{cases} \sum_{\ell=0}^{n} \widetilde{H}_\ell(z)\widetilde{H}_\ell(z^{-1}) = 1, \\[2mm] \sum_{\ell=0}^{n} \widetilde{H}_\ell(z)\widetilde{H}_\ell(-z^{-1}) = 0 \end{cases} \tag{11.73}$$

hold for a.e. $z \in \mathbb{C}$ for which $|z| = 1$.

Proof. Let us rewrite the terms appearing in (11.36):

$$T_{1/2} H_\ell(\gamma) = \widetilde{H}_\ell(e^{-2\pi i(\gamma - 1/2)}) = \widetilde{H}_\ell(-e^{-2\pi i \gamma}),$$

and, because the coefficients $h_{k,\ell}$ are assumed to be real,

$$\overline{H_\ell(\gamma)} = \sum_{k \in \mathbb{Z}} h_{k,\ell} e^{-2\pi i k \gamma} = \widetilde{H}_\ell(e^{2\pi i \gamma}).$$

Thus (11.36) is equivalent to the conditions

$$\begin{cases} \sum_{\ell=0}^{n} \widetilde{H}_\ell(e^{-2\pi i \gamma})\widetilde{H}_\ell(e^{2\pi i \gamma}) = 1, \\[2mm] \sum_{\ell=0}^{n} \widetilde{H}_\ell(e^{2\pi i \gamma})\widetilde{H}_\ell(-e^{-2\pi i \gamma}) = 0. \end{cases}$$

Putting $z = e^{2\pi i \gamma}$ now leads to the result. $\square$

Very often, conditions involving filters are formulated in terms of the so-called *polyphase decomposition* of the Z-transform. In order to introduce that, note that we can decompose a sequence $\{h_k\}_{k \in \mathbb{Z}}$ into "even" and "odd" parts:

$$\begin{aligned} (\ldots, h_{-2}, h_{-1}, h_0, h_1, h_2, \ldots) &= (\ldots, h_{-2}, 0, h_0, 0, h_2, \ldots) \\ &+ (\ldots, 0, h_{-1}, 0, h_1, 0, \ldots). \end{aligned}$$

By linearity, this decomposition implies that the Z-transformation of $\{h_k\}_{k\in\mathbb{Z}}$ can be written as

$$
\begin{aligned}
\tilde{H}(z) &= \left[\cdots + h_{-2}z^2 + h_0 + h_2 z^{-2} + \cdots\right] \\
&\quad + \left[\cdots + h_{-1}z + h_1 z^{-1} + h_3 z^{-3} + \cdots\right] \\
&= \left[\cdots + h_{-2}z^2 + h_0 + h_2 z^{-2} + \cdots\right] \\
&\quad + z^{-1}\left[\cdots + h_{-1}z^2 + h_1 + h_3 z^{-2} + \cdots\right] \\
&= \sum_{k\in\mathbb{Z}} h_{2k} z^{-2k} + z^{-1}\sum_{k\in\mathbb{Z}} h_{2k+1} z^{-2k}. \quad (11.74)
\end{aligned}
$$

The *polyphase components* of $\tilde{H}(z)$ are now defined as the two functions

$$
\widetilde{H_0}(z) := \sum_{k\in\mathbb{Z}} h_{2k} z^{-k}, \quad \widetilde{H_1}(z) = \sum_{k\in\mathbb{Z}} h_{2k+1} z^{-k};
$$

thus, via (11.74), the Z-transformation has the *polyphase decomposition*

$$
\tilde{H}(z) = \widetilde{H_0}(z^2) + z^{-1}\widetilde{H_1}(z^2).
$$

Consider now a given sequence of 1-periodic functions H_ℓ, $\ell = 0,\ldots,n$, or, equivalently, a sequence of filters $\{h_{k,\ell}\}_{k\in\mathbb{Z}}$, $\ell = 0,\ldots n$. Associated with the filter $\{h_{k,\ell}\}_{k\in\mathbb{Z}}$, we denote the polyphase components of $\widetilde{H_\ell}$ by $\widetilde{H_{\ell,0}}$ and $\widetilde{H_{\ell,1}}$. Define the $(n+1)\times 2$ matrix of polyphase components H_p by

$$
H_p(z) = \begin{pmatrix} \widetilde{H_{0,0}}(z) & \widetilde{H_{0,1}}(z) \\ \widetilde{H_{1,0}}(z) & \widetilde{H_{1,1}}(z) \\ \cdot & \cdot \\ \cdot & \cdot \\ \widetilde{H_{n,0}}(z) & \widetilde{H_{n,1}}(z) \end{pmatrix}. \quad (11.75)
$$

We will now formulate Theorem 11.6.1 in terms of the matrix $H_p(z)$ and its transpose $H_p^T(z)$.

Theorem 11.6.2 *Assume that the functions H_ℓ, $\ell = 0,\ldots,n$, have real Fourier coefficients $h_{k,\ell}$, $k\in\mathbb{Z}$. Then the condition (11.73) is satisfied if and only if*

$$
H_p^T(z^{-1})H_p(z) = \frac{1}{2}I \quad (11.76)
$$

for almost all $z\in\mathbb{C}$ with $|z| = 1$.

Proof. Note that $\widetilde{H_{\ell,k}}(z^{-1}) = \overline{\widetilde{H_{\ell,k}}(z)}$ for $\ell = 0,\ldots,n$, $k = 0,1$; this implies that $H_p^T(z^{-1}) = \overline{H_p^T}(z)$. In terms of the entries of the matrix $H_p(z)$, the condition (11.76) means that for almost all $z \in \mathbb{C}$ with $|z| = 1$,

$$
\begin{cases}
\displaystyle\sum_{\ell=0}^{n} \left| \widetilde{H_{\ell,0}}(z) \right|^2 = \frac{1}{2}, \\[2ex]
\displaystyle\sum_{\ell=0}^{n} \left| \widetilde{H_{\ell,1}}(z) \right|^2 = \frac{1}{2}, \\[2ex]
\displaystyle\sum_{\ell=0}^{n} \widetilde{H_{\ell,0}}(z^{-1}) \widetilde{H_{\ell,1}}(z) = 0, \\[2ex]
\displaystyle\sum_{\ell=0}^{n} \widetilde{H_{\ell,1}}(z^{-1}) \widetilde{H_{\ell,0}}(z) = 0.
\end{cases}
\tag{11.77}
$$

On the other hand, in terms of the polyphase decomposition, the two terms in (11.73) can be written as

$$
\sum_{\ell=0}^{n} \widetilde{H_\ell}(z) \widetilde{H_\ell}(z^{-1})
\tag{11.78}
$$

$$
= \sum_{\ell=0}^{n} \left(\widetilde{H_{\ell,0}}(z^2) + z^{-1} \widetilde{H_{\ell,1}}(z^2) \right) \left(\widetilde{H_{\ell,0}}(z^{-2}) + z \widetilde{H_{\ell,1}}(z^{-2}) \right)
$$

$$
= \sum_{\ell=0}^{n} \left| \widetilde{H_{\ell,0}}(z^2) \right|^2 + \sum_{\ell=0}^{n} \left| \widetilde{H_{\ell,1}}(z^2) \right|^2
$$

$$
+ z \sum_{\ell=0}^{n} \widetilde{H_{\ell,0}}(z^2) \widetilde{H_{\ell,1}}(z^{-2}) + z^{-1} \sum_{\ell=0}^{n} \widetilde{H_{\ell,0}}(z^{-2}) \widetilde{H_{\ell,1}}(z^2),
$$

respectively,

$$
\sum_{\ell=0}^{n} \widetilde{H_\ell}(z) \widetilde{H_\ell}(-z^{-1})
\tag{11.79}
$$

$$
= \sum_{\ell=0}^{n} \left(\widetilde{H_{\ell,0}}(z^2) + z^{-1} \widetilde{H_{\ell,1}}(z^2) \right) \left(\widetilde{H_{\ell,0}}(z^{-2}) - z \widetilde{H_{\ell,1}}(z^{-2}) \right)
$$

$$
= \sum_{\ell=0}^{n} \left| \widetilde{H_{\ell,0}}(z^2) \right|^2 - \sum_{\ell=0}^{n} \left| \widetilde{H_{\ell,1}}(z^2) \right|^2
$$

$$
- z \sum_{\ell=0}^{n} \widetilde{H_{\ell,0}}(z^2) \widetilde{H_{\ell,1}}(z^{-2}) + z^{-1} \sum_{\ell=0}^{n} \widetilde{H_{\ell,0}}(z^{-2}) \widetilde{H_{\ell,1}}(z^2).
$$

From here, it follows that if (11.77) is satisfied, then the conditions in (11.73) are satisfied as well.

Now assume that (11.73) holds. Adding, respectively, subtracting, the two equations in (11.73), and using the expressions derived in (11.78) and

(11.79) leads to the equations

$$\begin{cases} 2\sum_{\ell=0}^{n}\left|\widetilde{H_{\ell,0}}(z^2)\right|^2 + 2z^{-1}\sum_{\ell=0}^{n}\widetilde{H_{\ell,0}}(z^{-2})\widetilde{H_{\ell,1}}(z^2) = 1, \\ 2\sum_{\ell=0}^{n}\left|\widetilde{H_{\ell,1}}(z^2)\right|^2 + 2z\sum_{\ell=0}^{n}\widetilde{H_{\ell,0}}(z^2)\widetilde{H_{\ell,1}}(z^{-2}) = 1. \end{cases} \qquad (11.80)$$

The terms $z^{-1}\sum_{\ell=0}^{n}\widetilde{H_{\ell,0}}(z^{-2})\widetilde{H_{\ell,1}}(z^2)$ and $z\sum_{\ell=0}^{n}\widetilde{H_{\ell,0}}(z^2)\widetilde{H_{\ell,1}}(z^{-2})$ are the complex conjugated of each other, but by (11.80) they are also real; thus,

$$z^{-1}\sum_{\ell=0}^{n}\widetilde{H_{\ell,0}}(z^{-2})\widetilde{H_{\ell,1}}(z^2) = z\sum_{\ell=0}^{n}\widetilde{H_{\ell,0}}(z^2)\widetilde{H_{\ell,1}}(z^{-2}) \in \mathbb{R}. \qquad (11.81)$$

Finally, applying the first equation in (11.73) with z replaced by $-z$ leads to

$$\begin{aligned} 1 &= \sum_{\ell=0}^{n}\widetilde{H_\ell}(-z)\widetilde{H_\ell}(-z^{-1}) \\ &= \sum_{\ell=0}^{n}\left(\widetilde{H_{\ell,0}}(z^2) - z^{-1}\widetilde{H_{\ell,1}}(z^2)\right)\left(\widetilde{H_{\ell,0}}(z^{-2}) - z\widetilde{H_{\ell,1}}(z^{-2})\right) \\ &= \sum_{\ell=0}^{n}\left|\widetilde{H_{\ell,0}}(z^2)\right|^2 + \sum_{\ell=0}^{n}\left|\widetilde{H_{\ell,1}}(z^2)\right|^2 \\ &\quad - z\sum_{\ell=0}^{n}\widetilde{H_{\ell,0}}(z^2)\widetilde{H_{\ell,1}}(z^{-2}) - z^{-1}\sum_{\ell=0}^{n}\widetilde{H_{\ell,0}}(z^{-2})\widetilde{H_{\ell,1}}(z^2). \end{aligned}$$

Again by addition and subtraction with the equation in (11.79), this leads to

$$\begin{cases} 2\sum_{\ell=0}^{n}\left|\widetilde{H_{\ell,0}}(z^2)\right|^2 - 2z\sum_{\ell=0}^{n}\widetilde{H_{\ell,0}}(z^2)\widetilde{H_{\ell,1}}(z^{-2})) = 1, \\ 2\sum_{\ell=0}^{n}\left|\widetilde{H_{\ell,1}}(z^2)\right|^2 - 2z^{-1}\sum_{\ell=0}^{n}\widetilde{H_{\ell,0}}(z^{-2})\widetilde{H_{\ell,1}}(z^2) = 1. \end{cases} \qquad (11.82)$$

Combining (11.82) with (11.80) and (11.81) finally leads to (11.77).   □

It turns out that the condition (11.76) in Theorem 11.6.2 is well-known in the context of *filter banks*. In the rest of this section, we discuss this connection.

Intuitively, a filter bank is some kind of "black box," which performs operations on an incoming signal (i.e., a sequence of numbers). Typically, a filter bank splits the incoming signal into certain subsignals, which contain particular information about the signal. For this reason, filter banks of that type are called *analysis filter banks*. After processing the subsequences coming out of the analysis filter bank, engineers usually wish to get back to the original input sequence. Therefore, it is essential that an analysis filter bank is followed by another filter bank, which reconstructs the original signal from the subsignals; such a filter bank is called a *synthesis filter bank*. In that case, the entire system consisting of the two filter banks is said to have the *perfect reconstruction* property.

The filter banks considered here will contain three operations on the incoming sequence $\{x_k\}_{k\in\mathbb{Z}}$:

- **Convolution with a sequence** $\{h_k\}_{k\in\mathbb{Z}}$: The outcome is a new sequence, whose k-th coordinate is given by $\sum_{n\in\mathbb{Z}} h_n x_{k-n}$.

- **Downsampling:** The outcome is the sequence

$$\downarrow \{x_k\}_{k\in\mathbb{Z}} := (\cdots x_{-2}, x_0, x_2, \cdots).$$

  Thus, downsampling removes each second element in the sequence.

- **Upsampling:** The outcome is the sequence

$$\uparrow \{x_k\}_{k\in\mathbb{Z}} := (\cdots x_{-1}, 0, x_0, 0, x_1, \cdots).$$

  Thus, upsampling inserts zeroes between the elements in the sequence.

Note that downsampling is the left-inverse of upsampling, but not the right-inverse.

We will now describe a particular filter bank. The analysis filter bank will split the incoming signal $\{x_k\}_{k\in\mathbb{Z}}$ into $n+1$ subsignals: each of these signals is obtained by convolving $\{x_k\}_{k\in\mathbb{Z}}$ with a sequence $h_{k,\ell}$, $\ell = 0, \ldots, n$, followed by a downsampling. The synthesis filter bank first upsamples each of the incoming $n + 1$ subsignals, then convolves the resulting sequences with sequences $\{g_{k,\ell}\}_{k\in\mathbb{Z}}$, $\ell = 0, \ldots, n$, and finally add the outcoming $n+1$ signals; see Figure 11.6. We will assume that the sequences $\{h_{k,\ell}\}_{k\in\mathbb{Z}}$ and $\{g_{k,\ell}\}_{k\in\mathbb{Z}}, \ell = 0, \ldots, n$, are related by

$$g_{k,\ell} = h_{-k,\ell}, \ k \in \mathbb{Z}, \ell = 0, \ldots, n.$$

For the above system consisting of the analysis filter bank followed by the synthesis filter bank, the perfect reconstruction property can be formulated in terms of the polyphase components associated with the filters $\{h_{k,\ell}\}_{k\in\mathbb{Z}}$:

Theorem 11.6.3 *For the considered filter bank, the perfect reconstruction property is equivalent to the condition*

$$H_p^T(z^{-1})H_p(z) = I \ for \ z \in \mathbb{C} \ with \ |z| = 1. \tag{11.83}$$

A proof of Theorem 11.6.3 can be found in [6]. Note that the conditions in (11.83) and (11.76) are really "identical:" if one of these conditions is satisfied, the other will be satisfied if the filter sequences $\{h_{k,\ell}\}_{k\in\mathbb{Z}}$ are

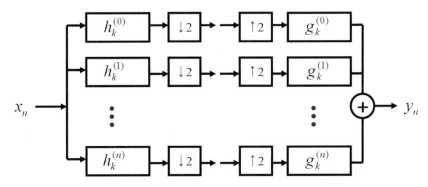

Figure 11.6. A filter bank consisting of an analysis filter bank composed with a synthesis filter bank.

either multiplied or divided by $\sqrt{2}$. In other words: if the condition (11.76) (and the general setup for the unitary extension principle) are satisfied, then the functions

$$\psi_\ell = \sqrt{2} \sum_{k\in\mathbb{Z}} h_{k,\ell} DT_{-k}\psi_0, \ \ell = 1, \ldots, n,$$

generate a tight frame with frame bound 1; if (11.83) is satisfied, the functions

$$\psi_\ell = \sum_{k\in\mathbb{Z}} h_{k,\ell} DT_{-k}\psi_0, \ \ell = 1, \ldots, n,$$

generate a tight frame with frame bound 1.

Thus, the conditions in Theorem 11.6.2 for construction of a tight wavelet frame are equivalent to the perfect reconstruction property for the above filter bank.

11.7   A survey on general wavelet frames

In this section, we will give a short description of wavelet frames with general dilation parameters and translation parameters. The reader might observe that many of the results are parallel to results obtained for Gabor frames in Chapter 9.

We will consider the functions $\psi^{a,b}$ in (11.1), with the assumption that the points (a, b) are restricted to discrete sets of the type $\{(a^j, kba^j)\}_{j,k\in\mathbb{Z}}$, where $a > 1, b > 0$; a is the *dilation parameter* or *scaling parameter* and b is the *translation parameter*. We hereby obtain the functions

$$(T_{kba^j} D_{a^j}\psi)(x) = (D_{a^j} T_{kb}\psi)(x) = \frac{1}{a^{j/2}}\psi(\frac{x}{a^j} - kb), \ \ j, k \in \mathbb{Z}.$$

Re-indexing (i.e., replacing j by $-j$), we see that

$$\{T_{kba^j} D_{a^j} \psi\}_{j,k\in\mathbb{Z}} = \{a^{j/2}\psi(a^j x - kb)\}_{j,k\in\mathbb{Z}}. \tag{11.84}$$

Definition 11.7.1 *Let $a > 1, b > 0$ and $\psi \in L^2(\mathbb{R})$. A frame for $L^2(\mathbb{R})$ of the form $\{a^{j/2}\psi(a^j x - kb)\}_{j,k\subset\mathbb{Z}}$ is called a wavelet frame.*

We first present a necessary condition for $\{a^{j/2}\psi(a^j x - kb)\}_{j,k\in\mathbb{Z}}$ to be a frame, due to Chui and Shi [20]. It plays the same role in wavelet analysis as Proposition 9.1.2 does for Gabor frames.

Proposition 11.7.2 *Let $a > 1, b > 0$ and $\psi \in L^2(\mathbb{R})$ be given. If $\{a^{j/2}\psi(a^j x - kb)\}_{j,k\in\mathbb{Z}}$ is a frame with frame bounds A, B, then*

$$bA \le \sum_{j\in\mathbb{Z}} \left|\hat{\psi}(a^j\gamma)\right|^2 \le bB, \ a.e. \ \gamma \in \mathbb{R}.$$

A sufficient condition for the wavelet system in (11.84) being a frame was obtained by Daubechies [25]. Casazza and Christensen obtained the following slight improvement, see [8] and [11]:

Theorem 11.7.3 *Let $a > 1, b > 0$ and $\psi \in L^2(\mathbb{R})$ be given. Suppose that*

$$B := \frac{1}{b} \sup_{|\gamma|\in[1,a]} \sum_{j,k\in\mathbb{Z}} \left|\hat{\psi}(a^j\gamma)\hat{\psi}(a^j\gamma + k/b)\right| < \infty. \tag{11.85}$$

Then $\{a^{j/2}\psi(a^j x - kb)\}_{j,k\in\mathbb{Z}}$ is a Bessel sequence with bound B, and for all functions $f \in L^2(\mathbb{R})$ for which $\hat{f} \in C_c(\mathbb{R})$,

$$\sum_{j,k\in\mathbb{Z}} |\langle f, D_{a^j} T_{kb}\psi\rangle|^2 = \frac{1}{b} \int_{-\infty}^{\infty} |\hat{f}(\gamma)|^2 \sum_{j\in\mathbb{Z}} |\hat{\psi}(a^j\gamma)|^2 d\gamma \tag{11.86}$$

$$+\frac{1}{b}\sum_{k\neq 0}\sum_{j\in\mathbb{Z}} \int_{-\infty}^{\infty} \hat{f}(\gamma)\overline{\hat{f}(\gamma - a^j k/b)}\hat{\psi}(a^{-j}\gamma)\overline{\hat{\psi}(a^{-j}\gamma - k/b)}\, d\gamma.$$

If furthermore

$$A := \frac{1}{b} \inf_{|\gamma|\in[1,a]} \left(\sum_{j\in\mathbb{Z}} \left|\hat{\psi}(a^j\gamma)\right|^2 - \sum_{k\neq 0}\sum_{j\in\mathbb{Z}} \left|\hat{\psi}(a^j\gamma)\hat{\psi}(a^j\gamma + k/b)\right| \right) > 0,$$

$$\tag{11.87}$$

then $\{a^{j/2}\psi(a^j x - kb)\}_{j,k\in\mathbb{Z}}$ is a frame for $L^2(\mathbb{R})$ with bounds A, B.

Nothing guarantees that a wavelet frame $\{a^{j/2}\psi(a^j x - kb)\}_{j,k\in\mathbb{Z}}$ constructed via Theorem 11.7.3 has a convenient dual frame: the canonical dual frame might not even have the wavelet structure. A construction of a pair of dual wavelet frames that is similar to the results for Gabor frames presented in Section 9.4 can be found in [51].

It is clear from the definition of a frame that a wavelet system Ψ containing a frame $\{a^{j/2}\psi(a^j x - kb)\}_{j,k\in\mathbb{Z}}$ is itself a frame if and only if Ψ is a Bessel sequence. An example of a wavelet system that contains $\{a^{j/2}\psi(a^j x - kb)\}_{j,k\in\mathbb{Z}}$ is

$$\{a^{j/2}\psi(a^j x - kb/n)\}_{j,k\in\mathbb{Z}}, \tag{11.88}$$

where $n \in \mathbb{N}$. We say that the wavelet system in (11.88) is obtained via *oversampling with factor n* of $\{a^{j/2}\psi(a^j x - kb)\}_{j,k\in\mathbb{Z}}$.

Oversampling will in general change the frame bounds, and for a tight wavelet frame it might happen that the oversampled frame is no longer tight. A positive result was obtained in [21], where the given conditions imply that $\{a^{j/2}\psi(a^j x - kb/n)\}_{j,k\in\mathbb{Z}}$ is tight if $\{a^{j/2}\psi(a^j x - kb)\}_{j,k\in\mathbb{Z}}$ is tight:

Theorem 11.7.4 *Let $a \geq 2$ be a positive integer and $b > 0$. Suppose that $\{a^{j/2}\psi(a^j x - kb)\}_{j,k\in\mathbb{Z}}$ is a frame for $L^2(\mathbb{R})$ with bounds A, B. Then, for any positive integer n that is relatively prime to a, the family in (11.88) is a frame for $L^2(\mathbb{R})$ with bounds nA, nB.*

In the special case $a = 2$, we see that tightness is preserved if n is odd. There exists examples, showing that tightness might not be preserved if n is even, cf. [22].

We saw in Example 11.1.2 that the canonical dual of a wavelet frame might not have the wavelet structure. However, there are cases where one can find another dual, which is also a wavelet system; such cases are characterized in Theorem 11.1.5. If the frame $\{a^{j/2}\psi(a^j x - kb)\}_{j,k\in\mathbb{Z}}$ has a wavelet dual $\{a^{j/2}\tilde{\psi}(a^j x - kb)\}_{j,k\in\mathbb{Z}}$ and n is a positive integer that is relatively prime to a, then the oversampled system (11.88) also has a dual with the wavelet structure, namely $\{\frac{1}{n}a^{j/2}\tilde{\psi}(a^j x - kb/n)\}_{j,k\in\mathbb{Z}}$. We refer to [21] for a proof.

Theorem 11.7.3 gives a sufficient condition for a function $\psi \in L^2(\mathbb{R})$ to generate a wavelet frame $\{\psi_{j,k}\}_{j,k\in\mathbb{Z}}$, expressed in terms of the Fourier transform $\hat{\psi}$. For special classes of functions ψ, we can give simpler conditions for ψ generating a wavelet frame. We will now consider functions ψ for which $\hat{\psi}$ is a characteristic function for a Lebesgue measurable set K in $\mathbb{R}$. In order for χ_K to belong to $L^2(\mathbb{R})$, we assume that K has finite Lebesgue measure. Further, for convenience, we only consider the case where the translation parameter is $b = 1$ and the dilation parameter is $a = 2$.

Definition 11.7.5 *A Lebesgue measurable set K in $\mathbb{R}$ is called a frame wavelet set if $|K| < \infty$ and the function ψ defined by $\hat{\psi} = \chi_K$ generates a wavelet frame $\{D^j T_k \psi\}_{j,k\in\mathbb{Z}}$ for $L^2(\mathbb{R})$.*

We will now present conditions for a set $K \subset \mathbb{R}$ being a frame set. We begin with some definitions:

Definition 11.7.6 *Let K be a measurable set in $\mathbb{R}$ with finite measure. We say that*

(i) *$x, y \in \mathbb{R}$ are δ-equivalent if there is an $j \in \mathbb{Z}$ such that*

$$x = 2^j y.$$

For $x \in K$, the number of elements $y \in K$ which belong to its δ-equivalence class is denoted by $\delta_K(x)$. Finally, let

$$K(\delta, k) := \{x \in K : \delta_K(x) = k\}, \ k \in \mathbb{N}.$$

(ii) *$x, y \in \mathbb{R}$ are τ-equivalent if there is an $k \in \mathbb{Z}$ such that*

$$x = y + k.$$

For $x \in K$, the number of elements $y \in K$ which belong to its τ-equivalence class is denoted by $\tau_K(x)$. Finally, let

$$K(\tau, k) := \{x \in K : \tau_K(x) = k\}, \ k \in \mathbb{N}.$$

Using the above notation, Dai et al. [24] were almost able to characterize frame wavelet sets:

Theorem 11.7.7 *Let K be a Lebesgue measurable set in $\mathbb{R}$ with finite measure. Then the following holds:*

(i) *K is a frame wavelet set if $\cup_{j \in \mathbb{Z}} 2^j K(\tau, 1) = \mathbb{R}$ (up to a null set) and there exists $M \in \mathbb{N}$ such that $K(\delta, m)$ and $K(\tau, m)$ are null sets for $m > M$; in this case, one is a lower frame bound for $\{D^j T_k \psi\}_{j,k \in \mathbb{Z}}$ and $M^{5/2}$ is an upper frame bound.*

(ii) *If K is a frame wavelet set, then $\cup_{j \in \mathbb{Z}} 2^j K = \mathbb{R}$ (up to a null set) and there exists $M \in \mathbb{N}$ such that $K(\delta, m)$ and $K(\tau, m)$ are null sets for $m > M$.*

For frame wavelet sets generating a tight frame, a complete characterization is obtained:

Theorem 11.7.8 *A Lebesgue measurable set K in $\mathbb{R}$ with finite measure is a frame wavelet set generating a tight frame if and only if the following conditions hold:*

(i) *$\cup_{j \in \mathbb{Z}} 2^j K = \mathbb{R}$ (up to a null set);*

(ii) *for some $m \geq 1$ we have $K = K(\tau, 1) = K(\delta, m)$.*

In case (i) and (ii) are satisfied, the frame bound for $\{D^j T_k \psi\}_{j,k \in \mathbb{Z}}$ is equal to m.

Let us show how the conditions in Theorem 11.7.8 can be reformulated. The condition $K = K(\tau, 1)$ means exactly that for $\gamma \in \mathbb{R}$, the point $\gamma + k$ belongs to K for at most one value of $k \in \mathbb{Z}$; or, expressed differently, that

$$\sum_{k \in \mathbb{Z}} \chi_K(\gamma + k) \leq 1, \quad a.e. \ \gamma \in \mathbb{R}. \tag{11.89}$$

Now assume that

$$\bigcup_{j \in \mathbb{Z}} 2^j K = \mathbb{R}, \text{ and for some } m \in \mathbb{N}, \ K = K(\delta, m). \tag{11.90}$$

Then, given $\gamma \in \mathbb{R}$ there exists $j' \in \mathbb{Z}$ such that $2^{-j'}\gamma \in K$. The δ-equivalence class of $2^{-j'}\gamma$ contains exactly m elements, so

$$\sum_{j \in \mathbb{Z}} \chi_{2^j K}(\gamma) = m, \quad a.e. \ \gamma \in \mathbb{R}. \tag{11.91}$$

Similarly, one proves that if (11.91) holds for some $m \in \mathbb{N}$, then (11.90) holds. Thus we have obtained an equivalent formulation of Theorem 11.7.8:

Theorem 11.7.9 *A Lebesgue measurable set K in $\mathbb{R}$ with finite measure is a frame wavelet set generating a tight frame if and only if (11.89) and (11.91) are satisfied for some $m \geq 1$.*

We illustrate the use of Theorem 11.7.9 with some examples:

Example 11.7.10 (i) Let $K = [-\frac{1}{2}, -\frac{1}{4}[\cup [\frac{1}{4}, \frac{1}{2}]$. Then, for any $\gamma \neq 0$ there exists exactly one value of $j \in \mathbb{Z}$ such that $\gamma \in 2^j K$, so (11.91) is satisfied with $m = 1$. Equation (11.89) is also satisfied, so K is a frame wavelet set, which generates a tight frame with frame bound one.

(ii) Similarly, for $n = 1, 2, \ldots$, the set $K = [-\frac{1}{2}, -\frac{1}{2^{n+1}}[\cup [\frac{1}{2^{n+1}}, \frac{1}{2}]$ is a frame wavelet set, which generates a tight frame with frame bound n.

(iii) Let $K = [-\frac{3}{4}, -\frac{1}{4}[\cup [\frac{1}{8}, \frac{1}{2}[$. Then

$$K(\tau, 1) = [-\frac{1}{2}, -\frac{1}{4}[\cup [\frac{1}{8}, \frac{1}{4}[,$$

and $\cup_{j \in \mathbb{Z}} 2^j K = \mathbb{R}$ up to a null set. Also, for $m \geq 2$ we have

$$K(\delta, m) = K(\tau, m) = 0.$$

Thus, by Theorem 11.7.7, K is a frame wavelet set.   □

11.8 The continuous wavelet transform

So far, we have concentrated on series expansions in terms of wavelet systems. We will now give an introduction to the continuous wavelet transform, which delivers wavelet-type expansions in terms of integrals rather than discrete sums. Let $\psi \in L^2(\mathbb{R})$. We say that ψ satisfies the *admissibility condition* if

$$C_\psi := \int_{-\infty}^{\infty} \frac{|\hat{\psi}(\gamma)|^2}{|\gamma|} d\gamma < \infty. \tag{11.92}$$

We also say that ψ is *admissible*. Note that if $\hat{\psi}$ is continuous in 0, which is, e.g., the case if $\psi \in L^1(\mathbb{R})$, then (11.92) can only be satisfied if $\hat{\psi}(0) = 0$, i.e., if $\int_{-\infty}^{\infty} \psi(x)dx = 0$. But if this condition is satisfied, weak decay conditions on $\hat{\psi}(\gamma)$ for $\gamma \to \pm\infty$ imply that (11.92) is satisfied.

Given an admissible function $\psi \in L^2(\mathbb{R})$, we define the *continuous wavelet transform* with respect to ψ of the function $f \in L^2(\mathbb{R})$ as the function $W_\psi(f)$ of two variables given by

$$\begin{aligned} W_\psi(f)(a,b) &= \langle f, \psi^{a,b} \rangle \\ &= \int_{-\infty}^{\infty} f(x) \frac{1}{|a|^{1/2}} \overline{\psi(\frac{x-b}{a})} \, dx. \end{aligned}$$

Proposition 11.8.1 *Assume that ψ is admissible. Then, for all functions $f, g \in L^2(\mathbb{R})$,*

$$\int_{-\infty}^{\infty} \int_{-\infty}^{\infty} W_\psi(f)(a,b) \overline{W_\psi(g)(a,b)} \frac{dadb}{a^2} = C_\psi \langle f, g \rangle. \tag{11.93}$$

Proof. Using the commutator relations for the Fourier transform and the operators T_b, D_a in (2.23),

$$\begin{aligned} W_\psi(f)(a,b) &= \langle f, \psi^{a,b} \rangle \\ &= \langle \mathcal{F}f, \mathcal{F}T_b D_a \psi \rangle \\ &= \langle \hat{f}, E_{-b} D_{1/a} \hat{\psi} \rangle \\ &= \int_{-\infty}^{\infty} \hat{f}(\gamma) e^{2\pi i b \gamma} |a|^{1/2} \overline{\hat{\psi}(a\gamma)} \, d\gamma. \end{aligned}$$

Consider for a moment a fixed value for a; then, this expression is the Fourier transform of the function

$$F_a(\gamma) = \hat{f}(\gamma) |a|^{1/2} \overline{\hat{\psi}(a\gamma)},$$

calculated at the point $-b$. Letting

$$G_a(\gamma) = \hat{g}(\gamma) |a|^{1/2} \overline{\hat{\psi}(a\gamma)},$$

it follows that

$$
\begin{aligned}
\int_{-\infty}^{\infty} W_\psi(f)(a,b)\overline{W_\psi(g)(a,b)}\, db
&= \int_{-\infty}^{\infty} \widehat{F_a}(-b)\overline{\widehat{G_a}(-b)}\, db \\
&= \langle \widehat{F_a}, \widehat{G_a} \rangle \\
&= \langle F_a, G_a \rangle \\
&= \int_{-\infty}^{\infty} \hat{f}(\gamma)\overline{\hat{g}(\gamma)}\, |a|\, |\hat{\psi}(a\gamma)|^2 d\gamma.
\end{aligned}
$$

Inserting this expression in the left-hand side of (11.93) and using Fubini's theorem gives

$$
\begin{aligned}
& \int_{-\infty}^{\infty}\int_{-\infty}^{\infty} W_\psi(f)(a,b)\overline{W_\psi(g)(a,b)}\, \frac{da\,db}{a^2} \\
&= \int_{-\infty}^{\infty}\int_{-\infty}^{\infty} \hat{f}(\gamma)\overline{\hat{g}(\gamma)}\, |a|\, |\hat{\psi}(a\gamma)|^2 d\gamma\, \frac{da}{a^2} \\
&= \int_{-\infty}^{\infty} \left(\int_{-\infty}^{\infty} \frac{1}{|a|}|\hat{\psi}(a\gamma)|^2 da \right) \hat{f}(\gamma)\overline{\hat{g}(\gamma)}\, d\gamma.
\end{aligned}
$$

By a change of variable,

$$
\int_{-\infty}^{\infty} \frac{1}{|a|}|\hat{\psi}(a\gamma)|^2 da = \int_{-\infty}^{\infty} \frac{1}{|a|}|\hat{\psi}(a)|^2 da = C_\psi;
$$

thus

$$
\begin{aligned}
\int_{-\infty}^{\infty}\int_{-\infty}^{\infty} W_\psi(f)(a,b)\overline{W_\psi(g)(a,b)}\frac{da\,db}{a^2}
&= C_\psi \langle \hat{f}, \hat{g} \rangle \\
&= C_\psi \langle f, g \rangle.
\end{aligned}
$$

$\square$

As in the Gabor case, we write the result in (11.8.1) as

$$
f = \int_{-\infty}^{\infty}\int_{-\infty}^{\infty} W_\psi(f)(a,b)\psi^{a,b}\frac{da\,db}{a^2}, \quad f \in L^2(\mathbb{R}), \tag{11.94}
$$

where the integral is understood in the weak sense.

Connecting with the theory for continuous frames in Section 5.8, we have the following:

Corollary 11.8.2 *If $\psi \in L^2(\mathbb{R})$ is admissible, then $\{\psi^{a,b}\}_{a\neq 0, b\in\mathbb{R}}$ is a continuous frame for $L^2(\mathbb{R})$ with respect to $\mathbb{R}\times\mathbb{R}\backslash\{0\}$ equipped with the measure $\frac{1}{a^2}da\,db$.*

11.9   Exercises

11.1 Let $\{D^j T_k \psi\}_{j,k\in\mathbb{Z}}$ be a frame with frame operator S. Prove that S commutes with the dilation operator D, and thereby that

$$\{S^{-1} D^j T_k \psi\}_{j,k\in\mathbb{Z}} = \{D^j S^{-1} T_k \psi\}_{j,k\in\mathbb{Z}}.$$

11.2 Prove Theorem 11.1.6 via Theorem 11.1.5.

11.3 Show how to modify the proof of Lemma 11.2.3 in order to prove (11.23).

11.4 Verify that (11.30) and (11.36) are equivalent.

11.5 Complete the proof of Corollary 11.5.2 by showing that (11.68) holds for $\nu = 1/2$.

11.6 Derive the expression (11.46) for the function ψ_2.

11.7 Prove (11.47) and provide the missing details in Example 11.2.12.

11.8 Prove Corollary 11.3.4 (Hint: The proof is similar to the proof of Corollary 11.3.3, except that one has to replace the function θ in the oblique extension principle by $\theta - \eta$.)

11.9 Derive the expressions in (11.62), (11.63), and (11.65).

11.10 Consider the B-spline B_2.
    (i) Use the results in Example 11.2.9 and Example 11.2.11 to calculate the Z-transforms $\widetilde{H}_\ell$, $\ell = 0, 1, 2$, and verify that the conditions in Theorem 11.6.1 are satisfied.
    (ii) Calculate the polyphase components for the Z-transforms $\widetilde{H}_\ell$, $\ell = 0, 1, 2$ and verify that the matrix $H_p^T(z)$ satisfies the conditions in Theorem 11.6.2.

11.11 Calculate the coefficients c_j in Theorem 11.3.5 for $m = 4$, $M = 2$.

11.12 Assume that K is a frame wavelet set, and let $\theta \in L^2(\mathbb{R})$ be a function with support on K. Assume that there exist constants $C, D > 0$ such that $C \le |\theta| \le D$. Prove that the function $\psi \in L^2(\mathbb{R})$ defined by $\hat{\psi} = \theta \chi_K$ generates a wavelet frame.

List of Symbols

$\mathbb{R}$:   The real numbers.

$\mathbb{R}^+$:   The strictly positive real numbers.

$\mathbb{N}$:   The natural numbers: 1,2,3,....

$\mathbb{Z}$:   The integers.

$\mathbb{Q}$:   The rational numbers.

$\mathbb{C}$:   The complex numbers.

$gcd(p,q)$:   The largest common divisor for $p, q \in \mathbb{N}$.

$\lfloor x \rfloor$:   The integer part of $x \in \mathbb{R}$, i.e., the largest integer not exceeding x.

$\overline{x}$:   The complex conjugated of $x \in \mathbb{C}$.

X, Y :   Banach spaces.

$\mathcal{H}, \mathcal{K}$:   Hilbert spaces.

$\oplus$:   Orthogonal direct sum.

$L^p(\mathbb{R})$:   The space of measurable functions $f : \mathbb{R} \mapsto \mathbb{C}$ for which $\int_{\mathbb{R}} |f(x)|^p dx < \infty$.

$C^k(\mathbb{R})$:   The space of k times differentiable functions with a continuous k-th derivative.

$C(a,b)$:   The space of continuous functions $f :]a, b[\to \mathbb{C}$.

$\mathcal{F}f(\gamma) = \hat{f}(\gamma)$:   The Fourier transform, for $f \in L^1(\mathbb{R})$ given by $\hat{f}(\gamma) = \int_{\mathbb{R}} f(x)e^{-2\pi ix\gamma} dx$.

$\ell^2(I)$:   The space of square summable sequences on I.

$|I|$:   The Lebesgue measure of a Borel set I, or when I is discrete, the number of elements in I.

χ_A :   The indicator function for a set A, $\chi_A(x) = 1$ if $x \in A$, otherwise 0.

$\overline{A}$:   The closure of a set A.

$A^\perp$:   The orthogonal complement of a subset A in a Hilbert space.

$\operatorname{supp} f :$ The support of the function f: $\operatorname{supp} f = \overline{\{x \in \mathbb{R} : f(x) \neq 0\}}$.

$\delta_{k,j} :$ The Kronecker delta: $\delta_{k,j} = 1$ if $k = j$, $\delta_{k,j} = 0$ if $k \neq j$.

$T_a :$ The translation operator $(T_a f)(x) = f(x - a)$.

$E_b :$ The modulation operator $(E_b f)(x) = e^{2\pi i b x} f(x)$.

$D_a :$ The dilation operator $(D_a f)(x) = \frac{1}{\sqrt{a}} f(\frac{x}{a})$, $\quad a > 0$.

$D :$ The dilation operator $(D f)(x) = 2^{1/2} f(2x)$.

$S :$ The frame operator.

$T :$ The pre-frame operator.

$U^\dagger :$ The pseudo-inverse of the operator U.

$\mathcal{N}_U :$ The kernel of the operator U.

$\mathcal{R}_U :$ The range of the operator U.

$\text{MSE} :$ The mean-square error.

References

[1] Aldroubi, A. and Gröchenig, K.: *Nonuniform sampling and reconstruction in shift-invariant spaces.* SIAM Review **43** no. 4 (2001), 585–620.

[2] Bachman, G., Narici, L., and Beckenstein, E.: *Fourier and wavelet analysis.* Springer, New York, 2000.

[3] Benedetto, J. and Treiber, O.: *Wavelet frames: multiresolution analysis and extension principles.* In "Wavelet transforms and time-frequency signal analysis," 1–36, (ed. Debnath, L.). Birkhäuser, Boston, 2001.

[4] de Boor, C.: *A practical guide to splines.* Springer, New York, 2001.

[5] de Boor, C., DeVore, R. A., and Ron, A.: *On the construction of multivariate (pre)wavelets.* Constr. Approx. **9** (1993), 123–166.

[6] Burrus, C. S., Gopinath, R. A., and Guo, H.: *Wavelets and wavelet transforms.* Prentice Hall, Englewood Cliffs, New Jersey, 1998.

[7] Bölcskei, H. and Janssen, A. J. E. M.: *Gabor frames, unimodularity, and window decay.* J. Fourier Anal. Appl. **6** no. 3 (2000), 255–276.

[8] Casazza, P. G. and Christensen, O.: *Weyl-Heisenberg frames for subspaces of $L^2(\mathbb{R})$.* Proc. Amer. Math. Soc. **129** (2001), 145–154.

[9] Casazza, P. G. and Christensen, O.: *Frames and Schauder bases.* In "Approximation Theory: In Memory of A. K. Varma," 133–139, (eds. Govil, N. K., Mohapatra, R. N., Nashed, Z., Sharma, A., Szabados, J.). Marcel Dekker, New York, 1998.

[10] Christensen, O.: *An introduction to frames and Riesz bases.* Birkhäuser, Boston, 2003.

[11] Christensen, O.: *Frames, bases, and discrete Gabor/wavelet expansions.* Bull. Amer. Math. Soc. **38** no. 3 (2001), 273–291.

[12] Christensen, O.: *A Paley–Wiener theorem for frames*. Proc. Amer. Math. Soc. **123** (1995), 2199–2202.

[13] Christensen, O.: *Pairs of dual Gabor frames with compact support and desired frequency localization*. Appl. Comp. Harm. Anal. **20** (2006), 403–410.

[14] Christensen, O. and Eldar, Y.: *Oblique dual frames and shift-invariant spaces*. Appl. Comp. Harm. Anal. 2004 **17** no. 1 (2004), 48–68 (special frame issue).

[15] Christensen, O., Kim, O. H., Kim, R. Y., and Lim J. K.: *Riesz sequences of translates and their generalized duals*. J. Geometric Analysis **16** no. 4 (2006), 585–596.

[16] Christensen, O. and Kim, R. Y.: *On dual Gabor frame pairs generated by polynomials*. Preprint, 2007.

[17] Chui, C.: *Wavelets - a tutorial in theory and practice*. Academic Press, San Diego, 1992.

[18] Chui, C.: *Multivariate splines*. SIAM, Philadelphia, 1988.

[19] Chui, C., He, W., and Stöckler, J.: *Compactly supported tight and sibling frames with maximum vanishing moments*. Appl. Comp. Harm. Anal. **13** no. 3 (2002), 226-262.

[20] Chui, C. and Shi, X.: *Inequalities of Littlewood-Paley type for frames and wavelets*. SIAM J. Math. Anal. **24** no. 1 (1993), 263–277.

[21] Chui, C. and Shi, X.: *Bessel sequences and affine frames*. Appl. Comp. Harm. Anal. **1** (1993), 29–49.

[22] Chui, C. and Shi, X.: *N× Oversampling preserves any tight affine frame for odd N*. Proc. Amer. Math. Soc. **121** no. 2 (1994), 511–517.

[23] Cvetković, Z. and Vetterli, M.: *Tight Weyl-Heisenberg frames*. IEEE Trans. Signal. Proc. **46** no. 5 (1998), 1256–1259.

[24] Dai, X., Diao, Y., and Gu, Q.: *Frame wavelet sets in* $\mathbb{R}$. Proc. Amer. Math. Soc. **129** no. 7 (2000), 2045–2055.

[25] Daubechies, I.: *The wavelet transformation, time-frequency localization and signal analysis*. IEEE Trans. Inform. Theory **36** (1990), 961–1005.

[26] Daubechies, I.: *Ten lectures on wavelets*. SIAM, Philadelphia, 1992.

[27] Daubechies, I., Grossmann, A., and Meyer, Y.: *Painless nonorthogonal expansions*. J. Math. Phys. **27** (1986), 1271–1283.

[28] Daubechies, I., Han, B., Ron, A., and Shen, Z.: *Framelets: MRA-based constructions of wavelet frames*. Appl. Comp. Harm. Anal. **14** no. 1 (2003), 1-46. 2001.

[29] Daubechies, I., Landau, H. J., and Landau, Z.: *Gabor time-frequency lattices and the Wexler-Raz identity*. J. Fourier Anal. Appl. **1** (1995), 437–478.

[30] Duffin, R. J. and Schaeffer, A. C.: *A class of nonharmonic Fourier series*. Trans. Amer. Math. Soc. **72** (1952), 341–366.

[31] Feichtinger, H. G.: *Modulation spaces: looking back and ahead*. Sampling Theory in Signal and Image Processing **5** no. 2 (2006), 109–140.

[32] Feichtinger, H. G. and Strohmer, T. (eds.): *Gabor analysis and algorithms: Theory and applications*. Birkhäuser, Boston, 1998.

[33] Feichtinger, H. G. and Strohmer, T. (eds.): *Advances in Gabor analysis*. Birkhäuser, Boston, 2002.

[34] Frazier, M., Garrigos, G., Wang, K., and Weiss, G.: *A characterization of functions that generate wavelet and related expansion*. J. Fourier Anal. Appl. **3** (1997), 883–906.

[35] Gopinath, R.: *Phaselets of framelets*. IEEE Trans. Signal Process. **53** no. 5 (2005), 1794–1806.

[36] Goyal, V. K., Kovačević, J., and Kelner, A. J.: *Quantized frame expansions with erasures*. Appl. Comp. Harm. Anal. **10** no. 3 (2000), 203–233.

[37] Gröchenig, K.: *Foundations of time-frequency analysis*. Birkhäuser, Boston, 2000.

[38] Gröchenig, K.: *Localization of frames, Banach frames, and the invertibility of the frame operator*. J. Fourier Anal. Appl. **10** no. 2 (2004), 105-132.

[39] Gröchenig, K., Janssen, A. J. E. M., Kaiblinger, N., and Pfander, G.: *Note on B-splines, wavelet scaling functions, and Gabor frames*. IEEE Trans. Inf. Theory **49** no. 12 (2003), 3318–3320.

[40] Heil, C. and Walnut, D.: *Continuous and discrete wavelet transforms*. SIAM Review **31** (1989), 628–666.

[41] Hernandez, E. and Weiss, G.: *A first course on wavelets*. CRC Press, Boca Raton, 1996.

[42] Hernandez, E., Labate, D., and Weiss, G.: A unified characterization of reproducing systems generated by a finite family II. J. Geom. Anal. **12** no. 4 (2002), 615–662.

[43] Heuser, H.: *Functional analysis*. John Wiley, New York, 1982.

[44] Janssen, A. J. E. M.: *The duality condition for Weyl-Heisenberg frames*. In "Gabor analysis: theory and application," (eds. Feichtinger, H. G. and Strohmer, T.). Birkhäuser, Boston, 1998.

[45] Janssen, A. J. E. M.: *Representations of Gabor frame operators*. In "Twentieth century harmonic analysis–a celebration," 73–101, NATO Sci. Ser. II Math. Phys. Chem., **33**, Kluwer Acad. Publ., Dordrecht, 2001.

[46] Janssen, A. J. E. M.: *On rationally oversampled Weyl-Heisenberg frames*. Signal Processing **47** (1995), 239–245.

[47] Janssen, A. J. E. M.: *Duality and biorthogonality for Weyl-Heisenberg frames*. J. Fourier Anal. Appl. **1** no. 4 (1995), 403–436.

[48] Janssen, A. J. E. M.: *From continuous to discrete Weyl-Heisenberg frames through sampling*. J. Fourier Anal. Appl. **3** no. 5 (1997), 583–596.

[49] Janssen, A. J. E. M.: *Zak transforms with few zeros and the tie*. In "Advances in Gabor analysis," (eds. Feichtinger, H. G. and Strohmer, T.). Birkhäuser, Boston, 2002

[50] Kim, J. M. and Kwon, K. H.: *Frames by integer translates*. To appear in J. Korean Society for Industrial and Applied Mathematics, 2007.

[51] Lemvig, J.: *Constructing pairs of dual bandlimited framelets with desired time localization*. Preprint, 2007.

[52] Lindenstrauss, J. and Tzafriri, L.: *Classical Banach spaces 1.* Springer, New York, 1977.

[53] Mallat, S.: *A wavelet tour of signal processing.* Academic Press, San Diego, 1999.

[54] Ron, A. and Shen, Z.: *Weyl-Heisenberg systems and Riesz bases in $L^2(\mathbb{R}^d)$.* Duke Math. J. **89** (1997), 237–282.

[55] Ron, A. and Shen, Z.: *Affine systems in $L_2(\mathbb{R}^d)$: the analysis of the analysis operator.* J. Funct. Anal. **148** (1997) 408–447.

[56] Ron, A. and Shen, Z.: *Affine systems in $L_2(R^d)$ II: dual systems.* J. Fourier Anal. Appl. **3** (1997), 617–637.

[57] Ron, A. and Shen, Z.: *Compactly supported tight affine spline frames in $L_2(R^d)$.* Math. Comp. **67** (1998), 191–207.

[58] Ron, A. and Shen, Z.: *Frames and stable bases for shift-invariant subspaces of $L^2(\mathbb{R}^d)$.* Canad. J. Math. **47** no. 5 (1995), 1051–1094.

[59] Rudin, W.: *Real and complex analysis.* McGraw-Hill, New York, 1986.

[60] Rudin, W.: *Functional analysis.* McGraw-Hill, New York, 1973.

[61] Rudin, W.: *Fourier analysis on groups.* Interscience Publishers, New York, 1962.

[62] Young, R.: *An introduction to nonharmonic Fourier series.* Academic Press, New York, 1980 (revised first edition 2001).

[63] Vetterli, M. and Kovačević, J.: *Wavelets and subband coding.* Prentice-Hall, Englewood Cliffs, New Jersey, 1995.

[64] Walnut, D.: *An introduction to wavelet analysis.* Birkhäuser, Boston, 2001.

[65] Wojtaszczyk, P.: *A mathematical introduction to wavelets.* Cambridge University Press, Cambridge, 1999.

[66] Zibulski, M. and Zeevi, Y. Y.: *Oversampling in the Gabor scheme.* IEEE Trans. SP **41** no. 8 (1993), 2679–2687.

Index

Applied and Numerical Harmonic Analysis

J.M. Cooper: *Introduction to Partial Differential Equations with MATLAB* (ISBN 0-8176-3967-5)

C.E. D'Attellis and E.M. Fernández-Berdaguer: *Wavelet Theory and Harmonic Analysis in Applied Sciences* (ISBN 0-8176-3953-5)

H.G. Feichtinger and T. Strohmer: *Gabor Analysis and Algorithms* (ISBN 0-8176-3959-4)

T.M. Peters, J.H.T. Bates, G.B. Pike, P. Munger, and J.C. Williams: *The Fourier Transform in Biomedical Engineering* (ISBN 0-8176-3941-1)

A.I. Saichev and W.A. Woyczyński: *Distributions in the Physical and Engineering Sciences* (ISBN 0-8176-3924-1)

R. Tolimieri and M. An: *Time-Frequency Representations* (ISBN 0-8176-3918-7)

G.T. Herman: *Geometry of Digital Spaces* (ISBN 0-8176-3897-0)

A. Procházka, J. Uhlíř, P.J.W. Rayner, and N.G. Kingsbury: *Signal Analysis and Prediction* (ISBN 0-8176-4042-8)

J. Ramanathan: *Methods of Applied Fourier Analysis* (ISBN 0-8176-3963-2)

A. Teolis: *Computational Signal Processing with Wavelets* (ISBN 0-8176-3909-8)

W.O. Bray and Č.V. Stanojević: *Analysis of Divergence* (ISBN 0-8176-4058-4)

G.T Herman and A. Kuba: *Discrete Tomography* (ISBN 0-8176-4101-7)

J.J. Benedetto and P.J.S.G. Ferreira: *Modern Sampling Theory* (ISBN 0-8176-4023-1)

A. Abbate, C.M. DeCusatis, and P.K. Das: *Wavelets and Subbands* (ISBN 0-8176-4136-X)

L. Debnath: *Wavelet Transforms and Time-Frequency Signal Analysis* (ISBN 0-8176-4104-1)

K. Gröchenig: *Foundations of Time-Frequency Analysis* (ISBN 0-8176-4022-3)

D.F. Walnut: *An Introduction to Wavelet Analysis* (ISBN 0-8176-3962-4)

O. Bratteli and P. Jorgensen: *Wavelets through a Looking Glass* (ISBN 0-8176-4280-3)

H.G. Feichtinger and T. Strohmer: *Advances in Gabor Analysis* (ISBN 0-8176-4239-0)

O. Christensen: *An Introduction to Frames and Riesz Bases* (ISBN 0-8176-4295-1)

L. Debnath: *Wavelets and Signal Processing* (ISBN 0-8176-4235-8)

J. Davis: *Methods of Applied Mathematics with a MATLAB Overview* (ISBN 0-8176-4331-1)

G. Bi and Y. Zeng: *Transforms and Fast Algorithms for Signal Analysis and Representations* (ISBN 0-8176-4279-X)

J.J. Benedetto and A. Zayed: *Sampling, Wavelets, and Tomography* (ISBN 0-8176-4304-4)

E. Prestini: *The Evolution of Applied Harmonic Analysis* (ISBN 0-8176-4125-4)

O. Christensen and K.L. Christensen: *Approximation Theory* (ISBN 0-8176-3600-5)

L. Brandolini, L. Colzani, A. Iosevich, and G. Travaglini: *Fourier Analysis and Convexity* (ISBN 0-8176-3263-8)

W. Freeden and V. Michel: *Multiscale Potential Theory* (ISBN 0-8176-4105-X)

O. Calin and D.-C. Chang: *Geometric Mechanics on Riemannian Manifolds* (ISBN 0-8176-4354-0)

J.A. Hogan and J.D. Lakey: *Time-Frequency and Time-Scale Methods* (ISBN 0-8176-4276-5)

C. Heil: *Harmonic Analysis and Applications* (ISBN 0-8176-3778-8)

K. Borre, D.M. Akos, N. Bertelsen, P. Rinder, and S.H. Jensen: *A Software-Defined GPS and Galileo Receiver* (ISBN 0-8176-4390-7)

T. Qian, V. Mang I, and Y. Xu: *Wavelet Analysis and Applications* (ISBN 3-7643-7777-1)

G.T. Herman and A. Kuba: *Advances in Discrete Tomography and Its Applications* (ISBN 0-8176-3614-5)

M.C. Fu, R.A. Jarrow, J.-Y. J. Yen, and R.J. Elliott: *Advances in Mathematical Finance* (ISBN 0-8176-4544-6)

O. Christensen: *Frames and Bases* (ISBN 0-8176-4677-6)